Ring-Opening Polymerization

Takeo Saegusa, EDITOR
Kyoto University

Eric Goethals, EDITOR
University of Ghent

An international symposium sponsored by the Division of Polymer Chemistry, Inc. at the 173rd Meeting of the American Chemical Society, New Orleans, La., March 21–23, 1977

ACS SYMPOSIUM SERIES **59**

AMERICAN CHEMICAL SOCIETY
WASHINGTON, D. C. 1977

Library of Congress CIP Data

Ring-opening polymerization.
  (ACS symposia series; 59 ISSN 0097-6156)

  Includes bibliographical references and index.

  1. Polymers and polymerization—Congresses. 2. Cyclic compounds—Congresses.
  I. Saegusa, Takeo, 1927-   . II. Goethals, Eric. III. American Chemical Society. Division of Polymer Chemistry. IV. Series: American Chemical Society. ACS symposium series; 59.

QD281.P6R56   547'.28   77-13631
ISBN 0-8412-0392-X   ACSMC8 59 1-352 1977

Copyright © 1977

American Chemical Society

All Rights Reserved. No part of this book may be reproduced or transmitted in any form or by any means—graphic, electronic, including photocopying, recording, taping, or information storage and retrieval systems—without written permission from the American Chemical Society.

PRINTED IN THE UNITED STATES OF AMERICA

ACS Symposium Series

Robert F. Gould, *Editor*

Advisory Board

Donald G. Crosby

Jeremiah P. Freeman

E. Desmond Goddard

Robert A. Hofstader

John L. Margrave

Nina I. McClelland

John B. Pfeiffer

Joseph V. Rodricks

Alan C. Sartorelli

Raymond B. Seymour

Roy L. Whistler

Aaron Wold

FOREWORD

The ACS SYMPOSIUM SERIES was founded in 1974 to provide a medium for publishing symposia quickly in book form. The format of the SERIES parallels that of its predecessor, ADVANCES IN CHEMISTRY SERIES, except that in order to save time the papers are not typeset but are reproduced as they are submitted by the authors in camera-ready form. As a further means of saving time, the papers are not edited or reviewed except by the symposium chairman, who becomes editor of the book. Papers published in the ACS SYMPOSIUM SERIES are original contributions not published elsewhere in whole or major part and include reports of research as well as reviews since symposia may embrace both types of presentation.

CONTENTS

Preface vii

1. Cationic Polymerization of Cyclic Amines 1
   E. J. Goethals, E. H. Schacht, P. Bruggeman, and P. Bossaer

2. New Aspects of the Chemistry of Living Tetrahydrofuran Polymers Initiated by Trifluoromethane Sulfonic Anhydride 13
   Samuel Smith, William J. Schultz, and Richard A. Newmark

3. New Developments in Graft Copolymerization by Oxonium Ion Mechanism 24
   K. I. Lee and P. Dreyfuss

4. Ring-Opening Polymerization with Expansion in Volume 38
   William J. Bailey, Robert L. Sun, Hirokazu Katsuki, Takeshi Endo, Hideaki Iwama, Rikio Tsushima, Kazuhide Saigou, and Michel M. Bitritto

5. Progress in Polymerization of Cyclic Acetals 60
   Stanisław Penczek and Przemysław Kubisa

6. Ring-Opening Polymerization of Macrocyclic Acetals 77
   Rolf C. Schulz, K. Albrecht, C. Rentsch, and Q. V. Tran Thi

7. Macrocyclic Formals 99
   Yuya Yamashita and Yuhsuke Kawakami

8. Stereoregularity as a Function of Side Chain Size in Perhaloacetaldehyde Polymerization 111
   D. W. Lipp and O. Vogl

9. Mechanism of the Cationic Polymerization of Lactams 129
   M. Rothe and G. Bertalan

10. Ring-Opening Copolymerization of Some Cyclic Compounds Containing Oxygen and Nitrogen Atoms 145
    H. L. Hsieh

11. Ring-Opening Polymerizations: Mechanism of Polymerization of ε-Caprolactone 152
    R. H. Young, M. Matzner, and L. A. Pilato

12. New Prospects in Homogeneous Ring-Opening Polymerization of Heterocyclic Monomers 165
    Ph. Teyssié, J. P. Bioul, A. Hamitou, J. Heuschen, L. Hocks, R. Jérôme, and T. Ouhadi

13. Optically Active Poly[oxy(1-alkyl)ethylene] 178
    Teiji Tsuruta

14. Stereoselective and Stereoelective Polymerization of Oxiranes and Thiiranes 191
    Nicolas Spassky

15. Rate and Stereochemistry of the Anionic Polymerization of α,α-Disubstituted-β-propiolactones 210
    Robert W. Lenz, Christian G. D'Hondt, and Ebrahim Bigdeli

16. Specific Interactions of Lithium Chloride in the Anionic Polymerization of Lactams 216
    Giorgio Bontá, Alberto Ciferri, and Saverio Russo

17. Isomerization Polymerization of Lactams 233
    H. K. Reimschuessel

18. Copolymeriaztion of ε-Caprolactam with β-(3,4-Diaminophenyl) Propionic Acid 251
    S. W. Shalaby and E. A. Turi

19. Anionic Polymerization of Fluorocarbon Epoxides 269
    James T. Hill and John P. Erdman

20. Ring-Opening Polymerization via C–C Bond Opening 285
    H. K. Hall, Jr., H. Tsuchiya, P. Ykman, J. Otton, S. C. Snider, and A. Deutschman, Jr.

21. New Polymers by Ring-Opening Polymerization of Norbornene Derivatives with Polar Substituents 303
    S. Matsumoto, K. Komatsu, and K. Igarashi

22. Polymerization of Aryl Cyclic Sulfonium Zwitterions 318
    D. L. Schmidt

23. Spontaneous Alternating Copolymerization of Cyclic Phosphorus Compounds via Phosphonium Zwitterion Intermediates 332
    Takeo Saegusa, Shiro Kobayashi, Yoshiharu Kimura, and Tsunenori Yokoyama

Index 345

PREFACE

The international symposium on which this volume is based consisted of invited lectures from the United States, Europe, and Japan and honors W. H. Bailey, of the University of Maryland, who received the ACS Witco Award in polymer chemistry. The full papers corresponding to all of the invited lectures cover a great variety of important aspects of ring-opening polymerization.

In polymer chemistry and in the associated industrial processes, polycondensation and vinyl polymerization have long played prominent roles. Ring-opening polymerization does not have a long history, but it has been extensively studied in the past two decades. Monomers suitable for ring-opening polymerization show a great variety of functional groups and ring sizes. Therefore the patterns of polymerization reactions are very diversified. Various polymerization catalysts with specific activities were discovered, and several of the polymerizations are now important to industry. Some of the commercially produced polymers obtained by ring-opening polymerization are nylon-6, polyacetal, poly(ethylene oxide), poly(propylene oxide) and their copolymers, poly(epichlorohydrin), poly(ethylenimine), poly(tetrahydrofuran), and poly(caprolactone).

The back-bone units of polymers made by ring-opening polymerization may contain one, two, or even three heteroatoms or no heteroatoms at all. These polymers exhibit a wide variety of properties. We believe that ring-opening polymerization has great possibilities for further progress, and we hope that this book will contribute to the understanding and the significance of this field.

At the symposium, we received a grant from the American Chemical Society (ACS-PRF Special Education Opportunities Grant) and donations from the following companies: Allied Chemical Co., Dow Chemical Co., Ethyl Corp., General Electric Co., Minnesota Mining and Manufacturing Co., Phillips Petroleum Co., and Tennessee Eastman Co. These funds were used mostly to subsidize the travel expenses of the speakers from the academic institutions outside the United States. We wish to express our hearty thanks to the American Chemical Society and to the above-named companies. We also thank W. J. Bailey and the other contributors for their excellent papers and for their cooperation in organizing, presenting, and publishing this symposium.

Kyoto University
Kyoto, Japan

TAKEO SAEGUSA

University of Ghent
Ghent, Belgium
July 21, 1977

ERIC GOETHALS

Cationic Polymerization of Cyclic Amines

E. J. GOETHALS, E. H. SCHACHT, P. BRUGGEMAN, and P. BOSSAER
Institute of Organic Chemistry, Rijksuniversiteit Gent, B-9000 Ghent, Belgium

It is now generally accepted that the propagation reaction in the cationic ring-opening polymerization of cyclic amines is a nucleophilic attack of the monomer nitrogen on a strained cyclic ammonium salt which is the active species of the growing macromolecule. The driving force of the polymerization is the relief of strain associated with the ring-opening of the active chain end. The resultant polymer molecules, however, also contain nucleophilic amino functions and therefore the polymer competes with monomer to react with the active species. This results in the formation of a (linear or cyclic) non-strained and therefore non-reactive ammonium salt.

With secondary cyclic amines (R = H), the proton on the terminated ammonium salt as well as the proton on the active species can be transferred to other amino groups present in the mixture including monomer. Dimers and other low oligomers are therefore the initial products, and the final products are branched polymers containing a distribution of primary, secondary and tertiary amino functions (1, 2).

With cyclic tertiary amines (R = alkyl), the formation of a non-strained ammonium salt is a real termination reaction. If the rate of this termination reaction is not negligably small compared with the rate of the propagation, in other words if the ratio k_p/k_t is not very high, the polymerization will stop before all monomer is consumed and low molecular weight polymers will be obtained. Therefore, it is interesting to know the factors which influence this termination. The main purpose of the investigations presented in this paper is to determine the relation between monomer structure and the ratio k_p/k_t. The initiator for all polymerizations was triethyloxonium tetrafluoroborate. This substance reacts very rapidly with amines so that it may be assumed that initiation is fast compared with propagation. Also, the counter ion $BF_4^\ominus$ is stable and has a low nucleophilicity so that termination reactions with counter ion may be neglected.

Methods for the Determination of k_p/k_t.

For fast polymerizations with values of k_p/k_t up to 100, a method based on the maximal conversions obtained with different initiator concentrations was used. The rate of polymerization R_p is given by:

$$R_p = -dm/dt = k_p\, m[P_n^+] \qquad [1]$$

where m = monomer concn. and $[P_n^+]$ = concn. of growing chains.

If termination is a first order reaction the rate R_t is given by:

$$R_t = -d[P_n^+]/dt = k_t[P_n^+] \qquad [2]$$

If termination occurs by reaction of the growing chains with any of the amino functions of polymer the rate R_t is given by the second order eqn.:

$$R_t = -d[P_n^+]/dt = k_t[P_n^+](m_o - m) \qquad [3]$$

where m_o = initial monomer concn.
Dividing eqn. [1] by eqn. [2] or [3] leads to:

$$\frac{dm}{d[P_n^+]} = \frac{k_p}{k_t} m \quad \text{or} \quad \frac{dm}{d[P_n^+]} = \frac{k_p}{k_t} \frac{m}{(m_o - m)}$$

Integration of these equations between the limits $m = m_o$, $[P_n^+] = fc_o$ and $m = m_f$, $[P_n^+] = 0$ leads to:

$$\ln \frac{m_o}{m_f} = \frac{k_p}{k_t} f\, c_o \qquad [4]$$

for a first order termination and to:

$$\ln \frac{m_o}{m_f} - \frac{m_o - m_f}{m_o} = \frac{k_p}{k_t} f \frac{c_o}{m_o} \qquad [5]$$

for a second order termination where f = efficiency factor of
initiation; c_o = initiator concentration, and m_f = monomer concentration at the end of the polymerization.

With triethyloxonium tetrafluoroborate, the initiation reaction is fast compared with propagation, so that $f \sim 1$. The type of termination reaction defines the functional form of m_o, m_f and c_o, i.e. eqn. [4] or [5]. Consequently the (grafical) solution of these equations permits to distinguish between first order and second order termination reaction and to determine the values of k_p/k_t.

For slow polymerizations, separate values for k_p and k_t can be derived from time-conversion curves by using eq. [6] or eq. [7] depending on whether termination occurs by a first order or a second order reaction.

$$\ln \frac{R_p}{mc_o} = \ln k_p - k_t \, t \qquad [6]$$

$$\ln \frac{R_p}{mc_o} = \ln k_p - k_t \int_o^t (m_o - m) dt \qquad [7]$$

R_p can be measured from the tangent at the time-conversion curves and $\int_o^t (m_o - m) dt$ is the area under a time-conversion curve up to time t when $(m_o - m)$ is used as the ordinate (3). If termination is slow compared with propagation, in other words if the ratio k_p/k_t has a high value, the k_p can be derived directly from first order plots of the polymerization. In that case the termination becomes significant only at nearly quantitative conversions and k_t can then be derived from experiments in which "dying" polymer solutions are used to initiate new polymerizations at regular intervals. These second polymerizations become slower as the initiating solutions become older. First order plots of these new polymerizations give straight lines the slopes of which are equal to $k_p[P_n^+]$. Since k_p is known, this method allows to measure the decrease of $[P_n^+]$ in the initiating polymer solution and hence to calculate k_t.

Results and Discussion.

1) N-Substituted Ethylenimines. These monomers polymerize very rapidly at 0°C and it was not possible to evaluate separate values for k_p and k_t at this temperature. Generally the polymerizations stop at limited conversions. It was found that eqn. [4] (and not eqn. [5]) leads to straight lines the slopes of which are equal to $f.k_p/k_t$. Some examples are shown in Figure 1. Consequently the termination reactions occur according to first order kinetics. This can be explained by assuming that the termination reaction is predominantly intramolecular which means that the terminated ammonium salts are cyclic. This is in agreement with the observation that a number of N-substituted aziridines form not

only polymer but also the cyclic dimer (disubstituted piperazine) or cyclic tetramer (4, 5). These cyclic oligomers are formed by degradation of the polymer. This is supported by the fact that these oligomers are formed mainly after the polymerization. Gel permeation chromatography analysis of the mixture shows that during the time oligomer is formed, polymer concentration decreases whereas residual monomer concentration remains unchanged. It thus seems that a reaction similar to termination continues to occur at the terminated chain ends; for example :

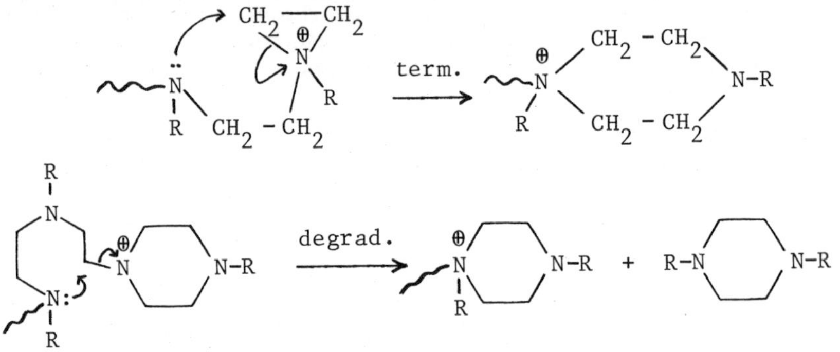

Table 1 gives a survey of values of $f \cdot k_p/k_t$ for different N-substituted ethylenimines together with the pK_b values of the monomers. It is clear that there is no simple relationship between k_p/k_t and the pK_b values. On the other hand, the trend for ethyl, isopropyl and tert.butyl aziridine strongly indicates that steric hindrance plays an important role.

Table 1: Values of $f \cdot k_p/k_t$ for the polymerization of N-substituted ethylenimines, 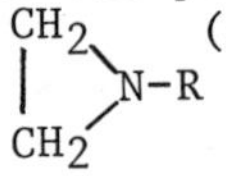 (a)

N-substituent (R)	$f \cdot k_p/k_t$ (b) (l.mol^{-1})	max.yield for m_o = 1 and c_o = 0.01 mol.l^{-1}	pK_b of monomer
$-C_2H_5$	6	2	6.09
$-CH(CH_3)_2$	21	15	6.23
$-C(CH_3)_3$	$\sim\infty$	100	5.44
$-CH_2C_6H_5$	85	55	7.24
$-CH_2CH_2C_6H_5$	14	12	6.75
$-CH_2CH_2CN$	82	55	8.67

(a) In CH_2Cl_2 at 0°C with Et_3OBF_4 as initiator.
(b) Efficiency factor $f \simeq 1$.

Of all monomers listed in Table 1 only tert.butylaziridine showed no evidence of a termination reaction. Even with small initiator concentrations (e.g. 10^{-3} mol.l^{-1}) did the polymerization go to completion. The absence of termination was further indicated by kinetic studies at low temperatures (-40 to -20°). As shown in Figure 2, the first order plots of the reaction gave perfect straight lines up to high conversion. The viscosity data given in Table 2, show that the molecular weight of the polymer can be controlled by the m_o/c_o ratio. This indicates that also transfer reactions are unimportant and is in agreement with the high living character for this polymerization.

Table 2: Intrinsic viscosities of poly(t.BA.HCl), as a function of m_o/c_o.

m_o (mol.l^{-1})	c_o (mol.l^{-1})	$\dfrac{m_o}{c_o}$	$[\eta]$ (a) (dl.g^{-1})
1.0	0.05	20	0.035
1.0	0.03	33	0.066
1.0	0.02	50	0.078
2.0	0.04	50	0.078
1.0	0.01	100	0.12
1.0	0.006	167	0.18
1.0	0.002	500	0.43
1.0	0.001	1000	0.84

(a) in aqueous 0.4 N.KCl at 25°C.

The absence of transfer reactions is confirmed by the possibility to produce block-copolymers of tert.butylaziridine by using "living" cationic polymers as initiator. This has been achieved with living poly(tetrahydrofurane) at 0°C (6) and with living polystyrene perchlorate, at -60°C (7). In both cases the formation of block-copolymers was demonstrated by the solubility properties of the obtained polymers which were different from those of homo poly-(tert.butylaziridine), and by gel permeation chromatography.

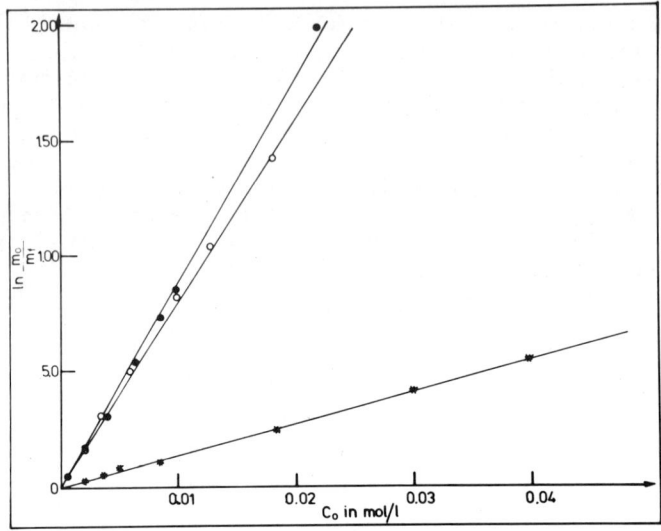

Figure 1. Determination of $f \cdot k_p/k_t$, according to Equation 4, for the polymerization of 1-benzyl aziridine (●), 1-(2-cyanoethyl)aziridine (○), and 1-(2-phenylethyl)aziridine (*) in CH_2Cl_2 at $0°C$ with triethyloxonium tetrafluoroborate as initiator, $m_o = 1.0 \; mol \cdot l^{-1}$

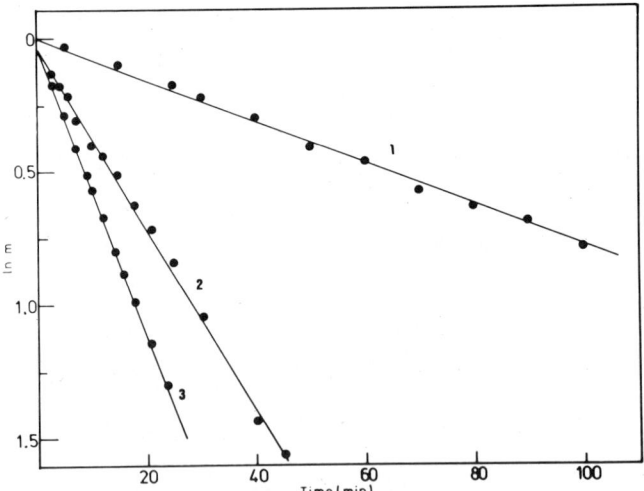

Figure 2. First-order plots of the polymerization of 1-tert-butylaziridine with triethyloxonium tetrafluoroborate at different temperatures. (1) $-40°C$; $c_o = 0.020 \; mol \cdot l^{-1}$. (2) $-30°C$; $c_o = 0.025 \; mol \cdot l^{-1}$. (3) $-20°C$; $c_o = 0.018 \; mol \cdot l^{-1}$. $m_o = 1.0 \; mol \cdot l^{-1}$.

When initiation is quantitative and in the absence of transfer and termination, the molecular weight at quantitative conversion is given by Mm_o/c_o (M = molecular weight of monomer). Accordingly, the last polymer in Table 2 (with $[\eta]$ = 0.84) has a molecular weight of 100,000.

The absence of a termination reaction in the polymerization of 1-tert.butylaziridine is explained by the high steric hindrance caused by the tert.butyl groups around the nitrogen atoms of the polymer chain. As a consequence a nucleophilic attack by these nitrogen atoms on the active species is no more possible. Poly(tert.butylaziridine) is a highly crystalline polymer with a m.p. of 142°C.

2) <u>N-substituted Propylenimines</u>. These monomers polymerize rather slowly at temperatures between 0 and 20°C and they lead to high conversions. As an example Figure 3 shows time-conversion curves for different monomers at 10°C. First order plots of these polymerizations generally give straight lines up to high conversions which indicates that termination reactions are not important during the course of polymerization. Values of k_p (derived from the first order plots) and of k_t (derived from second monomer additions as described above) are listed in Table 3. For these polymerizations propagation is clearly much slower than initiation and therefore the efficiency factor for initiation f, may be assumed to be equal to 1.

Table 3: Values of k_p and k_t for the polymerization of N-substituted propylenimines, $CH_3\text{-}CH\diagdown \atop CH_2 \diagup N\text{-}R$ (a)

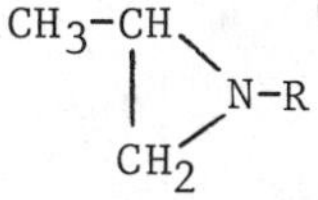

N-substituent (R)	$k_p \times 10^2$ (l.mol^{-1}sec^{-1})	$k_t \times 10^6$ (sec^{-1})	k_p/k_t (l.mol^{-1})	pK_b of monomer
$-CH_2C_6H_5$	1.27	11.6 (b)	1100 (b)	7.00
$-CH_2CH_2C_6H_5$	1.55	1.6	10,000	5.93
$-CH_2CH_2CN$	2.5-1.7	4.5	5,500 - 3,800	8.00

(a) In CH_2Cl_2 at 10°C with Et_3OBF_4 as initiator.
(b) Values obtained for polymerizations with m_o/c_o up to 40. For higher ratios k_t seems to increase markedly.

Comparison of the k_p/k_t values listed in Tables 1 and 3 clearly shows that the introduction of a methyl group in 2-position of the aziridine ring results in a dramatic increase of the living character of the polymerizations. With these monomers it is possible to prepare linear polyamines with a desired molecular weight by using the appropriate m_o/c_o ratio. In this way polymers with molecular weights up to 20.000 were obtained. If c_o is fur-

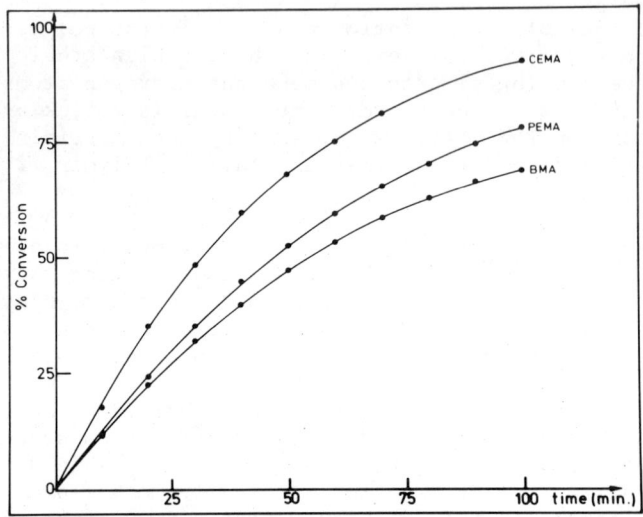

Figure 3. Time–conversion curves for the polymerization of 1-(2-cyanoethyl)-2-methyl aziridine (CEMA), 1-(2-phenylethyl)-2-methyl aziridine (PEMA), and 1-benzyl-2-methyl aziridine (BMA) in CH_2Cl_2 at $10°C$ with triethyloxonium tetrafluoroborate. $c_o = 0.015$ and $m_o = 0.80$ mol $\cdot$ l^{-1}.

ther decreased, the maximum yields become lower in accordance with eqn. [4] and consequently also the molecular weights level off. The different behaviour as far as termination is concerned between the aziridine monomers and their 2-methyl analogues is also well illustrated by the observation that a polymerized solution of 1-benzyl-2-methyl aziridine (BMA) is capable of initiating the polymerization of 1-benzyl aziridine (BA) but the reverse is not possible. This again proves that the former polymer (if not too old) still contains active chain-ends capable to initiate the polymerization of BA, whereas poly-BA is "dead" directly after the polymerization.

Even more striking is the behaviour of 1-benzyl-2,2-dimethyl aziridine (BDMA)

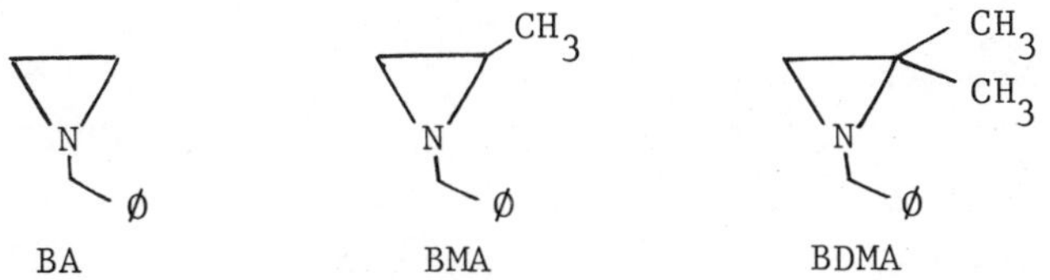

BA   BMA   BDMA

It was not possible to polymerise this monomer at temperatures between 0° and 120°C, in bulk or in solution. BDMA does react with triethyloxonium tetrafluoroborate to form the expected aziridinium salt but this salt does not give a ring-opening reaction with BDMA monomer. It is however an excellent initiator for the polymerization of BA :

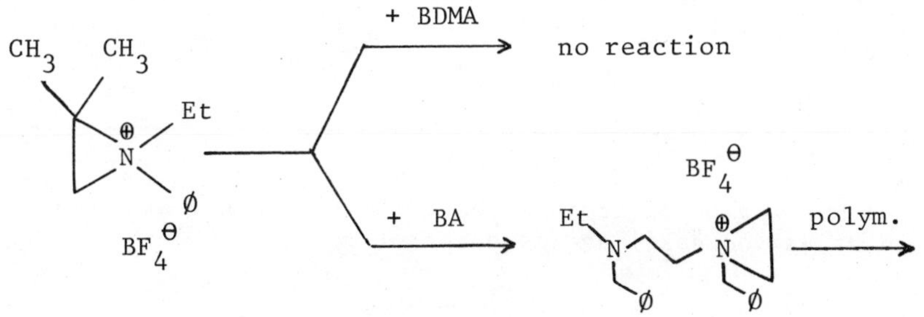

Also, BDMA does copolymerise with BA although the amount of BDMA incorporated in the copolymer is small. These observations lead to the conclusion that a propagation reaction between a highly substituted aziridinium salt and a highly substituted aziridine monomer is not possible, but that reaction between the highly sterically substituted aziridinium with a non-sterically hindered monomer or vice versa, is possible.

3) _Azetidines_. The polymerizations of these monomers are characterized by low rate constants of polymerization and by a high living character. Values of k_p and k_t for two azetidines are listed in Table 4. 1,3,3-Trimethylazetidine is a very sluggish

monomer even at 80°C. With this monomer, the active species for the polymerization can be seen in the NMR spectrum (8). The observed signals are in complete agreement with the azetidinium ion structure and they remain unchanged during and after the polymerization. At 80°C, 90% conversion is reached in a few hours but no termination could be observed after 10 days so that this system may be considered as a living polymerization (9). 1-Methylazetidine is more reactive than the trimethyl derivative but the k_t is still small compared with k_p and therefore, polymerization leads to high conversions.

Table 4: Rate constants in the polymerization of azetidines.

Monomer	▱N-CH₃	(CH₃)₂▱N-CH₃
Solvent	CH_2Cl_2	$C_6H_5NO_2$
Temp. (°C)	20	78
$k_p \times 10^4$ (l.mol^{-1}sec^{-1})	47	1.4
$k_t \times 10^4$ (l.mol^{-1}sec^{-1})	0.18	≈ 0
k_p / k_t	250	≈ ∞

The polymerization of another azetidine, conidine, has been described by Razvodovskii and coworkers (10).

conidine

The polymerization of this monomer initiated with borontrifluoride in methanol also proceeds via a living polymer mechanism.

In contrast with the aziridines, the termination reaction in the polymerization of 1-methylazetidine is of second order. This conclusion is based on the fact that eqn. [7], and not eqn. [6], leads to straight lines. Also, the viscosity of the polymerization mixture continues to increase when most of the monomer has been consumed. This is in accordance with the occurrence of a slow intermolecular termination reaction leading to branched structures. An analogous difference in termination behaviour between three- and four-membered heterocyclic monomers has been found with the sulfur analogues (thiiranes and thietanes)(11).

It has been postulated earlier (9) that the high living character of the polymerization of 1,3,3-trimethylazetidine is due to the high basicity of this amine (pK_b = 2.7) compared to linear tertiary amines (pK_b = 3,5-4,0). Assuming that the basicity of the polymeric amino groups is comparable to that of linear tertiary amines, this would result in a high preference of the monomer to react with the growing species compared with the polymer, which would lead to a high k_p/k_t ratio. In the case of 1-methylazetidine, the basicity of the monomer (pK_b = 3.6) is close to the basicity of linear tertiary amines and still k_p is 250 times higher than k_t. This leads to the conclusion that basicity is not the major factor that determines the rates of propagation and termination (although it may have some influence). Instead, steric hindrance around the amino functions is believed to be the major factor. Due to the cyclic structure of the monomers, the monomeric amino functions are relatively less hindered compared with the polymeric ones. In addition, amino groups in the polymer have twice the amount of substituents as the monomeric ones.

Experimental.

The azetidines were prepared by ring closure of the corresponding 3-amino propanol sulfates (14).

N-substituted aziridines were synthesized from the corresponding unsubstituted aziridine (12) or by ring closure of the corresponding 2-amino alcoholsulfate (13). The monomers were distilled from calcium hydride just before use. Triethyloxonium tetrafluoroborate was synthesized as described by Meerwein (15) and was purified by several reprecipitations with ether, from its methylene chloride. Time-conversion curves for the polymerization of the propylenimine monomers and 1-methylazetidine were obtained by dilatometry, those for 1,3,3-trimethylazetidine by NMR spectroscopy. Polymerizations were carried out in such a way that the reaction mixture was always under an atmosphere of dry nitrogen.

Literature cited.

(1) Dick C. and Ham G., J. Macromol.Sci.-Chem., (1970), 4, 1301.
(2) Schacht E.H. and Goethals E.J., Makromol.Chem., (1974), 175, 3447.
(3) Goethals E.J. and Drijvers W., Makromol.Chem., (1970), 136, 73.
(4) Schacht E.H., Bruggeman P. and Goethals E.J., Paper presented at the Int.Symp. on Cationic Polymerization, Rouen, 1973.
(5) Goethals E.J., Adv. Polym. Sci., (1977), 23, 121.
(6) Bucquoye M. and Goethals E.J., unpublished results (1976).
(7) Bossaer P.K., Goethals E., Hackett P., Pepper D.C., Europ.Polymer J., (1976), 12, in press.

(8) Goethals E.J. and Schacht E.H., J.Polym.Sci., Pol.Letters Edn. (1973), 11, 497.
(9) Schacht E.H. and Goethals E.J., Makromol.Chem., (1973), 167, 155.
(10) Razvodovskii E.F., Berlin A.A., Nekrazov A.V., Pushchaeva L.M., Puchkova N.G. and Enikolopyan N.S., Vysokomol.Soedin., Ser. A, (1973), 15, 2219 and 2233.
(11) Goethals E.J., Makromol. Chem., (1975), 175, 1309.
(12) Bestian H., Ann.Chem., (1950), 566, 210.
(13) Bottini A. and Roberts J.D., J.Am.Chem.Soc., (1952), 80, 5203.
(14) Anderson A.G. and Wills M.T., J.Org.Chem., (1968), 33, 2133.
(15) Meerwein H., Bottenburg E., Gold H., Pfeil E., Willfang G., J.Prakt.Chem., (1939), 154, 38; Org.Synth., (1973), Coll. Vol. V, 1080.

New Aspects of the Chemistry of Living Tetrahydrofuran Polymers Initiated by Trifluoromethane Sulfonic Anhydride

SAMUEL SMITH, WILLIAM J. SCHULTZ, and RICHARD A. NEWMARK
Central Research Laboratories, 3M Co., 3M Center, St. Paul, MN 55101

Smith and Hubin have described the polymerization of tetrahydrofuran (THF) using either $(CF_3SO_3)_2O$ or $(FSO_2)_2O$ as initiators (1). These anhydrides of the so-called "super" acids were found to yield living polymers of THF in which the end groups consisted of oxonium ions in equilibrium with covalently bonded esters. The nature of the reactions in the case of triflic anhydride initiation was postulated to be as follows (1).

Initiation:
$$\bigcirc\!\!\!\!O + (CF_3SO_2)_2O \longrightarrow \underset{I}{\bigcirc\!\!\!\!\overset{+}{O}-SO_2CF_3} \cdot CF_3SO_3^- \qquad (1)$$

Propagation:
$$I + n\,\bigcirc\!\!\!\!O \rightleftharpoons \underset{II}{CF_3SO_3\text{-}(C_4H_8O)_{n-1}C_4H_8\text{-}\overset{+}{\bigcirc\!\!\!\!O}} \cdot CF_3SO_3^- \qquad (2)$$

Equilibration Between Macroester and Macroion:
$$II \rightleftharpoons \underset{III}{\overset{+}{\bigcirc\!\!\!\!O}\text{-}(C_4H_8O)_{n-2}C_4H_8\text{-}\overset{+}{\bigcirc\!\!\!\!O}} \cdot 2CF_3SO_3^- \rightleftharpoons \underset{IV}{CF_3SO_3\text{-}(C_4H_8O)_{n+1}O_2SCF_3} \qquad (3)$$

Several papers have recently appeared in which ^1H, ^{19}F and ^{13}C nmr spectral analyses were used to investigate the nature of the macroester-macroion equilibrium which results when alkyl esters of the super acids are used as THF polymerization initiators (2-7). The exact determination of these equilibrium constants in various reaction solvents has been an especially noteworthy result (2b, 3). Very recently, the surprisingly dramatic effect of the overall concentration of poly-THF living end groups on the macroester-macroion equilibrium has been reported and attributed to ion aggregation effects which act to increase the ion/ester ratio (8).

Two important features distinguish THF polymerization initiated with super acid anhydrides from that initiated with the corresponding esters which have received so much study. Anhydride initiation is much more rapid than ester initiation and it leads to a polymer capable of growing at both ends, whereas ester initiation produces polymer growing at only one end.

These distinctions prompted us then to study in some detail the exact nature of the reaction of $(CF_3SO_2)_2O$ with THF.

EXPERIMENTAL

NMR spectra were obtained with a Varian XL-100 spectrometer. Chemical shifts were measured from tetramethylsilane (^1H) and $CFCl_3$ (^{19}F) reference internal standards and shifts are recorded here as positive when they are downfield from the reference. Samples used for ^{19}F nmr spectra were withdrawn by syringe through a septum cap closure from a reaction vessel which had been maintained at 0° during the mixing of reactants. All reactants had been distilled and great care was exercised to avoid moisture contamination. The ^{19}F nmr spectrum of the $(CF_3SO_2)_2O$ initiator used in this work indicated at least 95% purity, with CF_3SO_3H and $CF_3SO_3CF_3$ constituting the major impurities and being present in almost equal concentrations. The nmr spectra were determined at 25°, with the first spectrum obtained 3 minutes after initiation of the reaction. Mass spectra were obtained with a CEC 21-110C mass spectrometer and values are reported as molecular mass per unit charge, m/e.

Molecular weight distributions were obtained by gel permeation chromatography (GPC) (Waters Associates Chromatograph) using a set of six Styragel^R columns, each 122 x 0.63 cm, which were selected to achieve high resolution of low molecular weight fractions. The gels had rated pore sizes of 10^2 (3 columns), 10^3, 10^4, and 10^6 Å. The molecular weight distributions were determined in either chloroform or THF solutions at 25° using both standard differential refractive index and U.V. detectors. The latter was employed at a wave length of 2540 Å to detect the phenyl end groups of specially terminated polymeric intermediates. The phenyl groups were appended to reactive intermediates by terminating reactions with the addition of a 3-molar excess of sodium phenoxide in THF solution. Excess $NaOC_6H_5$ was ordinarily not removed since it did not interfere with either ^1H nmr or GPC spectra.

DISCUSSION OF RESULTS

^{19}F nmr. The addition of $(CF_3SO_2)_2O$ to a cyclohexane solution of THF immediately gave rise to the appearance of 3 distinct fluorine absorption peaks, as shown in Figure 1a. It is noteworthy that in every case studied the ^{19}F nmr indicated that the anhydride, which gives a sharp singlet peak at -73.2 ppm in cyclohexane solution, reacts essentially instantly on mixing with THF at 25°. (In one instance a known amount of reference trifluoromethyl benzene was added to the reactant solution and the ^{19}F nmr spectrum was integrated to prove that these three peaks accounted for all the fluorines derived from the anhydride). The ^{19}F nmr absorption peaks at -75.7, -75.8, and -78.6 ppm (upfield from $CFCl_3$) were assigned to tetramethylene bis(triflate) (V) (i.e. $CF_3SO_3\{CH_2\}_4O_3SCF_3$), macro-triflate ester IV and macro-triflate ion III, respectively, by the following considerations. (Strictly speaking, ^{19}F nmr observes only the macroester and macroion end groups and does not distinguish between either II and III or II and IV, since II contributes equally to the macroion and macroester peaks.)

The assignment of the peak at -78.6 ppm to macroion III was facilitated by studying the ^{19}F nmr spectra of solutions containing soluble triflate salts, e.g. $NH_4O_3SCF_3$ (chemical shift = -78.8 ppm). The assignments of the peaks at -75.7 and -75.8 ppm to V and macroester IV, respectively were made in a separate study in which an authentic sample of V, prepared using the procedure of reference ([1]), was added to a 10.2 molar solution of THF in cyclohexane and an initially strong peak at -75.7 ppm was observed. This peak slowly disappeared to give increasingly stronger peaks attributable to III and IV as the slow polymerization of THF initiated by V proceeded. (These ^{19}F nmr assignments for III and IV agree very well with the values previously reported for the corresponding living triflate end groups which had been characterized under very similar experimental conditions) ([6]).

As the polymerization reaction initiated by $(CF_3SO_2)_2O$ progressed at 25° the peak attributed to V slowly and steadily decreased, while the concentrations of both macroester and macroion increased, as illustrated in the kinetic data of Table I.

TABLE I. Relative Concentrations of Products by ^{19}F nmr Versus Polymerization Time at 25°. Reactant Concentrations: $(CF_3SO_2)_2O$ = 0.079 M; THF in Cyclohexane = 10.2 M.

Reaction Time (min.)	$CF_3SO_3(CH_2)_4O_3SCF_3$	Macroester	Macroion
3	36	58	7
10	33	59	8
30	16	70	11
74	10	77	13
280	3	79	18

It is interesting to note that the macroion/macroester ratio increased as V was depleted, a finding which seems to accord with the report that the equilibrium shown in Equation (3) shifts to produce more macroion as a consequence of the overall increase in the concentration of poly-THF living end groups ([8]).

An otherwise identical experiment to that shown in Table I was performed in which the very polar solvent, nitromethane, was substituted for the non-polar cyclohexane. In this case V was formed in almost the same proportion (32% after 4 minutes at 25°), but macroion concentration predominated over macroester, III = 47% and IV = 21% after 4 minutes. (This spectrum is shown in Figure 1b.) The increase in the dielectric constant of the medium would, of course, be expected to shift the equilibrium in the direction of macroion formation, a situation which had indeed been found previously ([3]).

Other solvents were investigated as diluents in the reaction of $(CF_3SO_2)_2O$ with THF. These included methylene chloride, carbon tetrachloride, nitrobenzene, toluene and o-dichlorobenzene. In each instance ^{19}F nmr showed that V formed to account for at least 30% of the total products formed early in the reaction. The effect of THF concentration on the formation of V was examined in one case. At THF concentrations of 10.2 and 5.6 molar in cyclohexane, V constituted 36% and 68%, respectively, of the total reaction products formed after 3 minutes at 25°.

GPC. The discovery that V was being formed and was acting as a very slow THF polymerization initiator, as indicated above and in reference (1), implied that further information concerning the progress of the polymerization could be gained by investigating the molecular weight distributions of products formed during the course of the reaction using reaction conditions identical to those described in Table 1. Toward that end, a flask reaction was run in which aliquot samples were terminated at various reaction stages by quenching with sodium phenoxide. This reaction is known to convert oxonium ion end groups to phenyl ethers (9) and we established in model reactions that it also converted triflate esters to phenyl ethers. The first sample quenched after 3 minutes of reaction (8.4% THF conversion) showed a bimodal molecular weight distribution by GPC with well resolved peaks at 21.5 Å and 75 Å end-to-end distance, as shown in Figure 2. A sample of the eluent at 21.5 Å was separated and dried and the mass spectrum of the product was run. One major peak was found at m/e 242, corresponding to the molecular ion of $C_6H_5O(CH_2)_4OC_6H_5$ (VI), and trace peaks were observed at m/e 314 and 386, corresponding to the THF-dimer and trimer diphenyl ether molecular ions, respectively. The U.V. trace of the GPC, which is indicative of the number-average molecular weight ($\overline{M}_n$), showed that the 21.5 Å peak constituted 35 mole % of the total product, in good agreement with the corresponding data shown for V in Table I. As the reaction progressed, GPC analyses showed that the distribution became increasingly unimodal. The polymer peak moved up-scale with the simultaneous appearance of a distinctive, increasingly low molecular weight tail, as V slowly disappeared by initiating new polymer chains growing at both ends. When equilibrium conversion of THF (69%) was reached after 90 minutes, the 21.5 Å peak was no longer discernible and the $\overline{M}_n$ and $\overline{M}_w$ values were 13,000 and 19,000, respectively (polydispersity = 1.5). (The Q factor for converting Å end-to-end distance to molecular weight for our calibrated column system was 29). (The distribution curve for the 90 minute reaction product is also shown in Figure 2). Prolongation of the reaction for an additional 3 hours caused the polydispersity to increase to 2.6, for reasons which have been investigated independently (10,11).

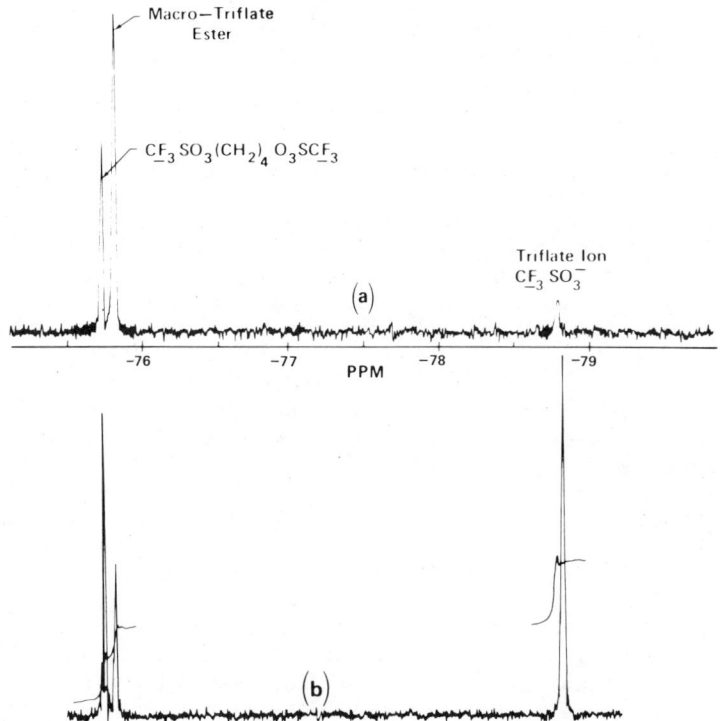

Figure 1. ^{19}F *NMR spectrum of products of reaction of 10.2M THF with 0.079M* $(CF_3SO_2)_2O$. *Conditions: 3 min at 25°. (a) in cyclohexane, (b) in nitromethane.*

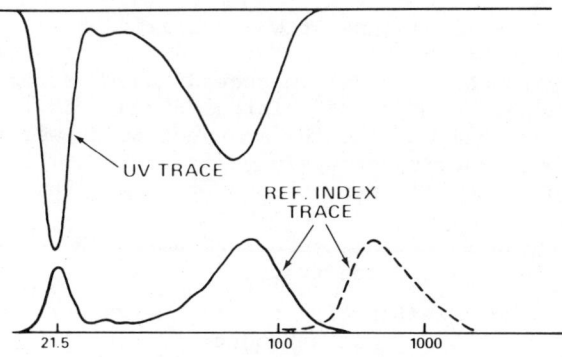

Figure 2. Mol wt distribution of phenyl ether-terminated products. [THF] *in cyclohexane* = 10.2M; $[(CF_3SO_2)_2O]$ = 0.079M. (———) *Product of 3-min reaction (8.4% THF conversion);* (- - -) *product of 90-min reaction (69% THF conversion).*

The first order dependence of THF polymerization rate on $(CF_3SO_2)_2O$ concentration has been reported (1). However, in the course of continued studies we have observed that as the relative concentration of anhydride is increased beyond about 2 mole percent of the THF concentration, then a precipitate occurs early in the reaction and lower than expected polymerization rates are observed. We investigated such a reaction at 25° in which THF concentration in cyclohexane was 9.4 M and anhydride concentration was 0.47 M. A white precipitate was observed to form at the beginning of the reaction. Aliquot samples of the stirred reaction mixture were withdrawn after 5, 15, and 120 minutes of reaction, quenched with sodium phenoxide and the resulting, now homogeneous solutions were examined by GPC. All molecular weight distributions were found to be trimodal as shown by the simultaneous U.V. and differential refractive index traces of the product formed in the 15-minute reaction (Figure 3). The mass spectrum of the dried total sample removed after 15 minutes of reaction was obtained and this is shown in Figure 4. Strong fragmentation peaks are seen at m/e values of 55, 77, 94, 107, 149 and 221, corresponding to the respective radical or molecular ions derived from the oligomeric diphenyl ethers: C_4H_7, C_6H_5, C_6H_5OH, $C_6H_5OCH_2$, $C_6H_5OC_4H_8$ and $C_6H_5(OC_4H_8)_2$. The high mass portion of the spectrum depended on the inlet temperature as expected for a mixture of oligomers. At relatively low temperature a significant peak at m/e 242 is detected for VI. At higher temperature a much stronger peak at m/e 386 is observed, corresponding to the trimer molecular ion $C_6H_5O(C_4H_8O)_3C_6H_5$ (VII). Only trace peaks were detected for m/e values corresponding to other oligomeric poly-THF diphenyl ethers. On this basis, assignments were made for the GPC peaks: 21.5 Å = VI; 35 Å = VII.

The trimer (VII) assignment was confirmed by an experiment in which the eluent of the GPC peak at 35 Å was collected and dried. The mass spectrum of this sample showed the expected intense molecular ion peak at m/e 386. Infrared analysis for hydroxy groups proved negative. The ^1H nmr spectrum showed the expected 1.5 tetramethylene oxide groups/phenoxy group. This spectrum and the various specific proton absorption assignments are shown in Figure 5.

The kinetic data of this reaction, based on the analysis of the UV traces of the GPC spectra, are summarized in Table II.

TABLE II. Product Distribution by GPC Analysis of the Polymerization Reaction at 25°. Reactant Concentrations: $(CF_3SO_2)_2O$ = 0.47 M; THF in Cyclohexane = 9.4 M.

Reaction Time (min.)	Relative Molar Concns. of Diphenyl Ethers			Polymer Mol. Wt.	
	VI	VII	Polymer	M_n	M_w
5	24	40	36	1,900	4,400
15	15	50	35	2,000	8,400
120	5	40	55	2,100	22,000

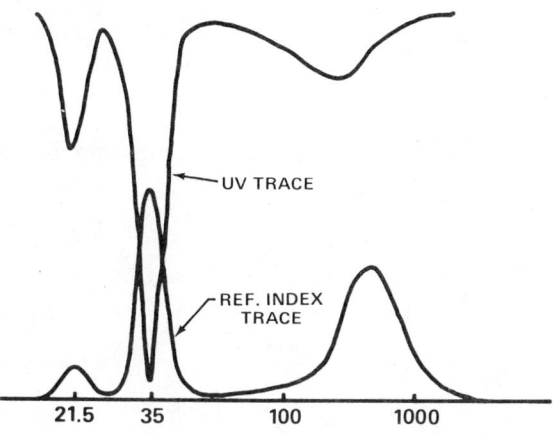

End-to-End Distance (Å)

Figure 3. Mol wt distribution of phenyl ether-terminated products. Reaction conditions: 15 min at 25°. THF in cyclohexane = 9.4M; $(CF_3SO_2)_2O - 0.47M$.

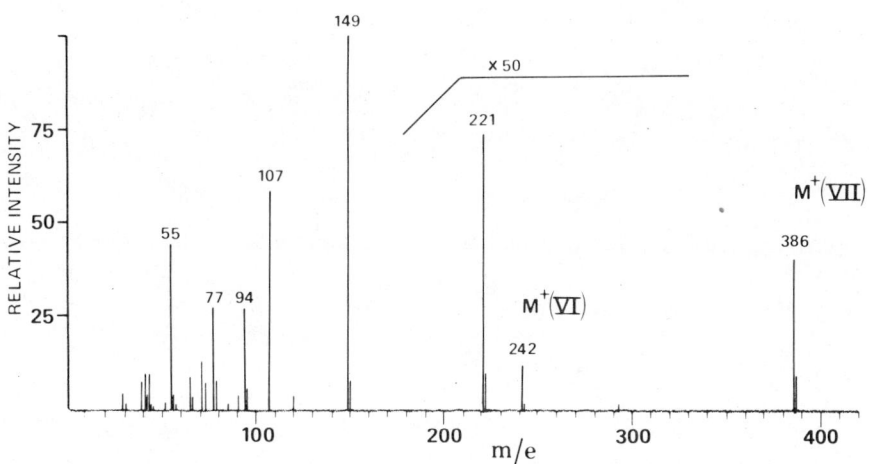

Figure 4. Mass spectrum of the total product, identical to that shown in Figure 3.

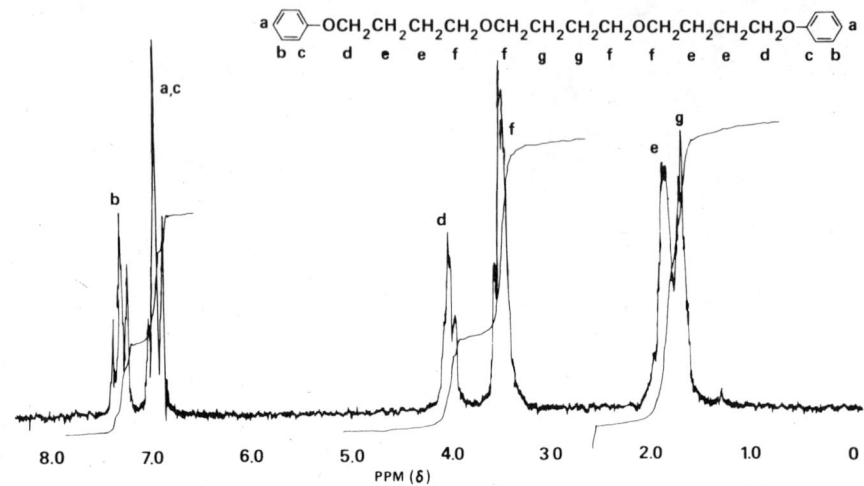

Figure 5. Proton NMR of the GPC eluent at 35 Å (see Figure 3).

It can be seen that V (the active intermediate which gives rise to VI) disappeared at a rate which is consistent with the previously discussed results. On the other hand, the reactive intermediate which gives rise to VII remained at a relatively constant concentration for the first 2 hours of reaction. These combined effects accounted for the unusually large disparity between $\bar{M}_n$ and $\bar{M}_w$ shown in Table II.

An authentic sample of VII was made in which 0.25 mole THF was added slowly at $-30°$ to a solution of 0.05 mole $(CF_3SO_2)_2O$ in 20 ml. CH_2Cl_2 while stirring, and then warmed to $25°$ after the addition was completed. The resulting dispersion was then diluted with 50 ml CH_2Cl_2, filtered and washed successively with CH_2Cl_2 and then THF. A white crystalline product, now known to be the THF-trimer bisoxonium salt,

$$[\text{structure}]\dot{O}(CH_2)_4\dot{O}[\text{structure}] \cdot 2CF_3SO_3^- \quad (VIII),$$

was obtained in quantitative yield, based on the starting anhydride. This salt was converted to the diphenyl ether VII by adding 5 g. to a solution of 10 g $NaOC_6H_5$ in a mixture of THF and ethanol, and then separated by precipitating it in water and purified by repeated water washing. Proton nmr and mass spectral analyses proved that the final product was VII in that they were identical to the spectra of the product separated by GPC, as discussed above.

The bisoxonium ion salt VIII decomposes on melting to give 1 mole of bisester V and 2 moles of THF (12). VIII is sparingly soluble in THF at $25°$ and very slowly disappears over several hours by initiating polymerization. It is very soluble in nitromethane, in which solvent it has been found to polymerize THF as rapidly as $(CF_3SO_2)_2O$ initiation.

Mechanisms of Formation of Monomer Bisester V and Trimer Bisoxonium Salt VIII. The initiation reaction shown in Equation (1) has been postulated to give rise to the oxonium ion salt I. (High electrical conductivity is manifest immediately following the addition of the anhydride to THF at $25°$). It is now believed that I disappears rapidly by following either of two reaction pathways having quite competitive rates. One involves nucleophilic addition of THF to I (Equation 2) and this rapidly leads to the formation of higher polymers. The alternative pathway is a cage reaction in which the $CF_3SO_3^-$ anion nucleophilically attacks the oxonium ion to open the ring to form V. This view is consistent with the facts that solvent polarity does not affect the relative yield of V (suggesting a cage reaction of the contact ion pair), and the relative yield of V increases as THF concentration is decreased, as discussed previously. V is a stable compound and ^{19}F nmr shows that it is not in equilibrium with I (1). V then disappears slowly as it initiates THF polymerization at a rate which appears to be similar to that of ethyl triflate (4).

The fact that V appears to form under all conditions of reacting $(CF_3SO_2)_2O$ with THF and then behaves as a slow polymerization initiator indicates that the previously reported narrow molecular weight distribution (polydispersity of 1.08) for a poly-THF prepared at -10° was probably in error (<u>1</u>). It is likely that the error might be attributed to the use of GPC columns in the earlier work which were not suited to resolve low molecular weight fractions.

As stated previously, VIII forms as a distinct species only when a relatively high concentration of anhydride is employed. It is important to note that the trimer is the lowest oligomer bis(triflate) which is capable of existing as a bisoxonium salt. Thus, in Equation (3) the equilibration between macrodication III and macrodiester IV can come into play only at the trimer stage. If VIII forms at a concentration greater than its very low saturation concentration, then it precipitates, thus driving the equilibrium essentially all the way toward bisoxonium ion triflate salt. As higher oligomer bis(triflates) slowly form, these are very soluble in THF and the normal equilibrium between macroester and macroion is re-established.

ABSTRACT

A detailed examination of the reaction of THF with $(CF_3SO_2)_2$ has been carried out and two prominently distinct oligomeric species have been found to be produced as intermediates during the polymerization which yields living products whose end groups consist of ions and esters in equilibrium. First, the ring-opened tetramethylene bis(triflate) ester is produced in all cases studied and its behavior as a relatively sluggish THF polymerization initiator causes the otherwise narrow molecular weight distribution to skew toward the low end. Second, at relatively high initial anhydride concentrations, the bisoxonium ion salt, $\left[\overset{+}{O}(CH_2)_4\overset{+}{O}\right] \cdot 2CF_3SO_3^-$, forms and separates as a pure crystalline precipitate. Reaction mechanisms are postulated to account for the surprising formation of these compounds during THF polymerization.

ACKNOWLEDGEMENT

We are indebted to Dr. Peter F. Cullen for providing the GPC analyses.

LITERATURE CITED

1. Smith, S., and Hubin, A.J., J. Macromol. Sci. - Chem., (1973), <u>A7</u>, 1399.
2. Kobayashi, S., Danda, H., and Saegusa, T., (a) Bull. Chem. Soc. of Japan (1973), <u>46</u>, 3214, (b) Macromol., (1974), <u>7</u>, 415.
3. Matyjaszewski, K., and Penczek, S., J. Polym. Sci. - Chem. Ed., (1974) <u>12</u>, 1905.

4. Matyjaszewski, K., Kubisa, P., and Penczek, S., J. Polym. Sci. Chem. Ed., (1975), 13, 763.
5. Matyjaszewski, K., Buyle, A.M., and Penczek, S., J. Polym. Sci. - Letters Ed., (1976), 14, 125.
6. Wu, T.K., and Pruckmayr, G., Macromol., (1975), 8, 77.
7. Pruckmayr, G., and Wu, T.K., Macromol., (1975) 8, 954.
8. Matyjaszewski, K. and Penczek, S., J. Polym. Sci. - Chem. Ed. (1977), 15, 247.
9. Saegusa, T., and Matsumoto, S., J. Polym Sci., (1968), A6, 1559.
10. Rosenberg, B.A., Ludvig, E.B., Gantmakher, A.R., and Medvedev, S.S., J. Polym. Sci., (1967), C16, 1917.
11. Croucher, T.G., and Wetton, R.E., Polymer, 17, (1976), 205.
12. Cash, D.J., personal communication.

3

New Developments in Graft Copolymerization by Oxonium Ion Mechanism

K. I. LEE and P. DREYFUSS
Institute of Polymer Science, The University of Akron, Akron, OH 44325

One important objective of our recent studies has been to find general ways to synthesize graft copolymers comprising hydrocarbon backbones with polar branches derived from cationically polymerizable heterocyclic monomers. Hitherto such polymers have not been available generally and are of interest because of the unique combination of properties that are potentially attainable.

We recently reported a new efficient method for the preparation of this type of graft copolymer (1). Our new method consists of initiating polymerization of the heterocycle from a hydrocarbon backbone containing a reactive halogen by adding a suitable salt as shown in equation 1, where X = halogen, Z = O, S,
$\overset{O}{\overset{\|}{O-C}}$, and MY is a salt of a metal (M) with a counterion (Y) capable of supporting onium ion polymerizations.

$$\text{Polymer-X} + \text{MY} + n Z \longrightarrow \text{Polymer-}[Z\sim]_n + \text{MX} \quad (1)$$

Not every halide, salt, and heterocycle can be used in our synthesis. We recently overviewed the scope of our discovery using studies with model halides, various heterocycles, a variety of salts, and numerous backbones to help define the limitations (2). Our model halide studies showed that reactive halides include allylic, tertiary, and benzylic chlorides, bromides, and iodides. Soluble silver salts with anions such as $SO_3CF_3^-$, BF_4^-, PF_6^-, AsF_6^-, SbF_6^-, and ClO_4^- are most suitable but $LiPF_6$ and $NaClO_4$ can also be used. We concluded that a suitable salt is one that is soluble in the polymerization medium, has an anion that leads to a stable onium salt, and leads to a product that is less soluble than the original salt. So far nine heterocycles have been polymerized by this method. These include tetrahydrofuran, 7-oxabicyclo-[2,2,1]heptane, oxetane, propylene oxide, styrene oxide, dioxolane, trioxane, ε-caprolactone, and thietane. Efforts to optimize conditions for selected monomers are in their early stages, and the preliminary data suggest that many of the expected side reactions are operative (3). Graft copolymers have been prepared from seven backbones: poly(vinyl chloride), polychloroprene, chlorinated EPDM, chlorobutyl rubber, bromobutyl rubber, chlorinated poly(butadiene), and chlorinated butadiene-styrene copolymer. In the most thoroughly examined cases with polytetrahydrofuran as the grafted copolymer and silver triflate as the inorganic salt, current data indicate that no unreacted backbone remains and no homopolymer forms. With some of the other monomers, studied only with $AgPF_6$ as the salt, homopolymer is formed in addition to graft. No evidence for cyclic oligomer formation was obtained with any of the monomers.

Efficiency of Initiation

Our goal is to prepare well-defined graft copolymers. The method described in this paper has the potential of leading to grafts with a controlled number of branches of known length. "Living" polymerizations of many of the heterocycles being studied are known and their rates of polymerization have been carefully determined (3,4). Thus preparation of graft copolymers with a predictable number and length of branches can be achieved if the efficiency of the initiation process is known and reproducible. Ideally the initiation should be instantaneous and 100% efficient. We therefore selected the study of the initiation process as our first in depth examination of our new process. In this paper we report data from the following experiments:

    1. Nmr studies of the products of reactions of model halides and silver salts in the presence of

the nonpolymerizable heterocycles, 2-methyltetrahydrofuran and tetrahydropyran.
2. Studies of the rate of formation of silver halide with model halides, different silver salts, and nonpolymerizable heterocycles.
3. Comparison of the rates of silver halide formation with rates of tetrahydrofuran polymerizations using model halides and different silver salts.
4. Application of the silver halide precipitation method to halogenated butyl rubbers in cyclic ether and comparison with results of graft copolymerization studies from the same backbones.

Possible Reaction Paths

One can imagine several different pathways that reaction of allyl halide, silver salt, and a heterocyclic ether might take. These are illustrated in the series of equations 2 where the reactants are allyl chloride, 2-methyltetrahydrofuran and silver hexafluorophosphate. The desired pathway is the addition of the allyl group to the heterocycle to form an oxonium ion (2-1). However, we know from previous work that when carbenium ions are possible intermediates, hydrogen abstraction (2-2) and elimination reactions (2-3) have to be considered (5,6). Reaction of the carbenium ion with the counterion to form allyl fluoride is a possibility (2-4). Reactions (2-2) to (2-3) are undesirable because the $H^{\oplus}PF_6^{\ominus}$ formed would initiate homopolymerization of the cyclic ether and pure graft copolymer would not be formed. Finally, as will be seen below, our data suggests that some type of coupling reaction may also be occurring occasionally and we include reaction (2-5) as a possibility. In analyzing our results, we looked for evidence of each of these pathways because we needed an explanation for some unexpected results.

Nmr Studies

We began our studies by examining the silver salt assisted reaction of 2-methyltetrahydrofuran with allyl chloride, allyl bromide, and allyl iodide.

Equation 2 - Possible Pathways

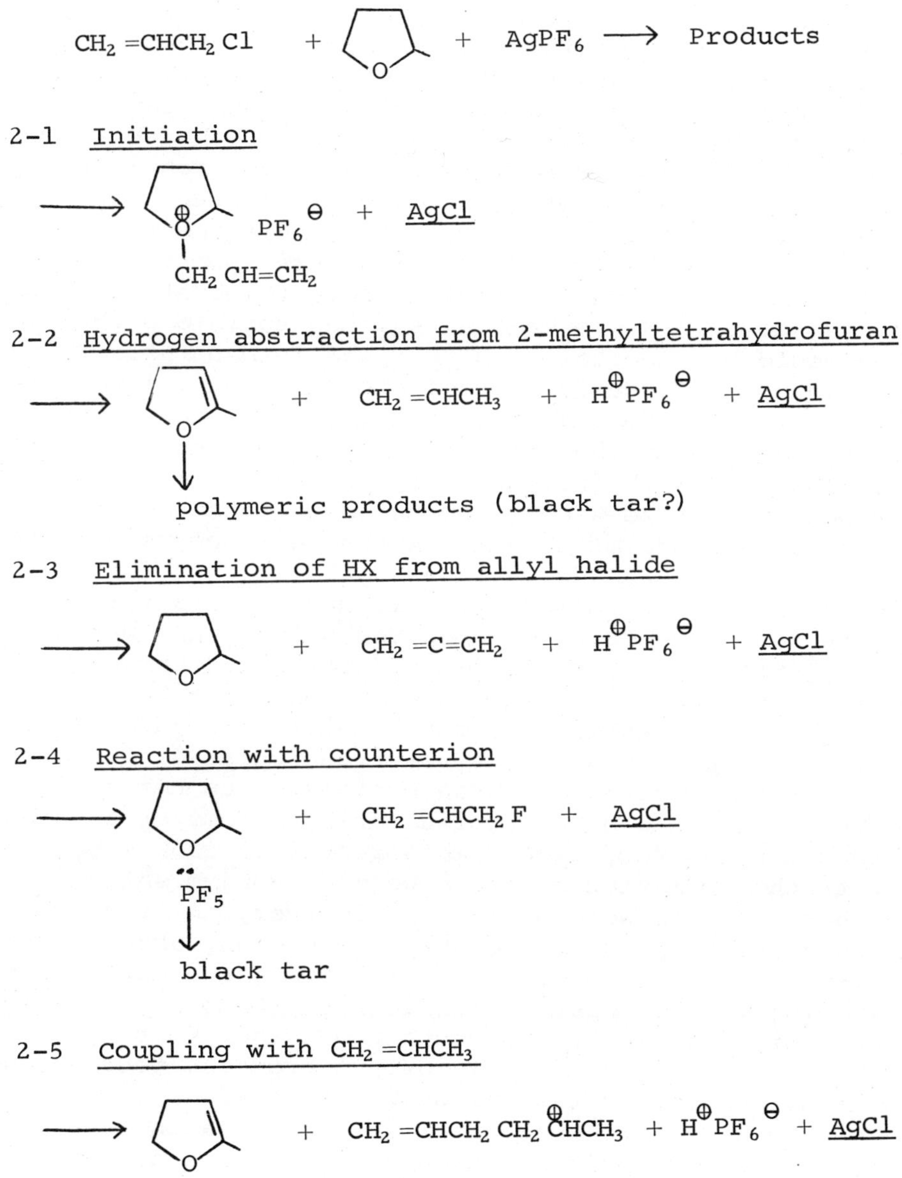

$$CH_2=CHCH_2Cl \quad + \quad \underset{O}{\bigcirc} \quad + \quad AgPF_6 \quad \longrightarrow \quad Products$$

2-1 Initiation

$$\longrightarrow \quad \underset{\underset{CH_2CH=CH_2}{|}}{\overset{\oplus}{O}}\!\!\bigcirc \quad PF_6^{\ominus} \quad + \quad \underline{AgCl}$$

2-2 Hydrogen abstraction from 2-methyltetrahydrofuran

$$\longrightarrow \quad \underset{O}{\bigcirc}\!\!= \quad + \quad CH_2=CHCH_3 \quad + \quad H^{\oplus}PF_6^{\ominus} \quad + \quad \underline{AgCl}$$

$$\downarrow$$

polymeric products (black tar?)

2-3 Elimination of HX from allyl halide

$$\longrightarrow \quad \underset{O}{\bigcirc}\!\text{-} \quad + \quad CH_2=C=CH_2 \quad + \quad H^{\oplus}PF_6^{\ominus} \quad + \quad \underline{AgCl}$$

2-4 Reaction with counterion

$$\longrightarrow \quad \underset{\overset{..}{O}}{\bigcirc}\!\text{-} \quad + \quad CH_2=CHCH_2F \quad + \quad \underline{AgCl}$$

$$PF_5$$
$$\downarrow$$
black tar

2-5 Coupling with $CH_2=CHCH_3$

$$\longrightarrow \quad \underset{O}{\bigcirc}\!\!= \quad + \quad CH_2=CHCH_2CH_2\overset{\oplus}{C}HCH_3 \quad + \quad H^{\oplus}PF_6^{\ominus} \quad + \quad \underline{AgCl}$$

Initially we had hoped to isolate both products of reaction 2-1. We found that although the silver halide precipitate could readily be isolated, attempts to isolate the oxonium ion salt gave only black tar and a gaseous product. This is consistent with the elusive character of allyl carbenium ions previously reported by Olah and Camisarov (7). The black tar could have been formed either from the furan in (2-2) or from the adduct of 2-methyltetrahydrofuran and PF_5, which is known to be unstable (8). Nmr spectra taken at various times on a T-60 Varian nmr Spectrometer after reactions carried out at room temperature showed a far downfield proton that shifted upfield with time as would be expected if $H^{\oplus}PF_6^{\ominus}$ had formed (5). We found no evidence for either propylene or allene, although we did demonstrate that under the conditions of our experiments the solubility of propylene would be so low in 2-methyltetrahydrofuran that it might well disappear before a spectrum could be taken. There was no evidence of a shift in the peaks corresponding to 2-methyltetrahydrofuran as would be expected if the oxonium ion had formed. The formation of allyl fluoride cannot be ruled out because allyl peaks persisted along with those of the 2-methyltetrahydrofuran even after all the silver chloride had precipitated.

If the reaction was carried out in liquid SO_2 at $-78°C$, it was possible to observe a shift of the allyl and 2-methyltetrahydrofuran protons in an nmr spectrum taken on a HR300 Varian nmr Spectrometer but the reaction was very slow. This result is consistent with the formation of the expected oxonium salt. No other products were apparent. However, the spectrum was quite complicated because of the many absorptions due to products and starting materials and we did not attempt to interpret it completely. Lambert and Johnson made similar comments about their nmr spectrum from a reaction of isopropyl bromide, tetrahydrofuran, and silver tetrafluoroborate (9).

We were more successful in demonstrating the formation of the addition product of initiation in an experiment using unsubstituted tetrahydrofuran and carrying out the polymerization with high allyl bromide-$AgSbF_6$ concentration and low enough conversion so that an oily product was formed. The nmr spectrum of a carefully purified sample of this polymer clearly showed the presence of allyl groups and the chemical shifts were consistent with the presence of an allyl ether linkage. Our previous graft copolymerization studies also indicate addition of the allyl group, since no homopolymer was formed during grafting. Apparently the allyl oxonium ion is unstable and decomposes rapidly. Side reactions occur if the heterocycle is not polymerizable but polymerization occurs when possible.

Overall, we concluded that in our case nmr studies are not a good method for studying the kinetics of the initiation process at room temperature. They were useful for gaining some insight into the products that might be formed.

Isolation of Silver Halide from Model Studies

Preliminary analysis of tetrahydrofuran polymerizations terminated by sodium phenoxide (10) indicates that the rate of formation of oxonium ion parallels the rate of precipitation of silver halide (11). Thus information about the rate and efficiency of formation of oxonium ions from silver salts, allyl halides and a cyclic ether was obtained by measuring the rate of formation of silver halide. If the ether was 2-methyl-tetrahydrofuran and the concentrations of allyl halides and $AgPF_6$ were 2×10^{-4} moles in 2 ml of cyclic ether (6.7×10^{-2} M), the formation of AgI and AgBr were much faster than that of AgCl. After 1 hr., the shortest time of observation, the percents of halide isolated were 3, 76 and 88 for AgCl, AgBr, and AgI, respectively (Table I, Figure 1). After 6 hrs, precipitations of AgCl, AgBr, and AgI were 10, 83, and 91% complete, respectively. At sufficiently long times 100% halide, within experimental error, was obtained in some cases.

To see if there is any influence of counterion other than $PF_6^\ominus$, the rate of formation of silver halide precipitation was measured using $AgSbF_6$ instead of $AgPF_6$. Figure 2 indicates that there is no essential difference between $PF_6^\ominus$ and $SbF_6^\ominus$.

The faster initiation by the bromide compared to the chloride in the polymerization of cyclic ethers was also demonstrated from the model reactions of allyl

Table I

Comparison of Rates of AgX Precipitation[a]

Salt	Halide[b]	Monomer	% AgX after 1 hr[c]
$AgPF_6$	Cl	2-MeTHF[d]	3
	Br	2-MeTHF	76
	I	2-MeTHF	89
$AgSbF_6$	Cl	2-MeTHF	2
	Br	2-MeTHF	78
	Cl	THP[e]	3
	Br	THP	74

[a] Comparisons are made at 1 hr because even though the rate of precipitation of AgCl had not yet reached a steady state, by this time precipitation of AgI was essentially complete. No measurements were made at less than 1 hr. R_p data was not calculated because under these circumstances the numbers could not be compared.

[b] All unsubstituted allyl halides. Halide concentration was approximately 10^{-4} moles in 2 ml cyclic ether in each case.

[c] % AgX = the actual amount of AgX isolated/ the expected amount of AgX based on the total amount of allyl halide charged.

[d] 2-Methyltetrahydrofuran.

[e] Tetrahydropyran.

halides and silver salts in the presence of tetrahydropyran (Figure 3). After 1 hr, about 74% of the expected amount of AgBr had precipitated whereas only 3% AgCl was isolated. After 6 hrs precipitations of AgBr and AgCl were 89% and 14% complete, respectively. These results indicate that the faster formation of oxonium ion from the bromide compared to the chloride is independent of the cyclic ether.

These results are in conformity with our expectation from the organic chemistry of small molecules. The easier departure of bromide compared to chloride by electrophilic assistance of the $Ag^{\oplus}$ would lead to more rapid formation of oxonium ion and hence more rapid conversion to polytetrahydrofuran.

Relative Rates of Tetrahydrofuran Polymerizations

Further evidence for the more rapid formation of active centers from the bromide compared to the chloride, for example, was obtained by examining the % conversion of tetrahydrofuran to polymer (Figure 4). After 1 hr the polymerization with allyl chloride had reached 2% conversion, while 65% conversion was obtained with allyl bromide in the same time interval. Concentrations of allyl halide and silver salt were comparable and all reactions were carried out at room temperature, where 75% is the thermodynamically expected maximum conversion. Also a sigmoidal conversion-time plot indicative of slow initiation was obtained with allyl chloride whereas a similar plot from allyl bromide was linear in the early stages. Again different counterions, $SbF_6^{\ominus}$, $PF_6^{\ominus}$, or $SO_3CF_3^{\ominus}$, made no great difference in the conclusions about relative rates of initiation with chloride, bromide, or iodide. Osmotic molecular weights of the resulting polytetrahydrofurans at equilibrium conversion were near those calculated from the amounts of halides and silver salts charged.

Isolation of Silver Halide after Reaction of Halogenated Polymers

These experiments were carried out using commercial chlorobutyl and bromobutyl rubbers dissolved in

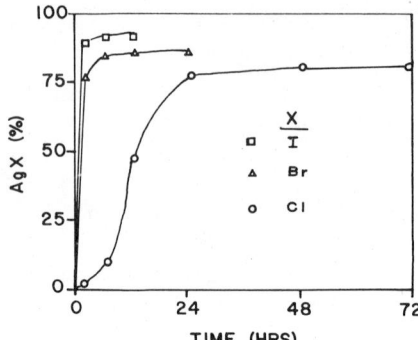

Figure 1. Effect of halide on formation rate of silver halide in reactions of 2-methyltetrahydrofuran with allyl halides and $AgPF_6$

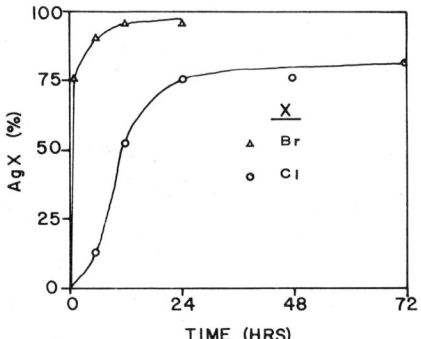

Figure 2. Effect of halide on formation rate of silver halide in reactions of 2-methyltetrahydrofuran with allyl halides and $AgSbF_6$

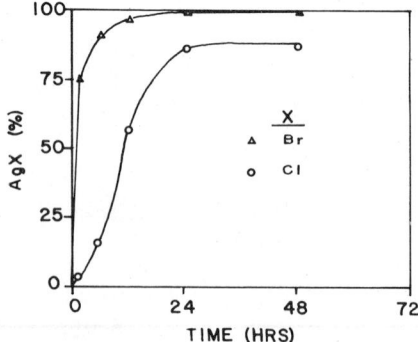

Figure 3. Effect of halide on formation rate of silver halide in reactions of tetrahydropyran with allyl halides and $AgSbF_6$

2-methyltetrahydrofuran and treated with $AgPF_6$. Samples containing approximately the same total percentage of halogen were used. The most probable structure and the percentage of allylic halogen in chlorobutyl rubber, whose structure is better known than that of bromobutyl rubber are shown below (12):

$$\pm CH_2 C(CH_3)_2 \pm \pm CH_2 C=CHCH_2 \pm \quad \text{or} \quad \pm CH_2 C(CH_3)_2 \pm \pm CH_2 CCHClCH_2 \pm$$
$$\qquad\qquad\qquad\qquad | \qquad\qquad\qquad\qquad\qquad\qquad\qquad\qquad ||$$
$$\qquad\qquad\qquad\qquad CH_2 Cl \qquad\qquad\qquad\qquad\qquad\qquad\qquad\qquad CH_2$$

$$98\% \qquad\qquad\qquad 2\% \qquad\qquad\qquad 98\% \qquad\qquad\qquad 2\%$$

As shown in Table II, again the rate of precipitation of AgBr was faster than that of AgCl. The amount of silver halide precipitated was similar in both cases and was considerably less than the total percentage of halogen in the polymer. This suggests that only about 16% of the halogen is active toward precipitation by our method and that either these halogens have a different, more reactive structure than that given above or that the percentage of halogens of the above structure is lower than previously supposed.

Table II

% Halide Precipitated from Halogenated Butyl Rubbers

Rubber	Time (hrs)	% Halide[a]
Chlorobutyl rubber	1	1.9
	6	7.6
	12	15.2
	24	15.2
Bromobutyl rubber	1	10
	6	16.7
	12	16.7
	24	16.7

[a] % halide = the actual amount of AgX isolated/the expected amount of AgX based on the total amount of halide in the rubber.

Unless the reaction with bromobutyl was carried out at high dilution (0.75%) compared to chlorobutyl rubber (2.35%), it was not possible to isolate the AgBr because the polymer gelled and trapped the salt. We suggest that a side reaction such as that illustrated in 2-5 occurs at higher concentrations. Alternately, since a reaction such as 2-3 could produce a polymer containing Diels-Alder active conjugated dienes (rather than the allene shown), the treated polymer might be self-curing by dimerization of these units. Data on this point was obtained by taking uv measurements of the original polymers dissolved in isooctane. The chlorobutyl rubber contained only an absorption at about 210 mµ, which could be associated with the olefin in the structure above, but the bromobutyl rubber also had additional absorptions between 220 and 240 mµ that would indicate a multiplicity of conjugated dienes even in the original polymer.

Grafting from Halogenated Butyl Rubbers

Further information about the relative reactivities of chlorobutyl and bromobutyl rubbers was obtained from grafting studies with polytetrahydrofuran as the branch. The results are summarized in Table III.

Table III

Polytetrahydrofuran Grafts from Chlorobutyl and Bromobutyl Rubber Backbones

Halide	Salt	PTHF Conversion after 24 hrs - %
Cl	$AgBF_4$	3
Cl	$AgPF_6$	13
Br	$AgBF_4$	0
Br	$AgPF_6$	5

It is noteworthy that after 24 hours the conversions to polytetrahydrofuran were all very low and that those from the bromobutyl rubber were lower than those from the chlorobutyl rubber. The differences between the results with $AgBF_4$ and $AgPF_6$ can be explained, since termination reactions are known to occur more readily with $BF_4^\ominus$ than with $PF_6^\ominus$ (4). The differences between the chloride and the bromide can be rationalized by assuming that reactions (2-2) and (2-3)

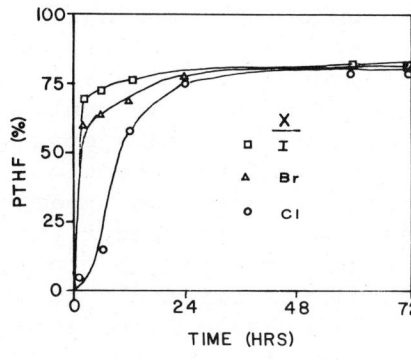

Figure 4. Effect of halide on conversion (%) to polytetrahydrofuran (PTHF) in reactions of tetrahydrofuran with allyl halides and $AgPF_6$

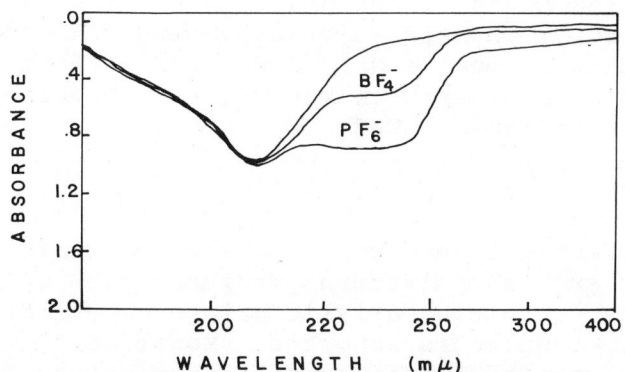

Figure 5. Uv absorption spectrum of chlorobutyl rubber in isooctane before and after preparation of the polytetrahydrofuran grafts described in Table III

or the Diels Alder dimerization are more important with the bromide than with the chloride, but that they can occur in both cases. Evidence that they do occur was obtained from uv spectroscopy. Figure 5 shows the uv spectra for treated and untreated chlorobutyl rubbers. A shoulder from 220 to 240 mμ appears in the polymer isolated after the grafting reaction. This shoulder is consistent with but not proof that dienes are being formed. Because the bromobutyl rubber uv spectrum already contained a similar absorption before the attempted grafting, it was difficult to obtain convincing evidence for an increased diene content after grafting. This was especially difficult since after the grafting reaction the polymer had a markedly increased tendency to gel and samples representative of the whole were not possible to obtain. The increased tendency to gel is consistent with increased diene content.

On the basis of our grafting studies with halogenated butyl rubbers we must conclude that although the bromide is more easily displaced to form AgX, as expected from organic chemistry, the chloride is more suitable for making grafts.

Summary

New insight into the chemistry of grafting polar branches such as polytetrahydrofuran from hydrocarbon backbones containing allylic halogens by adding suitable silver salts is reported. Model studies show that the rate of formation of silver halide can be used as an indication of the rate of formation of oxonium ions and that the rates of precipitation of AgX are in the expected order AgI>AgBr>AgCl. The results are independent of counterion and of heterocycle. Side reactions such as hydrogen abstraction, elimination of HX, and reaction with counterion are probable. It is concluded that although it is possible to prepare pure grafts by this method, precise control of the number and length of the branches may be difficult.

Literature Cited

1. Dreyfuss, P., and Kennedy, J. P., J. Polym. Sci., Polym. Letters Ed. (1976) **14**, 135.
2. Dreyfuss, P., and Kennedy, J. P., paper presented at 4th International Symposium on Cationic Polymerization, Akron, Ohio, June (1976). J. Polym. Sci., Polymer Symposia (in press).
3. Frisch, K. C., and Reegen, S. L., Eds., "Kinetics and Mechanisms of Polymerization Reactions, Vol 2, Ring-Opening Polymerization", Marcel Dekker, New York, N.Y., 1969.
4. Dreyfuss, P., and Dreyfuss, M. P., Chapter 4 in "Comprehensive Chemical Kinetics", Vol 15, p 259, Bamford, C. H., and Tipper, C. F. H., Eds., Elsevier Scientific Publishing Co., Amsterdam - W., Netherlands, 1976.
5. Dreyfuss, M. P., Westfahl, J. C., and Dreyfuss, P., Macromolecules, (1968) **1**, 437.
6. Pocker, Y. and Wong, W. H., J. Am. Chem. Soc. (1975) **97**, 7097.
7. Olah, G., Comisarov, M. B., J. Am. Chem. Soc. (1964) **86**, 5682.
8. Muetterties, L., Butler, T. A., Farlow, M. W., and Coffman, D. D., J. Inorg. Nucl. Chem. (1960) **16**, 52.
9. Lambert, J. B., and Johnson, D. H., J. Am. Chem. Soc. (1968) **90**, 1349.
10. Saegusa, T., and Matsumoto, S., J. Polym. Sci. (1968) **A1**, **6**, 1559.
11. Quirk, R. and Dreyfuss, P., unpublished results.
12. Baldwin, F. P., Gardner, I. J., Malatesta, A., and Rae, J. A., Paper No. 1 presented at 108th meeting Rubber Division of ACS, New Orleans, LA., Oct. 7, 1975.

Acknowledgment

Acknowledgment is made to the donors of the Petroleum Research Fund, administered by the American Chemical Society, for support of this research.

4

Ring-Opening Polymerization with Expansion in Volume

WILLIAM J. BAILEY, ROBERT L. SUN, HIROKAZU KATSUKI,
TAKESHI ENDO, HIDEAKI IWAMA, RIKIO TSUSHIMA,
KAZUHIDE SAIGOU, and MICHEL M. BITRITTO
Department of Chemistry, University of Maryland, College Park, MD 20742

For a number of industrial applications, such as strain-free composites, potting resins, high gloss coatings, binders for solid propellants, and impression materials, it appeared highly desirable to have monomers that will polymerize with near zero shrinkage. For other applications, such as precision castings, high strength adhesives, prestressed plastics, rock-cracking materials, elastomeric sealants, and dental fillings, it appeared highly desirable to have monomers that would undergo positive expansion on polymerization. For example, many composites involving high strength fibers in a polymeric matrix fail because of either poor adhesion between the matrix and the fibers or because of voids and microcracks in the matrix. Both of these problems are at least partially related to the fact that when available materials polymerize or cure, a pronounced shrinkage takes place. Examples are available from other fields to suggest that monomers that expand on polymerization would indeed produce strong adhesives. For example, when water freezes, it expands by 4%, and as a result ice will adhere to almost any surface, including Teflon which it does not even wet, by expanding into the various valleys and crevices of the irregular surface to promote strong micromechanical adhesion. For these reasons a research program was initiated to find monomers that would undergo either zero shrinkage or expansion upon polymerization.

Shrinkage that occurs during polymerization arises from a number of factors. One of the most important, however, is the fact that the monomer molecules are located at a van der Waals' distance from one another, while in the corresponding polymer the monomeric units move to within a covalent distance of one another. Thus, the atoms are much closer to one another in the polymer than they were in the original monomer. Smaller, but yet significant factors, are the change in entropy in going from monomer to the polymer, free volume in amorphous polymers, and how well the monomer and polymer pack if crystals are present in either phase.

In a condensation polymerization, in which a small molecule is eliminated, the shrinkage is partially related to the size of the molecule that is eliminated. For example, the shrinkage that takes place during the synthesis of nylon 6,6 can vary from 22% when hexamethylenediamine is condensed with adipic acid with the elimination of water to 66% when hexamethylenediamine is condensed with dioctyl adipate with the elimination of octyl alcohol. During addition polymerization, where no molecule is eliminated, the shrinkage, as indicated in Table I, can vary from 66% for the polymerization of ethylene to 6% for the polymerization of vinylpyrene. The shrinkage appears to correlate to a first approximation to the number of monomer molecules that are converted to polymer per unit volume. For example, styrene, which has approximately four times the molecular weight of ethylene, undergoes approximately only one-fourth the shrinkage that occurs during the polymerization of ethylene.

TABLE I. Calculated Shrinkages for Addition Polymerization

Monomer	Shrinkage, %
Ethylene	66.0
Propylene	39.0
Butadiene	36.0
Vinyl chloride	34.4
Acrylnitrile	31.0
Methyl methacrylate	21.2
Vinyl acetate	20.9
Styrene	14.5
Diallyl phthalate	11.8
N-Vinylcarbazole	7.5
1-Vinylpyrene	6.0

Ring-opening polymerization usually involves less shrinkage than simple addition polymerization. For example, Table II gives the calculated shrinkages for a selected number of ring-opening polymerizations. Ethylene oxide, which has a shrinkage of 23%, on the basis of its relative molecular weight with ethylene might have been expected to undergo a 40% shrinkage. One can rationalize the reduced shrinkage by noting that two processes are taking place during the polymerization of this monomer. First, the monomer units are moving from a van der Waals' distance to a covalent distance during polymerization, which should have resulted in a shrinkage of 40%, but at the same time the ring is opened and the oxygen atom moves from a covalent distance, with respect to the carbon atom, to a near van der Waals'distance with the recovery of about 17% of the shrinkage that occurred in the previous process. It is obvious from Table II, that the bigger the ring the closer to a true van der Waals' distance is approached during ring-opening and the smaller the shrinkage. One would predict that if the ring were large enough no shrinkage would be involved in the polymerization but the driving force for the polymerization would be quite small.

TABLE II. Calculated Shrinkages for Ring-Opening Polymerization

Monomer	Shrinkage, %
Ethylene oxide	23
Isobutylene oxide	20
Cyclobutene	18
Propylene oxide	17
Cyclopentene	15
Cyclopentane	12
Tetrahydrofuran	10
Cyclohexane	9
Styrene oxide	9
Cycloheptane	5
Cyclooctene	5
Bisphenol-A diglycidyl ether and diethylaminopropylamine	5
Cyclooctadiene	3
Cyclododecatriene	3
5-Oxa-1,2-dithiacycloheptane	3
Dimethylsilane oxide cyclic tetramer	2
Cyclooctane	2

It was reasoned from this study that if monomers were available in which at least two rings were opened for every new bond that was formed in the backbone, materials with either no change in volume during polymerization or slight expansion would be possible. It should be emphasized that this concept would eliminate from consideration the polymerization of a monomer, such as a diepoxide or a dianhydride, because, although two rings are opened during the polymerization, two new bonds are also formed at the same time.

It was shown that a variety of monomers would undergo such a polymerization. One of the first classes of compounds studied was the spiro ortho esters, of which the simplest example was 1,4,-6-trioxaspiro[4.4]nonane (I), which can be prepared from the condensation of butyrolactone with ethylene oxide in the presence of boron trifluoride in a 33% yield (1-3).

$$\begin{array}{c}CH_2-CH_2\\ \diagdown O \diagup\end{array} + \begin{array}{c}CH_2-CH_2\\ | \quad\quad\quad\diagdown\\ CH_2-CH_2 \diagup C=O\end{array} \xrightarrow[\text{13\% Shrinkage}]{BF_3} \begin{array}{c}CH_2-CH_2\\ | \quad\quad\quad\diagdown\\ CH_2-O\end{array} C \begin{array}{c}O-CH_2\\ \diagup \quad\quad |\\ O-CH_2\end{array}$$

d_4^{20} 0.869 d_4^{20} 1.11 d_4^{20} 1.16
                                                I

From a comparison of the densities of the materials involved, it is obvious that the spiro compound is a very compact monomer. When this monomer was polymerized in the dilatometer, placed in a constant temperature bath at 25°, in the presence of boron trifluoride, polymerization occurred over a 24-hr. period. During the polymerization the meniscus remained essentially at a constant level and indicated a slight increase in volume of 0.1%. Purification of the polymer by reprecipitation gave a 94% yield of a viscous liquid with a molecular weight of about 25,000. Although the polymer was difficult to purify its density indicated that the purified polymer is slightly more dense than the monomer (less than 0.1%).

The mechanism of the polymerization undoubtedly involves an oxonium ion and a stabilized carbonium ion:

Since the monomer contains two different types of oxygen atoms, attack can also occur at the other oxygen:

On this basis one would predict the formation of a polyester-ether containing both head-to-tail and head-to-head units. This fact was verified by the synthesis of the head-to-tail polymer by a direct transesterification procedure:

$$HO-CH_2-CH_2-O-CH_2-CH_2-CH_2-\overset{O}{\underset{\parallel}{C}}-OEt \longrightarrow \left[O-CH_2-CH_2-O-CH_2-CH_2-CH_2-\overset{O}{\underset{\parallel}{C}} \right]_x$$

Comparison of the NMR spectra of the two polymers gave strong evidence for the presence of head-to-head units (10 to 20%) in the ring-opened polymer.

The reason for this low shrinkage during polymerization can be rationalized by comparing the original monomer with the final ether-containing polyester. There are two processes which would lead to some contraction; one bond goes from a van der Waals' distance to a covalent distance, and one bond goes from a single bond to a double bond. For small atoms the covalent distance is only one-third the van der Waals' distance and this transformation, therefore, is much larger than the change from a single bond to a double bond. This shrinkage is counterbalanced by the two bonds that go from a covalent distance to a near van der Waals' distance in the final polymer. In this particular case, these processes seem to just about cancel one another.

Since this ether-containing polyester is a liquid, it is of little interest in its own right. However, if one desires a higher melting polymer from a spiro ortho ester, the introduction of cyclohexane rings gives a fairly large increase in the softening point. The introduction of one cyclohexane ring produces a monomer that gives a slight increase in volume upon polymerization (1%). The introduction of two cyclohexane rings gives a material with a softening point of over 100°C, but there is essentially no change in volume.

It was reasoned, therefore, that if a bifunctional spiro ortho ester could be prepared and a small amount of this bifunctional material were to be copolymerized with the 1,4,6-trioxaspiro[4.4]-nonane (I), a lightly cross-linked elastomeric material could be produced with essentially no change in volume. Model compounds showed that phenyl glycidyl ether would condense with γ-butyrolactone to produce a substituted spiro ortho ester in a 50% yield and that this material would indeed polymerize at $30°$ in the presence of boron trifluoride etherate.

By analogy with this synthesis a bifunctional spiro ortho ester was prepared by the condensation of hydroquinone diglycidyl ether and butyrolactone to produce in 30% yield a crystalline material with a melting point of $176°$. The synthesis was enhanced by the fact that the solid crystallized from the reaction mixture and was conveniently isolated by filtration (4).

On polymerization with boron trifluoride this monomer gave an insoluble, highly cross-linked resin. When a mixture of the trioxaspirononane I containing 10% of the bifunctional spiro ortho ester II was polymerized at $100°$ with boron trifluoride, a lightly cross-linked elastomer resulted. The high temperature was utilized to make sure that the mixture was homogeneous. The resulting elastomer had a swelling index of 12. A more tightly cross-linked elastomer was produced by polymerizing the spirononane with a mixture containing 30% of this spiro ortho ester to produce a material with essentially no change in volume that had a swelling index of 5 but was still somewhat elastomeric.

Although the adduct between an epoxy resin (bisphenol-A diglycidyl ether) and the butyrolactone did not give a material that could be isolated in the pure state, it was possible to use this reaction to prepare prepolymers which had a large shrinkage during the initial portion of the reaction when the material was liquid. When the prepolymer was further polymerized with boron trifluoride, a cross-linked material resulted with essentially no change in volume near the last part of the polymerization where the material becomes viscous and gels. This technique should allow the production of strain-free materials at a reasonable cost.

In order to demonstrate that expansion in volume would take place with bicyclic materials other than spiro derivatives, a ketal lactone was prepared by the method of Lange, Wamhoff, and Korte (5). Polymerization of this material with either boron trifluoride or a base produced the keto-containing polyester with essentially no change in volume.

Still another class of bicyclic materials that will polymerize with an increase in volume, are the 2,6,7-trioxabicyclo[2.2.2]octanes. For example, the monoethyl derivative III, which is a solid, will polymerize at 70° in the presence of boron trifluoride in about 10 minutes to produce the viscous liquid polymer IV with an increase in volume of 1.3% (6). When the polymerization was carried out from 0-5°, evidence was obtained from the infrared spectra (appearance of a strong transient band at 1600 cm^{-1}) that the polymerization took place stepwise to produce a stabilized carbenium ion which was converted to the final polymer. Introduction of

$$\text{Et-C}\begin{smallmatrix}\text{CH}_2\text{-O}\\ \text{CH}_2\text{-O}\\ \text{CH}_2\text{-O}\end{smallmatrix}\text{C-H} \xrightarrow[70\%]{BF_3 \cdot OEt_2} \left[\begin{smallmatrix}\text{-O-CH}_2 & \text{CH}_2\text{-}\\ \text{C} & \\ \text{Et} & \text{CH}_2\text{-O-C-H}\\ & \overset{O}{\|}\end{smallmatrix}\right]_x$$

     III   IV

bulky side groups gave similar bicyclo monomers, which produced high melting thermoplastic materials. Hall, DeBlauwe, and Pyriadi (7) had previously reported the polymerization of a 2,6,7-trioxabicyclo[2.2.2]octane at low temperatures under conditions that only one of the rings was opened.

    We have demonstrated, therefore, that polymerization with no change in volume or with expansion in volume is a general phenomena possible with a wide variety of liquid and crystalline cyclic and bicyclic monomers.

    Another very interesting class of compounds appeared to be the spiro ortho carbonates (8). Sakai, Kobayashi, and Ishii (9) recently described a method for synthesizing ortho carbonates using tin compounds with carbon disulfide. Using their method, we were able to synthesize a series of spiro ortho carbonates by the following set of reactions. This method worked well for 1,2-,1,3-, or 1,4- glycols to produce crystalline monomers (10).

$$\begin{smallmatrix}\text{CH}_2\text{OH}\\ \text{CH}_2\\ \text{CH}_2\text{OH}\end{smallmatrix} \xrightarrow[50\%]{Bu_2Sn=O} \text{CH}_2\begin{smallmatrix}\text{CH}_2\text{-O}\\ \\ \text{CH}_2\text{-O}\end{smallmatrix}\text{SnBu}_2 \xrightarrow{CS_2} \left[\text{CH}_2\begin{smallmatrix}\text{CH}_2\text{-O}\\ \\ \text{CH}_2\text{-O}\end{smallmatrix}\text{C=S}\right]$$

$$\xrightarrow{92\%} \text{CH}_2\begin{smallmatrix}\text{CH}_2\text{-O} & \text{O-CH}_2\\ & \text{C} \\ \text{CH}_2\text{-O} & \text{O-CH}_2\end{smallmatrix}\text{CH}_2 \xrightarrow[142°]{BF_3 \cdot OEt_2}$$

                                mp 141°

$$\left[\text{O-CH}_2\text{-CH}_2\text{-CH}_2\text{-O-}\overset{O}{\overset{\|}{C}}\text{-O-CH}_2\text{-CH}_2\text{-CH}_2\right]_x$$

    Since the spiro ortho carbonate was a highly crystalline material, initial polymerization studies were carried out above its melting point at 142°C. Although the polymerization could be carried out with a variety of cationic catalysts, such as boron trifluoride gas, boron trifluoride etherate, and aluminum chloride, boron trifluoride etherate proved to be the most convenient.

    Thus, when the polymerization of molten spiro ortho carbonate was carried out in bulk with boron trifluoride etherate at 142°C, a quantitative yield of polymer was obtained after several hours. [When polymerization was carried out at higher temperatures, the evolution of a gas (CO_2) was observed.] The polymer was purified by dissolution in chloroform followed by extraction of the solution with water. The structure of the polymeric material was proven not only by elemental analysis, but also by NMR and IR spectra. The polymer had an intrinsic viscosity of 0.26 in chloroform at 25°C. Although the relationship between molecular weight and intrinsic viscosity is unknown for this series of polymers, a reasonable assumption of the constants would indicate a molecular weight in excess of 100,000. These results tend to indicate that the strain inherent in the ortho carbonate structure provides a strong driving force for the polymerization. A very similar polymerization could be carried out at 100°C by addition of catalyst to the solid monomer.

When the polymerization was carried out in a dilatometer in which the bath was held at a constant temperature (142°C), the meniscus, instead of falling as is the usual case during polymerization, actually rose quite substantially. A calculation of the extent of change in volume indicates an expansion in excess of 2%. This compares very favorably with the very slight increase (0.14%) in volume reported earlier for the polymerization of a spiro ortho ester. This example, then, represents the first reported case in which a substantial amount of expansion in volume occurs during polymerization.

An even more remarkable relationship was discovered when the densities of the monomer and polymer were determined as a function of temperature. Table III lists the densities of the two materials at 25, 100, 130 and 142°C.

TABLE III. Calculation of Expansion During Polymerization

Temperature °C	Density of monomer, g/cc	Density of polymer, g/cc	Expansion in volume, %
25	1.31	1.20	9
100	1.30	1.14	14
130	1.30	1.11	17
142	1.12	1.10	2

The density of the amorphous liquid polycarbonate varied quite regularly and smoothly with changes in temperature from 1.20 g/cc at 25°C to 1.10 g/cc at 142°C. The density of the monomer, however, changed quite abruptly when it went from the molten monomer at 142°C to the crystalline monomer at temperatures below its melting point. Obviously, this data shows that the crystalline monomer is considerably more dense than the molten monomer. Similarly, the crystalline monomer was much more dense than the liquid polycarbonate. Thus, when the expansion in volume is calculated from the density of the crystalline monomer, the expansion was 9% at 25°C up to 17% at 130°C. Under ideal conditions the expansion might even be somewhat larger since the density of the crystalline monomer was determined by measuring the volume of a given weight of a solidified molten monomer. Under these conditions it is almost impossible to avoid the presence of some voids or the inclusion of a small amount of amorphous material. By coincidence, however, the 17% expansion is quite close to the expansion already calculated for the conversion of adamantane to polycyclopentenomer (2). Figure 1 gives the plot of the densities of the monomer and polymer vs temperature.

It is obvious from the data that the conversion of a crystalline monomer to an amorphous polymer represents the ideal case for the large expansion in volume since in most cases the crystalline monomer would be expected to be considerably more dense than the corresponding liquid monomer. This is just the opposite of the case in which a liquid monomer is converted to a crystalline polymer. For example, when liquid ethylene monomer is converted to crystalline polyethylene at 5^0C, a 66% shrinkage occurs. appears that the conversion of a liquid monomer to a crystalline polymer represents the ideal case to get the largest shrinkage during polymerization (3).

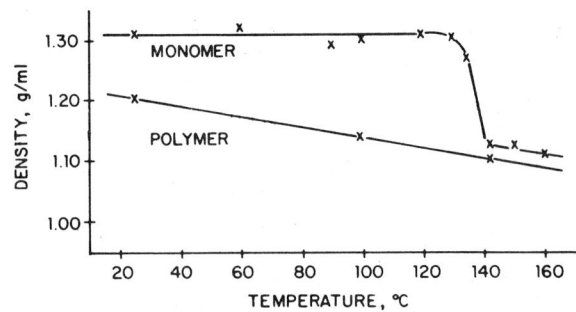

Figure 1. *Densities of the monomeric spiro ortho carbonate and related polyoxycarbonate vs. temperature*

An inspection of Fig. 1 indicates that the densities of the monomeric spiro ortho carbonate and the polymer appear to cross above 200^0. At that point, one would expect no change in volume during polymerization since the two materials have the same density. Above this critical temperature, one would expect to get shrinkage during the polymerization. Unfortunately, the polymerization cannot be carried out conveniently in this temperature range with the catalysts now available since carbon dioxide is liberated and the polycarbonate is not obtained in a pure form. At the lower end of the temperature scale the two lines appear to intersect again, but one would expect below the glass transition of the polymer that the density line would become more nearly horizontal and become essentially parallel to the line of the density of the monomer.

While at first it appeared difficult to find a polymerization procedure that would take full advantage of this large expansion, since the solid monomer is hard to introduce into a mold or a composite, it was found possible to make a slurry of this crystalline material in liquid epoxy monomer and copolymerize the two with controlled shrinkage. Expansion, contraction, or zero change in volume could be obtained depending on the concentration of the crystalline monomer in the slurry. The large volume increase also suggests that the polymerization may be used to replace explosives in cracking rocks in a quarry or for excavations.

A variety of analogs of this spiro ortho carbonate can be prepared and polymerized. For example, a higher softening polymer in this same series could be prepared from the trispiro analog, which has a melting point of $112°C$. Polymerization at room temperature produced a material with an increase in volume of 4%. Just below the melting point it gave a high melting polymer with an expansion of 7%. The spiro ortho carbonates appear to be a very versatile class of compounds for polymerization with expansion in volume.

Although the literature contains a large number of examples of ring-opening polymerizations involving ionic intermediates, there are very few examples involving radical ring-opening polymerizations. The few examples that exist in the literature involve the polymerization of vinylcyclopropane derivatives, such as 1,1-dichloro-2-vinylcyclopropane and 1-carbethoxy-2-vinylcyclopropane, or spiro-o-xylylene. Since these examples all contain a highly strained ring, it appeared possible that a number of other strained ring systems containing unsaturation either in or adjacent to the ring could also undergo ring-opening or double ring-opening polymerization by a radical mechanism. For this reason we undertook the synthesis of 3,9-dimethylene-1,5,7,11-tetraoxaspiro[5.5]undecane (VI) by the following set of reactions (11).

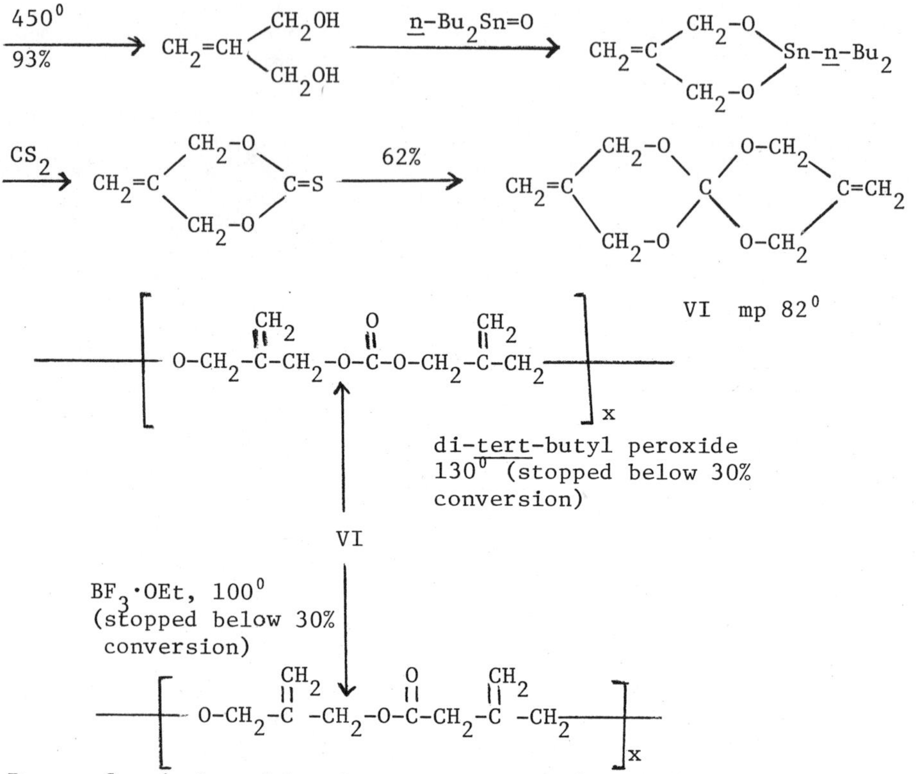

$\xrightarrow[93\%]{450°}$ $CH_2=CH{\overset{\displaystyle CH_2OH}{\underset{\displaystyle CH_2OH}{\diagdown}}}$ $\xrightarrow{\underline{n}\text{-}Bu_2Sn=O}$ $CH_2=C{\overset{\displaystyle CH_2-O}{\underset{\displaystyle CH_2-O}{\diagdown}}}Sn\text{-}\underline{n}\text{-}Bu_2$

$\xrightarrow{CS_2}$ $CH_2=C{\overset{\displaystyle CH_2-O}{\underset{\displaystyle CH_2-O}{\diagdown}}}C=S$ $\xrightarrow{62\%}$ $CH_2=C{\overset{\displaystyle CH_2-O}{\underset{\displaystyle CH_2-O}{\diagdown}}}C{\overset{\displaystyle O-CH_2}{\underset{\displaystyle O-CH_2}{\diagdown}}}C=CH_2$

$$\left[-O-CH_2-\underset{\underset{\displaystyle CH_2}{\|}}{C}-CH_2-O-\underset{\underset{\displaystyle O}{\|}}{C}-O-CH_2-\underset{\underset{\displaystyle CH_2}{\|}}{C}-CH_2- \right]_x$$

VI mp 82°

di-<u>tert</u>-butyl peroxide
130° (stopped below 30% conversion)

VI

BF$_3$·OEt, 100°
(stopped below 30% conversion)

$$\left[-O-CH_2-\underset{\underset{\displaystyle CH_2}{\|}}{C}-CH_2-O-\underset{\underset{\displaystyle O}{\|}}{C}-CH_2-\underset{\underset{\displaystyle CH_2}{\|}}{C}-CH_2- \right]_x$$

It was found when this monomer was treated with di-<u>tert</u>-butyl peroxide at 130° and the reaction was stopped below 30% conversion, a soluble polymer was obtained having a structure of a polycarbonate with pendant methylene groups. The structure of the polymer was established by elemental analysis as well as infrared and NMR spectroscopy. A very similar polymer could be attained by treatment of the monomer with boron trifluoride etherate at low conversions. The mechanism of the polymerization appeared to involve a radical double ring-opening according to the following mechansim (<u>12</u>):

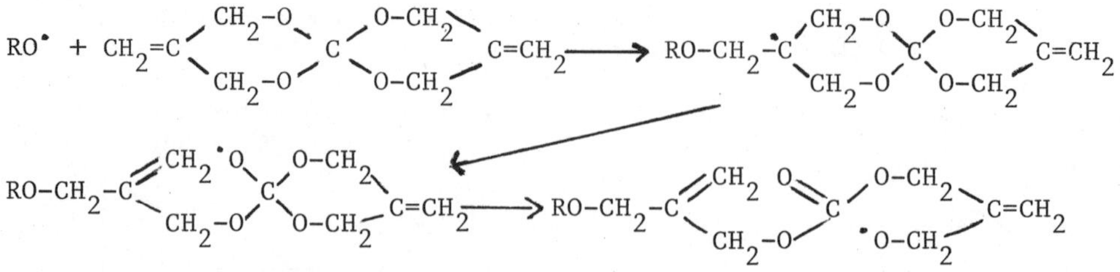

RO• + $CH_2=C{\overset{\displaystyle CH_2-O}{\underset{\displaystyle CH_2-O}{\diagdown}}}C{\overset{\displaystyle O-CH_2}{\underset{\displaystyle O-CH_2}{\diagdown}}}C=CH_2$ ⟶ $RO-CH_2-\overset{\bullet}{C}{\overset{\displaystyle CH_2-O}{\underset{\displaystyle CH_2-O}{\diagdown}}}C{\overset{\displaystyle O-CH_2}{\underset{\displaystyle O-CH_2}{\diagdown}}}C=CH_2$

$RO-CH_2-C{\overset{\displaystyle CH_2\overset{\displaystyle \bullet}{}O}{\underset{\displaystyle CH_2-O}{\diagdown}}}C{\overset{\displaystyle O-CH_2}{\underset{\displaystyle O-CH_2}{\diagdown}}}C=CH_2$ ⟵ $RO-CH_2-C{\overset{\displaystyle CH_2}{\underset{\displaystyle CH_2-O}{\diagdown}}}{\overset{\displaystyle O}{\|}}C{\overset{\displaystyle O-CH_2}{\underset{\displaystyle \bullet O-CH_2}{\diagdown}}}C=CH_2$

repeat
$$RO\left[CH_2-\overset{\overset{CH_2}{\|}}{C}-CH_2-O-\overset{\overset{O}{\|}}{C}-O-CH_2-\overset{\overset{CH_2}{\|}}{C}-CH_2-O\right]_x$$

The driving force for the double ring-opening polymerization apparently is the relief of the strain at the central spiro atom.

At high conversions this monomer produced a highly crosslinked resin, very similar in appearance to the material produced from the polymerization of diallyl carbonate. Furthermore, it was shown that this unsaturated spiro ortho carbonate would readily copolymerize with styrene, methyl methacrylate, and diallyl carbonate, but with slightly lower reactivity than these other monomers. As indicated in Figure 2 the volume change that occurred during homopolymerization was quite unusual. At room temperature, a 4.3% expansion in volume occurred, while just below its melting point at $70°$, a 7% expansion in volume occurred; at $85°$ a 2% expansion took place and the expansion decreased until at $115°$ no change in volume took place during polymerization; above $115°$ a slight shrinkage occurred. It is obvious from these data that the large expansion in volume that occurs below the melting point involves not only the increase in volume due to the double ring-opening, but also a change in volume of 3-6% due to the process of going from a crystalline monomer to a liquid monomer. Since the monomer is a crystalline solid, it is difficult to find examples of homopolymerization in which the full 7% expansion in volume can be utilized. However, in copolymerizations it is possible to use a slurry of the crystalline monomer in a liquid monomer so that as copolymerization progresses, the crystalline monomer dissolves with some expansion and also polymerizes with expansion.

A potential use of this monomer is in the area of dental fillings in which a slurry containing 20% of very fine crystals of the unsaturated spiro ortho carbonate VI in 60% of the adduct of methacrylic acid to bisphenol-$\underline{A}$ diglycidyl ether (Bis-GMA) plus 20% trimethylolpropane trimethacrylate produces on polymerization a material with essentially no change in volume. An investigation of a bubble test on tooth enamel showed that this copolymer had nearly double the adhesion to the tooth structure that the base resin had without the addition of the unsaturated spiro ortho carbonate. The copolymer also had improved impact strength but yet essentially the same modulus, and filled composites appeared to have somewhat improved abrasion resistance.

Since the synthesis of the spiro ortho carbonates through the tin compounds could be modified to produce unsymmetrical materials, we undertook the synthesis of the unsymmetrical 2-methylene-1,5,7,11-tetraoxaspiro[5.5]undecane by the following set of reactions:

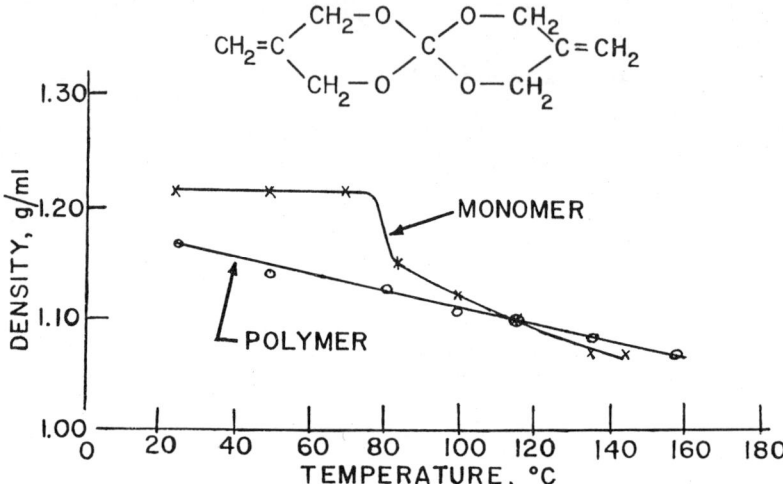

Figure 2. Densities of the monomeric unsaturated spiro ortho carbonate and related polyoxycarbonate vs. temperature

$$\text{CH}_2=\text{C}\begin{smallmatrix}\text{CH}_2-\text{OH}\\ \text{CH}_2-\text{OH}\end{smallmatrix} \qquad \text{HO-CH}_2-\text{CH}_2-\text{CH}_2-\text{OH}$$

$$\downarrow (\underline{n}\text{-Bu})_2\text{Sn=O} \qquad\qquad \downarrow \begin{array}{l}(1)\ \text{Na}\\ (2)\ (\underline{n}\text{-Bu})_3\text{SnCl}\end{array}$$

$$(\underline{n}\text{-Bu})_3\text{-Sn-O-CH}_2\text{-CH}_2\text{-CH}_2\text{-O-Sn}(\underline{n}\text{-Bu})_3$$

$$\downarrow \text{CS}_2$$

$$\text{CH}_2=\text{C}\begin{smallmatrix}\text{CH}_2-\text{O}\\ \text{CH}_2-\text{O}\end{smallmatrix}\text{Sn}(\underline{n}\text{-Bu})_2 \qquad \text{CH}_2\begin{smallmatrix}\text{CH}_2-\text{O}\\ \text{CH}_2-\text{O}\end{smallmatrix}\text{C=S}$$

$$\searrow\ 86\%\ \swarrow$$

$$\text{CH}_2=\text{C}\begin{smallmatrix}\text{CH}_2-\text{O}\\ \text{CH}_2-\text{O}\end{smallmatrix}\text{C}\begin{smallmatrix}\text{O-CH}_2\\ \text{O-CH}_2\end{smallmatrix}\text{CH}_2$$

mp 61-62°
VII

The resulting monomer was a crystalline solid with a melting point of 61-62°. When the polymerization was carried out in the presence of di-<u>tert</u>-butyl peroxide and the reaction was stopped at low conversion, a linear polycarbonate containing pendant methylene groups was obtained.

$$\text{VII} \xrightarrow[\substack{\text{di-}\underline{\text{tert}}\text{-butyl}\\ \text{peroxide}\\ 43\%}]{130°} \left[\text{CH}_2-\underset{\underset{\text{CH}_2}{\|}}{\text{C}}-\text{CH}_2-\text{O}-\underset{\underset{\text{O}}{\|}}{\text{C}}-\text{O-CH}_2-\text{CH}_2-\text{CH}_2\right]_x \quad [\eta]^{25°}_{\text{CHCl}_3} = 0.11$$

The structure of the polymer was established by elemental analysis as well as infrared and NMR spectroscopy. The structure of this material was very similar to the polymer that could be obtained by the ionic polymerization of this same monomer at low conversions. Bulk polymerization of VII with peroxide catalyst gave a material at 25° with an expansion of 4.5% and at 60° an expansion 5.5%; above the melting point of VII (61-62°) the expansion decreased until at 111°, the density of the monomer and the density of the polymer were the same.

When the 3-methylene derivative was mixed with an equal amount of styrene in the presence of di-<u>tert</u>-butyl peroxide and the reaction was stopped at below 30% conversion, a soluble copolymer was obtained containing 79% styrene and 21% of the linear polycarbonate units.

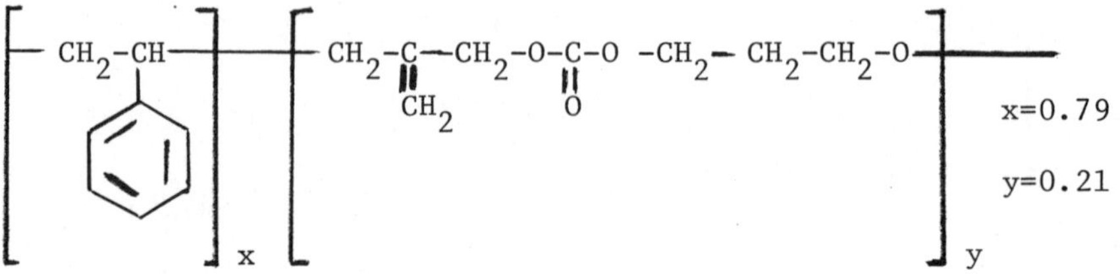

$$\left[-CH_2-CH(C_6H_5)- \right]_x \left[-CH_2-C(=CH_2)-CH_2-O-C(=O)-O-CH_2-CH_2-CH_2-O- \right]_y$$

x=0.79
y=0.21

By a very similar synthetic scheme, other unsaturated spiro ortho carbonates were prepared.

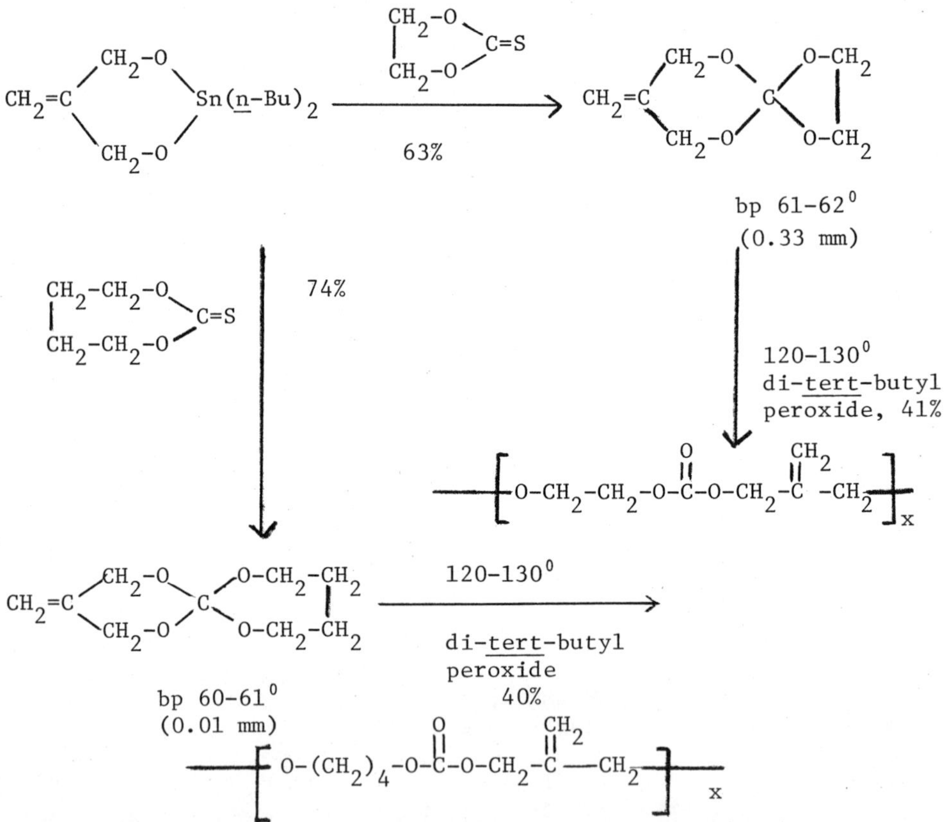

$$CH_2=C(CH_2-O)_2Sn(\underline{n}\text{-Bu})_2 + (CH_2-O)_2C=S \xrightarrow{63\%} CH_2=C(CH_2-O)_2C(O-CH_2)_2$$

bp 61-62° (0.33 mm)

$$(CH_2-CH_2-O)_2C=S \quad 74\%$$

120-130° di-tert-butyl peroxide, 41%

$$\left[-O-CH_2-CH_2-O-C(=O)-O-CH_2-C(=CH_2)-CH_2- \right]_x$$

$$CH_2=C(CH_2-O)_2C(O-CH_2-CH_2)_2 \xrightarrow{120-130°}$$

bp 60-61° (0.01 mm)

di-tert-butyl peroxide 40%

$$\left[-O-(CH_2)_4-O-C(=O)-O-CH_2-C(=CH_2)-CH_2- \right]_x$$

Bulk polymerization or solution polymerization in chlorobenzene gave soluble polymer if the reaction was stopped at low conversion. Both monomers gave cross-linked resins at high conversions.

Since several of the previously described ether-containing polycarbonates were low melting materials with a Tg below room temperature, it appeared highly desirable to synthesize a bis spiro ortho carbonate to be utilized as a cross-linking agent to convert these linear polymers into elastomers. By the method of Ishii (9) ethylene thiocarbonate was prepared in a 48% yield from ethylene carbonate through the use of tributyltin oxide and carbon disulfide. The intermediate thiocarbonate, which is a solid melting at 53°, was shown to be an excellent intermediate for producing a wide variety of spiro ortho carbonates. For example, when pentaerythritol was treated with the thiocarbonate in the presence of tributyltin oxide, a 20% yield of the bis spiro ortho carbonate, 1,4,6,10,12,15,16,19-octaoxatrispiro[4.2.2.4.2.2] nonadecane (VIII) melting point 215°, was obtained.

$$(\underline{n}\text{-Bu})_3\text{Sn-O-Sn}(\underline{n}\text{-Bu})_3 + \underset{\underset{\underset{O}{\|}}{C}}{\overset{CH_2\text{---}CH_2}{\underset{O\quad\quad O}{|\quad\quad|}}} \longrightarrow \begin{bmatrix} CH_2\text{-O-Sn}(\underline{n}\text{-Bu})_3 \\ CH_2\text{-O-Sn}(\underline{n}\text{-Bu})_3 \end{bmatrix} \xrightarrow[48\%]{CS_2}$$

$$\underset{\text{mp } 53°}{\begin{matrix}CH_2\text{-O}\\CH_2\text{-O}\end{matrix}\!\!>\!\!C=S} + \underset{HOCH_2}{\overset{HOCH_2}{\diagdown}}C\underset{CH_2OH}{\overset{CH_2OH}{\diagup}} \xrightarrow[20\%]{(\underline{n}\text{-Bu})_3\text{Sn-O-Sn}(\underline{n}\text{-Bu})_3}$$

$$\begin{matrix}CH_2\text{-O}\\CH_2\text{-O}\end{matrix}\!\!>\!\!C\!\!<\!\!\begin{matrix}O\text{-}CH_2\\O\text{-}CH_2\end{matrix}\!\!>\!\!C\!\!<\!\!\begin{matrix}CH_2\text{-O}\\CH_2\text{-O}\end{matrix}\!\!>\!\!C\!\!<\!\!\begin{matrix}O\text{-}CH_2\\O\text{-}CH_2\end{matrix}$$

VIII
mp 215°

When this material was treated with boron trifluoride at 150°, a hard, highly crossed-linked, insoluble resin, was obtained. On the other hand, when a homogeneous mixture containing 90% of the spiro ortho carbonate V and 10% of the bis spiro ortho carbonate VIII was treated with boron trifluoride at 145°, a 3% expansion occurred to produce a clear, solid elastomer with a swelling index of 10. When only 5% of the bis spiro ortho carbonate was used, a soluble polymer containing only a small amount of insoluble material was obtained. This would indicate that the reactivity of the bis spiro compound is less than that of V, so that a larger amount of VIII is required in order to produce an effective cross-linked network.

$$\underset{\underset{90\%}{V}}{\left[\begin{array}{c}CH_2-O\\CH_2-O\end{array}\right]C\left[\begin{array}{c}O-CH_2\\O-CH_2\end{array}\right]CH_2} + \underset{\underset{10\%}{VIII}}{\left[\begin{array}{c}CH_2-O\\CH_2-O\end{array}\right]C\left[\begin{array}{c}O-CH_2\\O-CH_2\end{array}\right]C\left[\begin{array}{c}CH_2-O\\CH_2-O\end{array}\right]C\left[\begin{array}{c}O-CH_2\\O-CH_2\end{array}\right]}$$

$145°$, BF_3, 3% expansion

$$\left[\begin{array}{c}-O-(CH_2)_3-O-\overset{O}{\overset{\|}{C}}-O-(CH_2)_3-\\-O-(CH_2)_3-O-\overset{O}{\overset{\|}{C}}-(CH_2)_3-\end{array}\right]_m^{} \quad \left[\begin{array}{c}-O-CH_2-CH_2-O-\overset{O}{\overset{\|}{C}}-O-CH_2\\-O-CH_2-CH_2-O-\overset{O}{\overset{\|}{C}}-O-CH_2\end{array}\right]_{}^{}C\begin{array}{c}CH_2\\CH_2\end{array}\right]_n$$

It was reasoned that if a bicyclic monomer could be prepared containing a functional group, that a large number of prepolymers could be utilized in producing materials with no change in volume on curing. Thus, in the synthesis of the bis spiro ortho carbonate just discussed, a small amount of a dihydroxy derivative was isolated. By optimizing the conditions a 16% yield of the dihydroxy spiro ortho carbonate was prepared.

$$\begin{array}{c}CH_2-O\\CH_2-O\end{array}C\begin{array}{c}O-CH_2\\O-CH_2\end{array}C\begin{array}{c}CH_2-OH\\CH_2-OH\end{array}$$

mp $155°$

A more versatile synthesis was developed according to the following equation:

$$CH_3-CH_2-C(CH_2-OH)_3 + (Bu_3Sn)_2O \xrightarrow[12 \text{ hr}]{\text{reflux in toluene}}$$

$$\begin{array}{c}HO-CH_2\\CH_3-CH_2\end{array}C\begin{array}{c}CH_2-O-SnBu_3\\CH_2-O-SnBu_3\end{array} \xrightarrow[\substack{100°\\18 \text{ hr}\\78\%}]{CS_2} \begin{array}{c}HO-CH_2\\CH_3-CH_2\end{array}C\begin{array}{c}CH_2-O\\CH_2-O\end{array}C\begin{array}{c}O-CH_2\\O-CH_2\end{array}C\begin{array}{c}CH_2-OH\\CH_2-CH_3\end{array}$$

By a similar reaction with ethylene thiocarbonate in place of carbon disulfide, a monohydroxy derivative was prepared.

$$\text{HO-CH}_2\text{-C(CH}_2\text{CH}_3\text{)(CH}_2\text{-O-SnBu}_3\text{)(CH}_2\text{-O-SnBu}_2\text{)} + \text{(CH}_2\text{-O)}_2\text{C=S} \xrightarrow[14\%]{22°, 48 \text{ hr, CHCl}_3}$$

spiro orthocarbonate product, mp 66°

These materials could be used to produce polyurethanes with no change in volume on curing. For example, the dihydroxy spiro ortho carbonate can be allowed to react with hexamethylene diisocyanate at room temperature to produce a linear polymer which on treatment with boron trifluoride gives a cross-linked resin with essentially no change in volume.

dihydroxy spiro orthocarbonate + $O=C=N-(CH_2)_6-N=C=O$

$\xrightarrow{\text{CHCl}_3, 30 \text{ hr. } 22°}$

linear polymer with $-O-\overset{O}{\underset{\|}{C}}-NH-(CH_2)_6-NH-\overset{O}{\underset{\|}{C}}-$ repeat unit, subscript x

$\xrightarrow{\text{BF}_3}$ cross-linked resin

Since rings are more compact than open chain analogs, it appeared possible to use thermal ring-opening to control shrinkage during polymerization. For example, cyclobutene is about 20% more dense than butadiene. While the polymerization of butadiene involves a shrinkage of 36%, if it were possible to convert cyclobutene to this same material, the process would involve a shrinkage of only 18%. Furthermore, most of the critical or damaging shrinkage that takes place during polymerization is that which occurs after the gel point in cross-linked materials or when the monomer-polymer mixture approaches the glass transition point in linear thermoplastic materials. When the monomer-polymer mixture is quite fluid, no strains are built up and the effect of the shrinkage can be partially overcome by the introduction of additional monomer. However, the shrinkage that takes place near the end of the polymerization results in the build up of strains, the formation of microcracks, and the introduction of poor adhesion. It was reasoned therefore, that if a polymeric material could be prepared containing rings that would open thermally during the critical portion of the polymerization, the effect of the shrinkage could be minimized. One of the materials that appears to meet this criteria is dimethyl cyclobutene-1,2-dicarboxylate (IX). When this material is mixed with methyl methacrylate and the mixture is treated with a peroxide catalyst, the double bond in the ring is fairly inert and doesn't take part to any large extent in the polymerization. However, if near the end of the polymerization the temperature is increased to 150^0, the cyclobutene ring opens to produce a diene X with an increase in volume of 5% (<u>17</u>).

$$\begin{array}{c} CH_3O_2-C \quad CO_2CH_3 \\ | \quad \quad | \\ C = C \\ | \quad \quad | \\ CH_2-CH_2 \end{array} \quad \xrightarrow{150^0} \quad \begin{array}{c} CH_3O_2C \diagdown \quad \diagup CO_2CH_3 \\ C-C \\ \diagup \quad \diagdown \\ CH_2 \quad \quad CH_2 \end{array}$$

$$\text{IX} \hspace{6cm} \text{X}$$

Since the double bonds in this compound are now reactive, part of the double bonds are incorporated into the polymer network to give a cross-linked material. The volume change can be controlled to some extent by the ratio of the cyclic ester added and the extent of polymerization that has taken place when the temperature is raised to 150^0. Theoretically a large number of ring compounds and polymers could be used to control shrinkage or to promote expansion of polymers on ring-opening.

A variety of ring systems containing sulfur, nitrogen and carbon are being investigated to produce multiple ring-openings to give polymers containing a variety of chemical structural units. It is hoped that these materials will find wide utility for the uses discussed earlier and will prove to be a very general solution to the problem of shrinkage during polymerization.

The authors are grateful to the Naval Air Systems Command and to the National Institute of Dental Research for support of this research.

Literature Cited

1. Bailey, W.J., and Sun, R.L., Amer. Chem. Soc., Div. Polym. Chem. Prepr., (1972), <u>13</u> (1), 400.
2. Bailey, W.J., J. Elastoplast., (1973), <u>5</u>, 142.
3. Bailey, W.J., J. Macrolmol. Sci.-Chem., (1975),<u>A9</u>(5), 849.
4. Bailey, W.J., Iwama, H., and Tsushima, R., J. Polymer Sci., Polym Symposia Edition, in press; Abstracts of the 4th International Symposium on Cationic Polymerization, Akron, Ohio, June

20-23 (1976).
5. Lange, C., Wamhoff, H., and Korte, T., Chem. Ber., (1967), 100, 2312.
6. Bailey, W.J., and Saigou, K., J. Polym. Sci., Polym. Letters Ed., in press.
7. Hall, H.K., Jr., DeBlauwe, Fr., and Pyriadi, T., J. Am. Chem. Soc., (1975), 97, 3854.
8. Bailey, W.J., and Katsuki, H., Amer. Chem. Soc., Div. Polym. Chem., Prepr., (1973), 14, 1679.
9. Sakai, S., Kobayashi, Y., and Ishii, Y., J. Org. Chem., (1971), 36, 1176.
10. Bailey, W.J., Katsuki, H., and Endo, T., Amer. Chem. Soc., Div. Polym. Chem., Prepr., (1973), 14, 1976.
11. Bailey, W.J., Katsuki, H., and Endo, T., Amer. Chem. Soc., Div. Polym. Chem., Prepr., (1974), 15, 445.
12. Endo, T., and Bailey, W.J., J. Polym. Sci., Polym. Letters Ed., (1975), 13, 193.
13. Endo, T., and Bailey, W.J., Makromol. Chem., (1975), 176, 2897.
14. Endo, T., and Bailey, W.J., J. Polym. Sci., Polym. Chem. Ed., (1975), 13, 2525.
15. Bailey, W.J., and Endo, T., J. Polym. Sci., Polym. Chem. Ed. (1976), 14, 1735.
16. Bailey, W.J., and Tsushima, R., J. Polymer Sci., in press.
17. Bailey, W.J., and Bitritto, M., J. Polymer Sci., in press.

5

Progress in Polymerization of Cyclic Acetals

STANISŁAW PENCZEK and PRZEMYSŁAW KUBISA
Polish Academy of Science, 90-362 Łódź, Poland

In our previous review paper, presented at the Rouen Symposium on Cationic Polymerization, we stressed some of the major differences between the cationic polymerization of cyclic ethers and cyclic acetals |1|. These differences are mainly caused by the much larger basicity (nucleophilicity) of cyclic ethers, than that of cyclic acetals; moreover, cyclic ethers are more basic (nucleophilic) than their polymers, whilst polyacetals seem to be more basic than their corresponding monomers.

Thus, in polymerization of cyclic ethers (or, at least in polymerization of THF), tertiary oxonium ions $\underline{1}$ are the only growing species |$\underline{2}$||$\underline{3}$||$\underline{4}$|:

$$CH_2OCH_2CH_2CH_2CH_2\!-\!\overset{+}{O}\!\diagup\!\!\!\!\diagdown\!\!\begin{array}{c}CH_2-CH_2\\ |\\ CH_2-CH_2\end{array}$$

$$\underline{1}$$

whereas in the polymerization of cyclic acetals, including 1,3-dioxolan (Diox), the equilibrium between the macroalkoxycarbenium ions $\underline{2}$ with their tertiary oxonium $\underline{3}$ counterparts is in our opinion the best representation of the active species :

(1)

$$\overset{+}{O}CH_2CH_2OCH_2 \; + \; \begin{array}{c}O\diagdown CH_2\!-\!\cdots\\ \diagup CH_2\\ O\diagup\\ \diagdown CH_2\!-\!\cdots\end{array} \rightleftharpoons OCH_2CH_2OCH_2\cdots\overset{+}{O}\diagup\!\!\!\!\diagdown\begin{array}{c}CH_2\\ CH_2\\ |\\ CH_2\\ \diagdown\cdots\end{array}$$

$\underline{2}$ polymer segment $\underline{3}$

Unfortunately, our knowledge of the carbenium-oxonium ion equilibria is very limited; some first quantitative data from our laboratory |5| were discussed in the Rouen paper |1|.

Properties of 2 and 3 in equilibrium (1) may depend very much on the polymerization conditions and structure of the cyclic acetal. At the sufficiently large excess of a polyacetal, 3 can become the predominant structure.

The other topic, which will be covered in this paper, was previously reviewed by Plesch at the IUPAC Symposium in Budapest |6|, and more recently at the I-st IUPAC Symposium on Ring-Opening Polymerization held in Jabłonna (1975) |7|. In this part of our paper the structure of the end-groups in poly-Diox is described, and the mechanistic consequences of the alleged macrocyclic or linear structures of the isolated dead macromolecules is discussed.

In 1975 Rosenberg, Irzhakh and Enikolopian published a book entitled "Interchain exchange in polymers" |8|, summarizing results of the Moscow group on the polymerization of cyclic acetals. Although some of the conclusions of this book would certainly be presented today differently in light of the new experimental data, the reader may find there an unorthodox solution of the majority of kinetic problems pertinent to the nonstationary polymerizations, including polymerization of cyclic acetals.

We shall start, however, this review of the progress in the polymerization of cyclic acetals from a brief description of the new polyacetals prepared, and from summarizing of the new data on the thermodynamics of polymerization of substituted 1,3-dioxolans.

Thermodynamics of Polymerization. New Polyacetals.

Ivin and Leonard |9| extended the thermodynamic treatment of the polymer-monomer equilibrium to the nonideal systems, accounting for the polymer-monomer interaction described by the Flory parameter χ_{mp}. For a bulk process, the following expression was obtained for the free energy change upon the conversion of one mole of pure monomer into one base-mole of amorphous polymer (ΔG_{1c}) :

$$\Delta G_{1c} = RT[\ln \phi_m + 1 + \chi_{mp}(\phi_p - \phi_m)] \tag{2}$$

where $\phi_m (= 1-\phi_p)$ is the equilibrium monomer volume fraction, computed from the experimentally determined equilibrium monomer concentration at the given temperature T. In this method the Flory parameter χ_{mp} is arbitrarly chosen (e.g. 0.4 for Diox-poly-Diox interaction) and assumed to be independent on temperature. Linearity of the plot of $\Delta G_{1c}/RT$ as a function of

$1/T$ indicates the reliability of these simplifications. More recently it has been observed |10|, that introduction of a term including the monomer-solvent and polymer-solvent interactions allows the polymer-monomer equilibria in solution to be uniformly treated. This term is related to the heat of mixing of the solvent used with monomer and polymer. It does not depend, however, very much on the solvent structure for Diox, and, therefore, $|Diox|_e$ is almost solvent independent |11| (although, only CH_2Cl_2, C_2H_5Cl and C_6H_6 were studied). It is worth noting, that in contrast to Diox, differences between the equilibrium monomer concentrations are much more pronounced in the polymerization of THF |12|. Indeed, $|THF|_e = 5.5$ mole·l^{-1} in CH_3NO_2 solvent and only 3.5 mole·l^{-1} in CCl_4 solvent ($|THF|_o = 7.0$ mole·l^{-1}, 25°).
These large differences reflect the higher basicity of THF and much stronger acid-base interactions between solvents and THF, than solvents and Diox.

The thermodynamic non-ideality of these systems are stressed, because some authors are still tending to determine what they are calling the thermodynamic quantities (like ΔH_p^o and ΔS_p^o) on the basis of simpler relationships, holding only for the ideal systems.

On the other hand, it has to be remembered, that in the polymerization kinetics, the proper value of $|monomer|_e$ has to be used, and that it changes with both $|monomer|_o$ and solvent structure. The combined results of bulk |13| and solution polymerization of Diox (taken largely from Reference |11|) allowed Leonard to calculate $\Delta H_{1c} = -4.0 \pm 0.1$ kcal·mole^{-1} and $\Delta S_{1c} = -11.0 \pm 0.3$ cal·mole^{-1}·deg^{-1}. These results agree well with values obtained from an equilibrium between gaseous monomer and amorphous polymer |14|.

Following the Ivin-Leonard's method, Okada determined recently the thermodynamic functions for the polymerization of 4-methyl-Diox |15| and, (assuming $\chi_{mp} = 0.3$), found $\Delta H_{1c} = -3.2 \pm 0.2$ kcal/mole and $\Delta S_{1c}^o = -12.7 \pm 0.8$ cal·mole^{-1}·deg^{-1}. Another work, performed in CH_2Cl_2 solvent for the some monomer, and not accounting for the discussed above interactions, gave the apparent values (ΔH_{ss}^{app} and ΔS_{ss}^{app}) depending, as it could be expected, on the starting monomer concentration |16|.

Theoretical Interpretation of the Polymerizability of Dioxolans.

Theoretical interpretation of the ring-chain equilibria, published by Jacobson and Stockmayer in 1950 |17| can only be applied to the case when chains or rings are so large that the configurational entropy is governed by the Gaussian function and the energy difference between the chain and ring forms is negligible. For a small ring, such as the five-membered ring of Diox, Jacobson-Stockmayer equation cannot be applied. In such case, as shown in many works, summarized recently by Hall for various cyclic monomers |18||19|, stability of the ring is connected only with the strain in the ring, caused mostly by the deviation in valency angles. Hall explicitly showed that the difference in strain energy between monomer and polymer equals the enthalpy of polymerization, provided that no substituents are present, or else conformational strains in the polymer may outweigh the strain in the ring.

It has already been observed in the polymerization of substituted ε-caprolactams |20||21|, that the substitution of hydrogen atoms decreases the polymerizability of monomers. The same phenomena were observed in the polymerization of 4,4-dimethyl-, cis-4,5-dimethyl and trans-4,5-dimethyl-Diox |15|.

These differences were interpreted in the polymerization of ε-caprolactams |20| from the change of thermodynamical properties caused by the existence of rotational isomers. In analysing polymerization of substituted dioxolans Okada took a slightly modified approach, comparing energetical differences between dioxolans and their polymers. Low - molecular weight acetals, e.g. dimethoxymethane and its homologues exist predominantly in the gauche form to avoid the rabbit--ear effects in the anti-form |22|:

gauche   anti

Polydioxolans are also assumed to exist in the gauche form, because the rabbit-ear effect in the anti-form is larger (1 kcal·mole^{-1}) than the gauche interaction of the methyl groups (usually considered to be from 0.6 to 0.9 kcal·mole^{-1}).

Substitution of the H atoms by CH_3 groups destabilizes monomers by replacing the cis- geminal C^4-H and C^5-H bonds opposition with a greater C^4-H and C^5-CH_3 opposition.

All of the Okada's calculations were based on the assumption, that the stable conformation of Diox is the "envelope" form, in which one of the carbon atoms of the ethylene group is located at the tip of the flap, giving the dihedral angle of the cis- neighboring hydrogens of the ethylene group equal to 35 degree.

Substitution in a polymer chain leads to the increased energy of the gauche-interactions and the difference between these two effects gives eventually a deviation in ΔH_{1c} ($\Delta\Delta H_{1c}$) for a substituted Diox from unsubstituted monomer. These calculations led Okada to the following estimated values of - ΔH_{1c} (given below in kcal·mole^{-1}) for various methyl substituted Diox :

[structures: 2.8, 1.4, 2.5, 1.4]

For 4-methyl-Diox there is a good agreement with the value determined experimentally (3.2 kcal·mole^{-1}).

Thus, these finding are in accordance with a general observation that in the polymerization of hetero - cyclic monomers substitution leads to decreased probability of chain formation. The extent of sensitivity of a given class of monomers to substitution is given by the ring strain of the parent, unsubstituted monomer.
Thus, even for highly substituted oxiranes (e.g. tetramethyl oxirane) complete polymerization can be achieved, because the ring strain overshadows any other effect.

Jedliński analysed in a series of papers the ^{1}H-NMR spectra of various substituted 1,3-dioxolans in order to understand the stereochemistry of these monomers. Then, following earlier work, described previously for the unsubstituted dioxolenium salts, studied the kinetics of H$^-$ transfer from these monomers to the triphenylmethylium cation |23|, as the first reaction, preceding the true initiation. This approach, is complementary to that of Okada, which gives a thermodynamic information about the polymerizability, while Jedliński tends to characterize the influence of structure (stereochemistry) on the rate of reactions pertinent to elementary reactions. There are till now, however, no quantitative informations available about the reactions related to the actual polymerization process (initiation, chain growth). Recently, Kops reported on the polymerization of bicyclic dioxolans, cyclic formals of trans- and cis- cyclohexanediols :

Only the trans- monomer polymerized, giving high molecular weight, solid polymer |24|. This result is in accordance with a more general phenomenon of the inc - reased strain in the trans-joined rings, due to the enhanced angular strain.

In the previous paragraph we discussed polymerization of 1,3-dioxolans substituted at C^4 and C^5. Information on the polymerization of dioxolans substituted at C^2 is very limited; we shall confine ourselves to the polymerization of 2-vinyl-dioxolans and 2-vinyl--dioxans.

Polymerization of the Unsaturated Cyclic Acetals.

Polymerization of 2-vinyl-1,3-dioxolan (4) |25| |26| 2-vinyl-1,3-dioxane (5) |27| |28| |29| and related mono - mers, substituted at C^2 :

$$\begin{array}{cc} \text{CH}_2\text{—CH}_2 & \text{CH}_2\text{—CH}_2\text{—CH}_2 \\ |\quad\quad| & |\quad\quad\quad\quad| \\ \text{O}\diagdown\diagup\text{O} & \text{O}\diagdown\quad\diagup\text{O} \\ \text{CH} & \text{CH} \\ | & | \\ \text{CH}_2\text{=CH} & \text{CH}_2\text{=CH} \\ \underline{4} & \underline{5} \end{array}$$

have been investigated during the last fifteen years in at least five laboratories. After the original discovery of Mukaiyama |25|, who found that 4 polymerizes, at least partially, to the linear polyester:

$$\{CH_2CH_2CH_2COCH_2\}\atop{\overset{\|}{O}}$$

Tada, Saegusa, and Furukawa |26| interpreted this result as a consequence of the hydride-shift polymerization, similar to that elaborated extensively by Kennedy |30| for branched α-olefins :

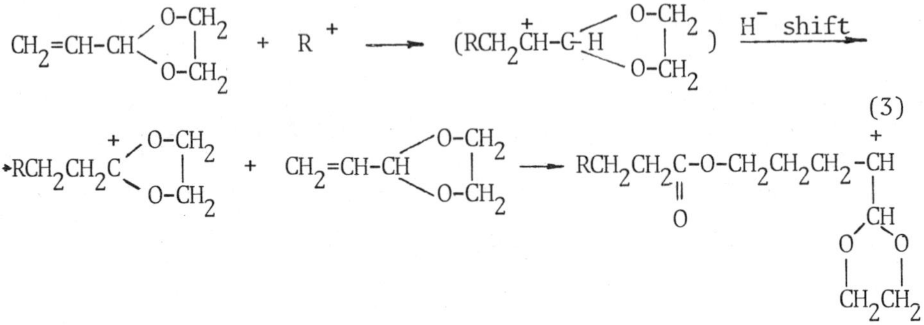

$$CH_2=CH-CH\begin{matrix}O-CH_2\\|\\O-CH_2\end{matrix} + R^+ \longrightarrow (RCH_2\overset{+}{CH}-C-H\begin{matrix}O-CH_2\\|\\O-CH_2\end{matrix}) \xrightarrow{H^- \text{ shift}}$$

$$\rightarrow RCH_2CH_2\overset{+}{C}\begin{matrix}O-CH_2\\|\\O-CH_2\end{matrix} + CH_2=CH-CH\begin{matrix}O-CH_2\\|\\O-CH_2\end{matrix} \longrightarrow RCH_2CH_2\underset{\underset{O}{\|}}{C}-O-CH_2CH_2CH_2-\overset{+}{CH}\begin{matrix}\\|\\CH\\O^{\diagup}{\diagdown}O\\|\quad|\\CH_2CH_2\end{matrix}$$

(3)

Although the most nucleophilic site of attack in 4 are rather the oxygen atoms, but, apparently the cationated 4, e.g.:

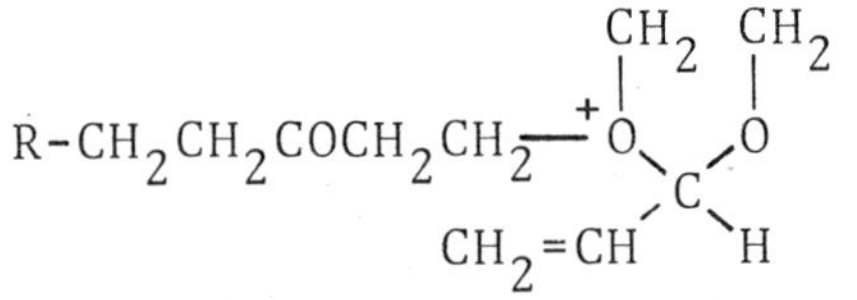

$$R-CH_2CH_2COCH_2CH_2 \!-\!\!\overset{+}{\underset{CH_2=CH}{O}}\!\!\underset{\diagdown C \diagup}{\overset{CH_2\ CH_2}{\underset{|\quad|}{\diagup\ O\ \diagdown}}}\!\!H$$

or its open-chain isomer, stabilized by the formation of the allylic - type carbenium ion, are not sufficiently reactive in the chain growth to compete with the H⁻ ion transfer processes.

More detailed analysis of polymerization of 4, and particularly an analysis of the ¹H-NMR spectra of poly-5 revealed |27|, that the complete structure of polymers is much more complex. Almost all of the repeating units that one could imagine were found, the most important ones being (for poly-5) as shown below:

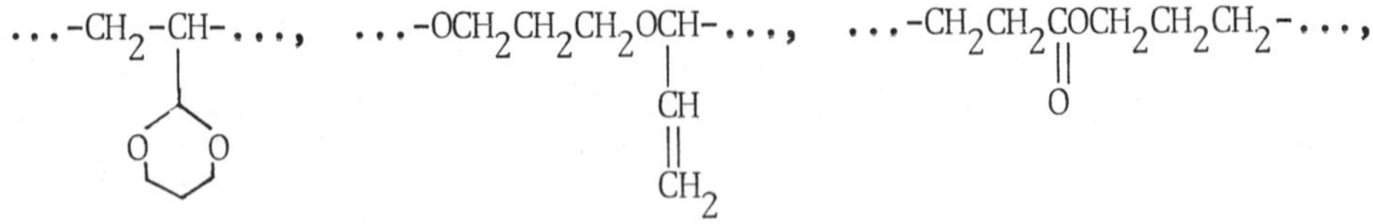

$$...-CH_2-CH-..., \quad ...-OCH_2CH_2CH_2OCH-..., \quad ...-CH_2CH_2\underset{\underset{O}{\|}}{C}OCH_2CH_2CH_2-...,$$
$$ \overset{\diagup O\ \ O \diagdown}{} \qquad\qquad\qquad \begin{matrix}|\\CH\\ \|\\CH_2\end{matrix}$$

The two first structures contain groups still reactive in the chain, and this is why these polymers are of interest for polymer chemists working in the polymer syntheses. The two-stage polymerization of these easily available monomers has been expected to provide a new group of reactive polymers. Free-radical polymerization, followed by the cationic formation of the network (and, in principle, vice versa) was also successfully applied by Minato to (1,3-dioxolan-4-yl) methyl acrylate |31|.

End-Groups in Polyacetals.

Poly-1,3,5-trioxan is known to contain hemiacetal -OH groups; some of these are formed because of the chain transfer to water |32|. Acetylation of the -OH groups greatly enhances thermal stability of the polyoxymethylene polymers, and is at the base of the commercialization of the first polyacetal (a homopolymer of CH_2O) |32a| known, in fact, many years ago from the classical works of Staudinger and Kern.

Jaacks a.o. discovered |32| formation of the methoxyl end groups in poly-1,3,5-trioxan, determined by the Zeisel method, and assumed, that these groups result from the H^- ion shift (intramolecularly) or transfer (intermolecularly).

Similar reaction was proposed by us more recently in the polymerization of Diox, conducted above 0°, to account for the methoxyl end-groups observed in the ^1H-NMR spectra |33|.

End-Groups in Poly-1,3-dioxolan.

Gresham polymerized Diox with mineral and Lewis acids and was unable to detect any end-groups |34|. Plesch confirmed Gresham's observation |35|, assumed that poly-Diox are mostly cyclic and on this basis proposed a mechanism of propagation with protonic acids ("ring-expansion"). Jaacks, in apparently similar conditions ($HClO_4$, CH_2Cl_2 solvent) found earlier, that polymers are rather linear, and quantitatively determined ethyl alcohol from the hydrolyzed end-groups, formed when a living-linear (on his opinion) polycation was killed with sodium ethylate |36|. These results were recently challenged by Plesch |37|.

In the polymerization of Diox initiated with triethyloxoniumhexafluorophosphate $((C_2H_5)_3O^+PF_6^-)$ Worsfold |38| claimed that he could not find any end-groups in poly-Diox formed, although a triplet from a CH_3CH_2O group is seen in the ^1H-NMR spectrum given in his paper. Ponomarenko a.o. |39|, by using $(C_2H_5)_3O^+SbCl_6^-$, labelled with ^{14}C in the ethyl group, concluded, that the number of moles of C_2H_5 groups, incorporated into the macromolecules, is close to the number of moles of the used initiator. Okada |40| in his study of oligomers isolated at low conversion (polymerization of Diox initiated with with $(C_2H_5)_3O^+BF_4^-$ and killed with CH_3ONa) observed linear oligomers with ethylate and methylate end-groups.

Semlyen |41| in a report describing polymerization initiated with $BF_3 \cdot O(C_2H_5)_2$ showed that gc-ms method gave evidence for the existance of cyclic oligomers formed in the process, which led to the mixture of linear and cyclic products with a distribution governed for larger macrocyclics by the Jacobson-Stockmayer theory.

This is an important result, because an agreement between the observed distribution of the cyclic oligomers with a distribution predicted by the Jacobson-Stockmayer theory (a slope equal to 2.5 for the plot of log K_x on log x, where K_x is the molar cyclization-equilibrium constant for macrocycles with a polymerization degree equal to x) strongly indicates that polymerization proceeds with a linear active species, forming macrocycles by back-biting and end-to-end closure.

Thus, because of the existing controversy, whether and when poly-Diox contain the end-groups, and because of the far reaching conclusions based on either macrocyclic or linear structures of the isolated poly-Diox, we reinvestigated recently this problem.

First of all we decided to use methods which would not involve destruction of the end-groups (like hydrolysis used by Jaacks |36| and then by Plesch |37| as a possible source of amibiguity. Secondly, we assumed, that both end-groups should be observed; the initial one, formed from an initiator, and the terminal one, coming from the killing agent. Thus, we initiated polymerization either by benzoilium hexafluoroantimonate $(C_6H_5CO^+SbF_6^-)$, assuming that the benzoate end-groups should be observable in UV, or with $(C_2H_5)_3O^+SbF_6^-$, assuming, that in the FPT-^{1}H-NMR spectra, $\underline{CH_3}CH_2O$ triplet from the end-group should be seen.

C_2H_5ONa, $N(CH_3)_3$ and $P(C_6H_5)_3$ killing agents were used and studied in FPT-^{1}H-NMR.

The benzonoate end-groups absorb at λ_{max}=230 nm (like the low molecular-weight benzoates); thus assuming that ε_{max} for ethyl benzoate is equal to ε_{max} of the benzoate end-groups the $\overline{DP}_n$ of poly-Diox were calculated. Table 1 summarizes some of these results where $\overline{DP}_n$(calcd.) are compared with $\overline{DP}_n$ (UV) and $\overline{DP}_n$ (osm.). The former values were calculated assuming, that the polymerization is a living one, i.e. that every molecule of initiator gives one macromolecule with no transfer, $\overline{DP}_n$(UV) was calculated as described above, and $\overline{DP}_n$(osm.) was measured by high-speed osmometry. Since polymers taken for measurements were isolated and purified by several dissolution/precipitation cycles, some amount of the lower molecular-weight material could be lost. Polymerization was conducted in CH_3NO_2 or in CH_2Cl_2 solvents at -15° in order to minimize the H$^-$ ion transfer (it has been shown in our laboratory that below -20°

the H⁻ ion transfer from e.g. dimethoxymethane to methoxycarbenium ion $-CH_3OCH_2SbF_6^-$ becomes immeasurably slow)

Table 1 |42|
Comparison of $\overline{DP}_n$(calcd.) $\overline{DP}_n$(UV) and $\overline{DP}_n$(osm.) of poly-1,3-dioxolans prepared with $C_6H_5CO^+SbF_6^-$ in CH_3NO_2 or CH_2Cl_2 solvents at $-15°$, and terminated with C_2H_5ONa. $|Diox|_o = 5.4$ mole·l^{-1}.

$|C_6H_5CO^+SbF_6^-|$ × 10³, mole·l⁻¹	$\overline{DP}_n$(calcd.) $\dfrac{|Diox|_o - |Diox|_e}{|C_6H_5CO^+SbF_6^-|_o} \cdot 10^{-3}$	$\overline{DP}_n$(UV) 10⁻³	$\overline{DP}_n$(osm.) 10⁻³
4.05	1.15	1.30	1.39
2.75	1.69	1.73	1.77
2.70	1.72	2.07	1.88
2.65	1.41	1.23	1.33
1.10	4.19	5.24	4.35
0.95	4.81	3.69	4.00

The second end-group, introduced upon a termination reaction, was observed by FPT-¹H-NMR for samples of poly--Diox, prepared from a perdeuterated Diox(Diox-d₆). This approach decresases an over-all number of protons in the sample and increases proportion of protons in the end-groups. Application of the Fourier-Pulse-Transform method for accumulation of the spectra enhanced the signal to noise ratio sufficiently to observe structure and concentration of the end-groups by FPT-¹H-NMR. Some of the pertinent results are shown in Table 2.
An agreement (within 20-25%) between $\overline{DP}_n$ calculated and measured by UV and/or ¹H-NMR methods indicates that practically all of the initiator used is present in the macromolecules. An agreement between $\overline{DP}_n$ found from the end-groups and $\overline{DP}_n$ measured osmometrically means, that the proportion of cyclic macromolecules is low, as it could be predicted, for instance, from the Jacobson-Stockmayer theory. This proportion, in principle, could be detected by comparing $\overline{DP}_n$ (end groups) and $\overline{DP}_n$(osmometry) but our accuracy of measurements is,

at least at present, not sufficiently high for these comparisons.

Table 2
Comparison of $\overline{DP}_n$(calcd.) and $\overline{DP}_n$(NMR) of poly--1,3-Diox, prepared from perdeuterated Diox(-d_6). Polymerization conditions: CH_3NO_2 solvent, -15⁰, 12 hrs.

Starting concn. of initiator 10^3 mole·1^{-1}	|Diox-d_6|$_o$ mole·1^{-1}	$\overline{DP}_n$(calcd.) $\dfrac{\|Diox\|_o - \|Diox\|_e}{\|initiator\|_o}$	$\overline{DP}_n$ found from ^{1}H-NMR	
			Initiator	Terminating agent
|$C_6H_5CO^+SbF_6^-$|$_o$ 4.7	4.05	700	$C_6H_5C(O)O$ 650	$-OC_2H_5$ 575
|$(C_2H_5)_3O^+SbF_6^-$|$_o$ 5.2	4.75	750	C_2H_5O- 1000	$-\overset{+}{P}(C_6H_5)_3$ 920

Thus, we can conclude, that poly-Diox, prepared with $C_6H_5CO^+SbF_6^-$ or $(C_2H_5)_3O^+SbF_6^-$ initiators are mostly linear, and macromolecules contain an initial end-group coming from an initiator and the terminal end-group coming from the terminating agent, e.g.:

$$C_6H_5\overset{O}{\overset{\|}{C}}(OCH_2CH_2OCH_2)_nOCH_2CH_3$$
$$CH_3CH_2(OCH_2CH_2OCH_2)_n\overset{+}{P}(C_6H_5)_3 \quad (4)$$

Polymerization degrees measured indicate also, that polymerization (at least in conditions given in Table 1 and 2) proceeds without an appreciable transfer affecting the polymerization degree.

Structure of Poly-Diox (cyclic vs linear) and Mechanism of Polymerization |43|.

As it will be shown in this paragraph, neither predominantly linear nor predominantly cyclic structures of the isolated, <u>killed</u> macromolecules are the straightforward arguments by themselves for the linear or cyclic growth of the <u>living</u> macromolecules. Indeed, let us consider an assumed equilibrium between living cyclic and living linear poly-Diox:

(5)

$$X-\overset{a}{\underset{b}{O}}\overset{\boxed{c}}{\underset{|}{O}}\overset{}{\underset{CH_2-O\sim\sim}{O}}\overset{}{\underset{}{O\sim\sim}} \rightleftharpoons X-O\underset{}{\overset{}{O}}\overset{a}{\underset{b|}{O_+^{\boxed{c}}}}\overset{}{\underset{CH_2-O\sim}{O\sim\sim}} \rightleftharpoons \text{linear living macromolecule}$$

7A   7B   7C

Equilibrium (5), describing the instantaneous state, should also be supplemented with the temporarily dead cyclic and linear (holding two ends coming from an initiator X) macromolecules. Cyclic living macromolecules 7A and 7B are results of the back-biting leading to the end-to-end closure (7A) or a back-biting to any of the oxygen atoms in the chain. The probability of the former process is enhanced very much, particularly at the early stages of polymerization, when the oxygen atom in the initial end-group (e.g. oxygen atom in the ether end-group) is much more nucleophilic than the oxygen atoms in the acetal bonds along the chain.

Let us now examine reaction of a killing agent with these living macromolecules. The linear living macromolecules will give their linear dead replica, but the cyclic-living ones may give either cyclic-dead or linear-dead macromolecules, depending on the initiator used, and therefore on a structure of X in 7A. In this structure there are three nonequivalent bonds: a, b, and c, that can be broken upon an attack of the killing agent.

If X=e.g. CH_3 or C_2H_5, then there are in 7A two rather stable bonds a and c, and one much less stable acetal bond b. Thus, even if cyclic 7A were a predominant structure at some stage of polymerization, then their reaction with killing agent would give mostly linear dead macromolecules. Thus, although it has been shown in the previous paragraph, that poly-Diox prepared with $C_6H_5CO^+SbF_6^-$ and $(C_2H_5)_3O^+SbF_6^-$ are linear, this is not sufficient to argue that the chain growth proceeds with a linear macrocation 7C.

In order to distinguish between the extreme structures 7A and 7C, as predominant during the chain growth, it is, therefore, necessary to observe directly the position of X; in 7A it is adjacent to the positively charged oxygen atom, in 7C it is a part of the ether chain end. If $X=C_2H_5$, then the differences between the chemical shifts in ^{1}H-NMR are as follows:

$\underline{CH_3}CH_2\overset{+}{O}$ $\underline{CH_3}CH_2$

δ 1.75(t)   δ 1.15(t)

thus, the difference between chemical shifts is sufficiently large and both structures can be independently observed by ^{1}H-NMR.

This is shown in Figure 1, taken from Reference |43|, and illustrating the change of the position of a triplet of the $\underline{CH_3}CH_2$-protons directly in the living polymerization system, consisting of a deuterated Diox(-d_6) and $(C_2H_5)_3O^+SbF_6^-$ initiator in CD_3NO_2 solvent. At the beginning of polymerization only the $\delta 1.75$ triplet is seen, while at equilibrium only the $\delta 1.15$ triplet; in the intermediate stages both triplets are observed. Additional splitting of the $\delta 1.15$ triplet into two triplets with a difference in chemical shifts equalling only 13 Hz (300 MHz spectrum) and the ratio of integrations 1:2, appears as a result of the simultaneous presence of two kinds of ethoxy groups, namely one from the polymer end-group, and the second one from ethyl ether, liberated from the initiating tertiary oxonium salt. Evacuation of the sample in high-vacuum removes completely ethyl ether, as it can be judged from the disappearance of its triplet from the spectrum.

Addition of the $N(CH_3)_3$ killing agent to the living system does not change the position of the $\delta 1.15$ triplet. The final spectrum of the killed system is shown in Figure 2 (also taken from Reference |43|. In this spectrum two singlets due to the $(\underline{CH_3})_3N$ and $(\underline{CH_3})_3N^+$ protons are observed, the ratio $|\underline{CH_3}CH_2O-|/|-N^+(\underline{CH_3})_3|$ (from the corresponding integrations) is equal to 1:1.1. Results reported in this paragraph, and based on the recent work from our laboratory, strongly indicate that polymerization of Diox, initiated by $(C_2H_5)_3O^+SbF_6^-$ proceeds, at least in CH_3NO_2 solvent, on the linear active species.

Systems with protonic acids initiators may behave differently, because if X=H in 7A, then the bond a (H-Ö⟨) in the secondary oxonium ion becomes the weakest one and this system is therefore much more susceptible to transfer (by transferring "H$^+$" from 7A) than systems initiated by stable cations. However, the thermodynamically controlled distribution of the polymerization degrees should be similar for both systems. This is, nevertheless, not so; for $(C_2H_5)_3O^+SbF_6^-$ and $C_6H_5CO^+SbF_6^-$ initiating systems described in References |42| |43|, $\overline{DP}_n$(calcd.) is equal to $\overline{DP}_n$(found) (within<25%) while in hands of Plesch |35|, working with $HClO_4$ initiator, $\overline{DP}_n$ found experimentally were always much lower than $\overline{DP}_n$(calcd.).

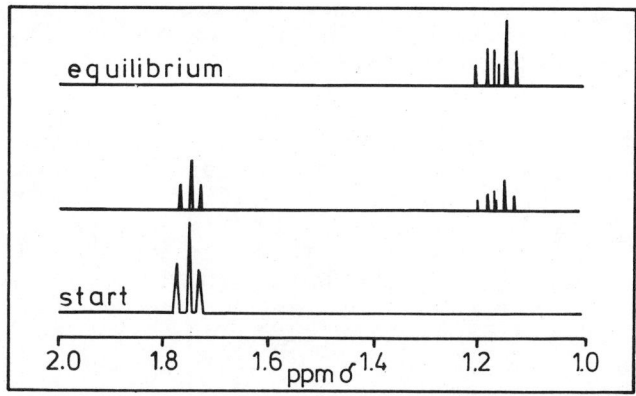

Figure 1. 300 MHz-^{1}H-NMR spectra of the absorption region from 1.0 to 2.0 δ ppm (C$\underline{H}_3$CH$_2$O groups, in ionic and covalent species) of the polymerization of Diox-d$_6$ (|Diox| = 4.0 mol · l^{-1}) initiated wtih $(C_2H_5)_3O^+SbF_6^-$ ($3.10^{-2°}$ mol · l^{-1}) in CD$_3$NO$_2$ solvent at 25° (43)

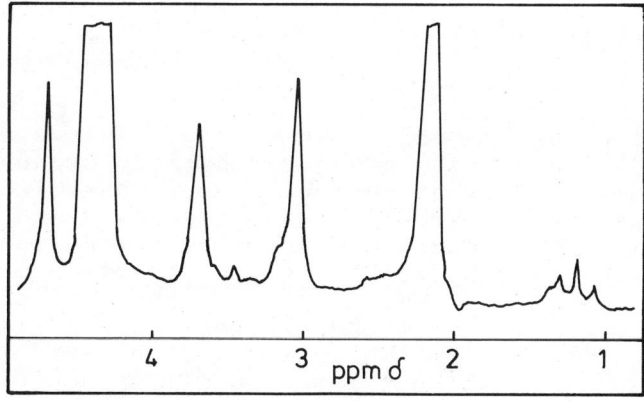

Figure 2. ^{1}H-NMR spectrum showing both end groups in poly (Diox-d$_6$): C$\underline{H}_3$CH$_2$O– at δ 1.15 and –N$^+$(CH$_3$)$_3$ at δ 3.05 (CH$_3$)$_3$N added in five-fold excess over growing species. Polymerization conditions as in Figure 1 (43)

Structure of the Active Species in the Polymerization of Cyclic Acetals.
Kinetics of Propagation.

In the previous paragraph a direct spectroscopic evidence eliminated the "ring-expansion" assumption as the supposed mechanism of propagation in the polymerization of Diox initiated with stable cations. This mechanism requires in fact, that the bond b in 7A is so strong, that during the whole polymerization process it does not brake in any other process but four-center inclusion of a monomer. The stability of bond b can be studied on the model compounds, for instance 8a bearing close resemblance to 7A :

$$CH_3OCH_2^+ \;+\; O\!\!<\!\!^{CH_3}_{CH_3} \;\rightleftharpoons\; CH_3\overset{a+c}{-\!O\!-}CH_3 \atop \underset{O-CH_3}{\underset{CH_2}{\overset{b|}{|}}} \tag{6}$$

(with SbF_6^- anion)

Szymański |44| has shown recently in our laboratory, that in reaction of 8 (at least in SO_2 solvent) with an excess of dimethylether :

$$CH_3\overset{a+c}{-\!O\!-}CH_3 \atop \underset{O-CH_3}{\underset{CH_2}{\overset{b|}{|}}} \;+\; O\!\!<\!\!^{CH_3}_{CH_3} \;\not\rightarrow\; 8 \;+\; O\!\!<\!\!^{CH_3}_{CH_3} \quad (CH_3)_3O^+ \;+\; (CH_3O)_2CH_2 \tag{7}$$

only bond b, brakes, as shown in eq.7. This exchange reaction is very fast (as judged by line broadening in NMR) inidicating, that bond b is very weak. Two equivalent bonds a and c are much more stable.

This latter conclusion is based on the fact, that the addition of $(CH_3)_3O^+SbF_6^-$ to the reaction mixture consisting of 8 and $(CH_3)_2O$ gives a <u>separate</u> singlet of $(CH_3)_3O^+$ (because of the slow (on the NMR time-scale) exchange with $(CH_3)_2O$))which was not observed in the reaction mixture without intentionally added $(CH_3)_3O^+SbF_6^-$. Thus, the trimethyloxonium salt is not formed in a system described by eq.7.

We previously proposed, that the linear growing macrocation in the polymerization of Diox is best described by the equilibrium between the macroalkoxycarbenium ion 2 with its tertiary oxonium counterpart 3, which, at the high ratio of a polymer to active species, becomes a predominant structure.

Our recent data |45| on the rate constant of propagation for Diox initiated by $C_6H_5CO^+SbF_6^-$, are in agreement with previously reported $k_p \approx 10^2$ l·mole^{-1}·s^{-1} at 25º in CH_2Cl_2 solvent. This is higher than k_p reported for polymerization initiated by $HClO_4$ |35|. Perhaps, in the polymerization of cyclic acetals the macroester-macroion-pair equilibrium, described for the polymerization of THF, also takes place, as proposed recently for the ClO_4^- anion |46|.

REFERENCES :

1. S.Penczek, Makromol.Chem. 175, 1217 (1974)
2. B.A.Rosenberg, E.B.Ludvig, A.R.Gantmacher and S.S. Miedwiediew, Vysokomol. Soed. 6, 2035 (1964)
3. K.Matyjaszewski, P.Kubisa and S.Penczek, J.Polymer Sci. A12, 1333 (1974)
4. T.K.Wu and G.Prukmayr , Macromolecules 7, 136 (1974)
5. S.Słomkowski and S.Penczek, J.Chem.Soc.Perkin II (1974), 1718
6. P.H.Plesch IUPAC International Symposium on Macromolecules Budapest 1969, Plenary Lecture, p.213
7. P.H.Plesch, I-st IUPAC International Symposium on Rings-Opening Polymerization, Jabłonna (Poland), 1975, Plenary Lecture, Pure & Appl.Chem., in press
8. B.A.Rosenberg, W.I.Irzak and W.S.Enikolopian "Interchain exchange in polymers" Chimia, Moscow, 1975 (in Russian)
9. K.Ivin and J.Leonard, European Pol.J. 6, 331 (1970)
10. J.Leonard, Macromolecules 2, 661 (1969)
11. L.I.Kozub, M.A.Markevich, A.A.Berlin and N.S. Enikolopian Vysokomol.Soed. 10, 2007 (1968)
12. S.Penczek and K.Matyjaszewski, submitted for publication
13. R.Binet and J.Leonard, Polymer 14, 355 (1973)
14. W.K.Busfield, R.M.Lee and D.Merigold, Makromol.Chem. 156, 183 (1972)
15. M.Okada, K.Mita and H.Sumimoto, Makromol.Chem. 176, 859 (1975)
16. Y.Firat and P.H.Plesch, Makromol.Chem. 176, 1179 (1975)
17. H.Jacobson and W.H.Stockmayer, J.Chem.Phys. 18, 1600 (1950)
18. H.K.Hall Jr., M.K.Brandt and R.M.Mason, J.Amer.Chem. Soc. 80, 6420 (1958)
19. H.K.Hall Jr. and J.H.Baldt, J.Amer.Chem.Soc. 93, 140 (1971)

20. Y.Yumoto, J.Chem.Phys. 29, 1234 (1958)
21. J.Šebenda, I-st IUPAC International Symposium on Ring-Opening Polymerization, Jabłonna (Poland), 1975, Plenary Lecture, Pure & Appl.Chem., in press
22. K.Pihlaja, Acta Chem.Scand. 25, 451 (1971)
23. Z.Jedliński, J.Łukaszczyk, J.Dudek,and M.Gibas, Macromolecules 9, 622 (1976) and references cited thereof
24. J.Kops and Spanggaard, Makromol.Chem. 175, 3077 (1974)
25. T.Mukaiyama, T.Fujisawa, H.Nohira,and T.Hyngaji, J.Org.Chem. 27, 3337)1962)
26. K.Tada, T.Saegusa,and J.Furukawa, Makromol.Chem. 95, 168 (1966)
27. M.Sumitomo, M.Okada,and H.Ito, J.Polymer Sci. A1, 6, 3182 (1968)
28. J.Martinez-Madrid and J.L.Mateo, Makromol.Chem. 136, 113 (1970)
29. Z.Jedliński, J.Maślińska-Solich, J.Polymer Sci.A1, 6, 3182 (1968)
30. J.P.Kennedy and A.L.Langer, Fortschr, Hochpolym. Forsch. 3, 508 (1964)
31. H.Minato and N.Muramatsu, Bull.Chem.Soc.Japan 42, 1146 (1969)
32. W.Kern, H.Deibing, A.Giefer, and V.Jaacks, Pure & Appl.Chem.12, 37 (1966)
32a.C.E.Schweitzer, R.N.Mc Donald, and J.O.Punderson, J.Appl.Polymer Sci. 1, 185 (1959)
33. A.Stolarczyk, P.Kubisa and S.Penczek, submitted for publication
34. W.S.Gresham, U.S.P. 2394910 (1946)
35. P.H.Plesch and P.H.Westermann, J.Polymer Sci C16 , 3837 (1968)
36. V.Jaacks, K.Boehlke, and E.Eberius, Makromol.Chem. 118, 354 (1968)
37. Y.Firat, F.R.Jones, P.H.Plesch, and P.H.Westermann, Makromol.Chem.Suppl. 1, 203 (1975)
38. E.J.Black and D.J.Worsfold, J.Macromol.Sci. A9, 1523 (1975)
39. Z.N.Nysenko, E.L.Berman, E.B.Ludvig, A.P.Klimow, W.A.Ponomarenko, and G.W.Isagulanz, Vysokomol. Soed. 18, 1696 (1976)
40. Y.Yamashita, M.Okada,and H.Kasahara, Makromol.Chem. 117, 256 (1968)
41. J.M.Andrews and J.A.Semlyen, Polymer 13, 142 (1972)
42. P.Kubisa and S.Penczek, submitted for publication
43. R.Szymanski in preparation
44. P.Kubisa, Bull.Acad.Pol.Sci.,in press

6

Ring-Opening Polymerization of Macrocyclic Acetals

ROLF C. SCHULZ, K. ALBRECHT, C. RENTSCH, and Q. V. TRAN THI

Institute of Organic Chemistry, University of Mainz, D-65 Mainz, West Germany

Numerous oxacyclic compounds are well known to form polymers in the presence of cationic initiators. In this way polyethers and polyacetals are obtained (1)-(11). Besides the parent compounds, listed in Table 1 many substituted oxacycles, furthermore bicyclic (12)-

Table I. Some polymerizable oxacyclic compounds

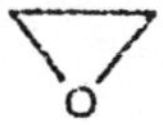

b.p.10,7°C (1)-(4)

b.p.47,7°C (1) (4) (5)

b.p.65°C (1) (4) (6)

b.p.121,5°C (7)

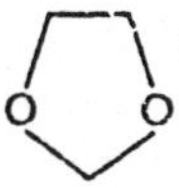

b.p.76°C (8)

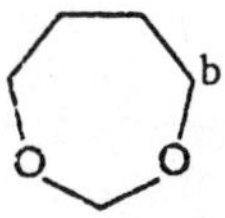

b.p.119°C (9)

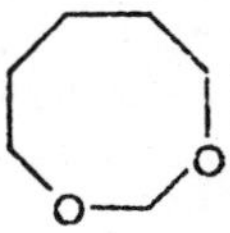

b.p.1,33,2°C (1o)

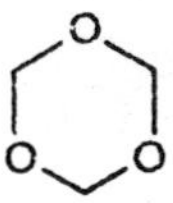

m.p.64°C
b.p.119°C (11)

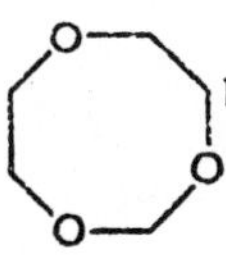

b.p.152°C (1o)

(15) as well as spirocyclic oxygen containing compounds (16);(17) are polymerizable. The polymerization mechanism does not only depend on the monomer, but also on the initiator and the experimental conditions.

In particular the polymerization of 1.2.- and 1.3-epoxides (1)-(5);(18);(19); tetrahydrofurane (1)-(4);(6);(2o); dioxolane (21);(22) and trioxane (11); (23)-(26) was thoroughly investigated. For reviews see (4);(8);(27);(28);(3o). It should be emphasized, that different oxacyclic monomers can also be copolymerized by cationic catalysts. Of great practical importance is e.g. the copolymerization of trioxane with ethylene oxide or dioxolane (31). Macromolecules with a statistic distribution of oxymethylene- and oxyethylene-units are formed in this way. On the other hand, however, the homopolymerization of dioxolane yields a polymer consisting of strictly alternating oxymethylene- and oxyethylene units (21);(32); therefore it can formally be considered as an alternating copolymer (eq.i).

$$\underset{O\diagdown O}{\bigcirc} \longrightarrow \left[-\underset{M}{CH_2O} - \underset{E}{CH_2CH_2O} - \right]_x \qquad (i)$$

It is not formed by a normal copolymerization starting from 2 different monomers, but since the monomer itself already contains both units in the ratio of 1 to 1. We wanted to investigate, whether it would be possible to prepare copolymers with other sequences from analogous monomers by homopolymerization. For this purpose one needs cyclic acetals, which contain the oxymethylene- and oxyethylene-units in the desired molar ratio. Of course during the polymerization of these monomers no elimination of formaldehyde or rearrangement may occur, since otherwise the regular sequence in the polymer is disturbed.

Monomers, which should be able to form sequenced copolymers according to the described principle, are the compounds /1/-/6/.

In the following, preparation and properties of these monomers and the corresponding polymers will be described.

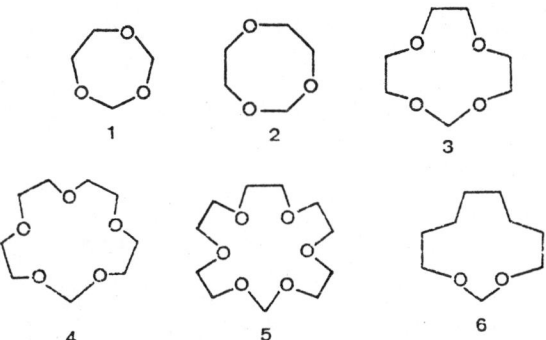

1.3.5-trioxacycloheptane.(trioxepane)/1/

Trioxepane /1/ is formed as a by-product during the copolymerization of trioxane or tetroxane with ethylene oxide or dioxolane (33)-(35). For its preparation a mixture of dioxolane, paraformaldehyde and sulphuric acid as catalyst is heated up to 100°C for 5 h. Afterwards one distils at 12 torr and 50°C (36). After repeated fractionational distillations from lithium-aluminium hydride a gaschromatographically pure monomer is obtained (b.p. 130°C). In the H-NMR spectrum only two sharp singlets appear (see Fig.1). The signal at δ = 4.92 ppm is assigned to the methylene protons (M) and the signal at δ = 3.82 ppm to the ethylene protons (E). The peak ratio is exactly 1 to 1. An addition of shift reagents (Eu(FOD)3) leads to a shift without splitting of the signals (37).

The monomer is easily polymerizable by cationic catalysts in solution and in bulk. Colourless, waxlike polymers are obtained. At the polymerization of /1/, either the bond between O1 and C2 or the bond between C2 and O3 can be cleaved. In both cases polymers with the same triad-sequence consisting of 2 oxymethylene and 1 oxyethylene-units (MME) would occur.

$$\text{structure} \longrightarrow -\underset{M}{CH_2O}-\underset{M}{CH_2O}-\underset{E}{CH_2\,CH_2O}- \longleftarrow \text{structure} \quad \text{(ii)}$$

A change of the ring opening mechanism or a transacetalization would of course lead to other sequences.

Whereas Gresham and Ball (36) assume, that the polymers of trioxepane have a regular structure with a ratio of 2 M to 1 E, Duke (38) concluded from IR- and NMR-measurements, that longer M-sequences must exist. In our own NMR-spectroscopic investigations we also found, that in the homopolymers of trioxepane also MMM-triads (δ = 4,89 ppm) and EME-triads (δ = 4,77 ppm) occur beside the expected MME-triads (see Fig.1). Furthermore from the 13C-NMR-spectra we were able to determine pentad-sequences (see Fig.2) and after addition of Eu(FOD)3 even heptad-sequences (37). Beside this we confirmed, that in the polymer the mole fraction of the M-units is evidently larger than the calculated value of 0,666. Therefore the polymer made from /1/ has neither the right overall composition nor the expected regular structure.

In order to clear up these anomalies the progress of the polymerization in dichloroethane with borontrifluoride at different temperatures was investigated. Hereto the decrease of the monomer has been determined by gas chromatography (39). An example of a time-conversion curve is shown in Fig.3. The polymerization proceeds rather quickly; the monomer concentration reaches a final state, which does not change over several hours. This concentration increases with increasing polymerization temperature (see Table 2). These facts lead us to conclude that it is an equilibrium polymerization. The plot of ln (M) against 1/T for temperatures between 0° and 60°C is shown in Fig.4. We calculated ΔS_{SS} = -18,9 J/Mol·K and ΔH_{SS} = -6,6 kJ/Mol. By extrapolation to a monomer concentration of (M) = 1 Mol/l in equilibrium, a formal ceiling temperature of 80°C results. In fact at 80°C and with a monomer concentration of 1 Mol/l no polymerization takes place.

But as we found, in the gas chromatogramm of the reaction mixture, dioxolane too is formed during polymerization (see Fig.3). This fact explains the NMR-spectroscopic statement, that the polymer does not have the same composition as the monomer, but contains an excess of M-units. The concentration of dioxolane also reaches a final value, which increases with rising polymerization temperature (see Table 2). But this means, that the composition of polymer depends on temperature and approaches the theoretical value only at low polymerization temperature. The described results show, that in the polymerization of trioxepane

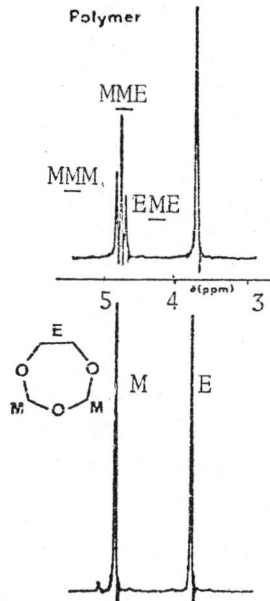

*Figure 1. H-NMR spectra of trioxepane /**1**/ and the polymer*

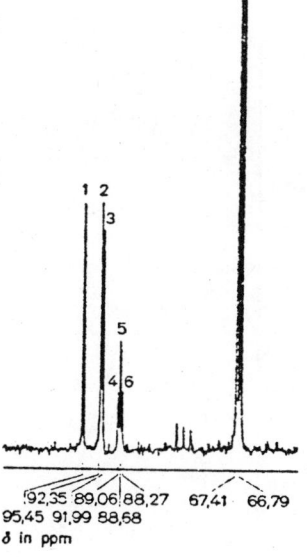

Figure 2. ^{13}C-NMR spectrum of a polymer of trioxepane ($CDCl_3$; 25, 2 MHz). (1) MEMEM; (2) MMMEM; (3) EMMEM; (4) MMMMM; (5) EMMMM; (6) EMMME; (7) MMEMM; (8) EMEME, EMEMM.

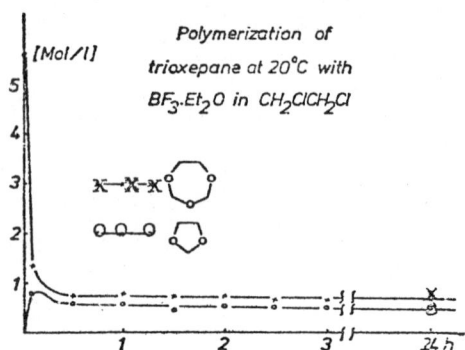

Figure 3. Time-conversion curve for the consumption of monomer (×) and formation of dioxolane (○) during the polymerization of trioxepane /1/

Table II. Equilibrium concentration of Dioxolane $[DOL]_e$ during polymerization of trioxepane /1/ with BF_3·etherate in dichloroethane

$[M]_0$ (Mol/1)	temp. (°C)	$[DOL]_e$ (Mol/1)
5,55	0	0,23
5,72	20	0,50
5,82	30	0,60
5,50	45	1,01
5,71	60	1,36

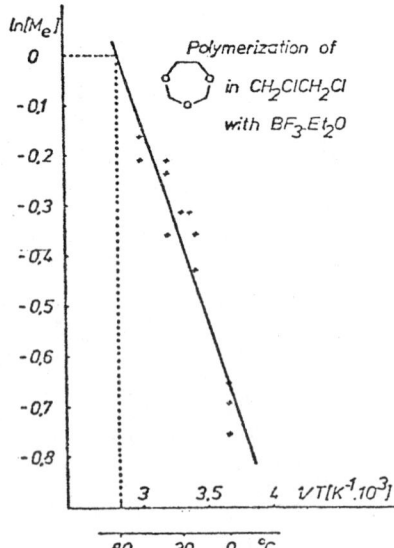

Figure 4. Monomer concentration at equilibrium in the polymerization of trioxepane /1/

several reactions occur simultaneously and different polymers and monomers are formed side by side. The following scheme comprises the observations.

```
                        (iv)
         [dioxolane]  +  CH₂O  ⇌  stat. copolymer

                        (VI)
(iii) ⇅              ╱

         [trioxepane]  ⇌  polymer + x [dioxolane]
              (V)
```

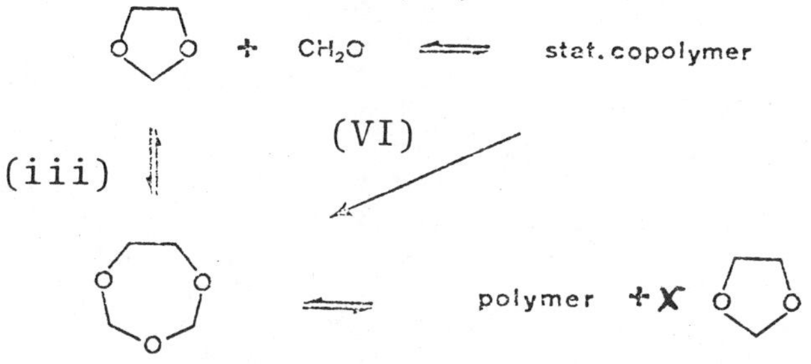

Reaction (iii) describes the syntheses of trioxepane from dioxolane and formaldehyde; during polymerization, of course, a cleavage into the components could occur again. But under the same conditions a copolymerization of dioxolane and formaldehyde is also possible (iv), leading to a statistic copolymer. Polymerization of trioxepane (V) leads to a polymer whereby, however, a certain part "$\underline{x}$" of dioxolane is formed. Therefore the polymer contains somewhat more than the calculated quantity of M-units. Only at low temperature, when "$\underline{x}$" becomes zero, do monomer and polymer have the same over all composition and one can suppose a real polymerization-depolymerization-equilibrium. Whether the sequence is hereby retained, depends on the possibilities of ring opening and transacetalisation, discussed above. Reaction (vi) describes the above mentioned formation of trioxepane as a by-product during copolymerization of trioxane and dioxolane (33)-(35).

1.3.6-trioxacyclooctane (trioxocane) /2/

According to Astle (4o) et al. trioxocane can be obtained by condensation of diethylene glycol with paraformaldehyde; it is easily polymerizable by cationic catalysts or by electrochemical initiation (41). Copolymerization with e.g. trioxane or dioxolane are possible, too (41). Kinetics, thermodynamics and mechanism of homopolymerization have been studied in detail by several authors. According to the analytic results of Weichert (42) the structure of the polymers of /2/

can only be explained by a ring opening mechanism between oxygen and carbon of the formal group. Later, Yamashita et al. (43) studied the kinetics of the polymerization of /2/ in dichloroethane using triethyl oxonium tetrafluoroborate. They found that the number average molecular weight of the polymer remains almost constant throughout the polymerization (DP $\sim$ 3o) indicating that the rate of initiation is considerably lower than that of propagation and that the polymerization is accompanied by some chain breaking reactions. The thermodynamic parameters of the equilibrium polymerization of /2/ (and some related cyclic formals) was studied in more detail by Yamashita et al. (22) and by Busfield and Lee (44). Furthermore, it was postulated that the polymer degraded exclusively to monomer in the presence of boron trifluoride.

In our own work we were occupied predominantly with the NMR-spectroscopic sequence analysis, in order to see, whether the concept for preparing sequenced copolymers, described at the beginning, could be validated (41). In the H-NMR-spectrum the monomer /2/ shows only two sharp singlets at δ = 3,8 ppm (E) and δ = 4,9 ppm (M) with a peak ratio of 4 to 1 (see Fig.5). Addition of Eu (DPM)3 effects a shift to lower field and a strong splitting of the ethylene signal, as the protons at C4 and C8 are not equivalent to the protons at C5 and C7. In the H-NMR-spectra of the homopolymer also only two peaks occur, having the same peak ratio as in the monomer (see Fig.5). From this, it can be concluded, that not only the overall composition but also the order of M- and E-units is the same in polymer and in monomer. Hence there is only one kind of ring opening and rearrangements or eliminations can be excluded. That means, in fact, that at the ring opening homopolymerization of trioxocane, a sequenced copolymer with a regular sequence of (MEE)-triads is formed (eq.vii).

$$\text{[trioxocane ring]} \longrightarrow \left[-\underset{M}{CH_2O} - \underset{E}{CH_2CH_2O} - \underset{E}{CH_2CH_2O} - \right]_x \quad (vii)$$

This finding agrees with the results of Weichert (42) who analysed the structure of polytrioxocane by acidic decomposition. Indications of endgroups have not been found, from which one should not conclude, that macrocyclic polymers are in hand.

1.3.6.9-tetraoxacycloundecane (triethylene glycol formal) /3/

This compound and its polymerizability was first mentioned by Carothers (45). It is prepared from triethylene glycol and paraformaldehyde; by addition of strong acids a prepolymer is produced. From this, the monomer /3/ is split off in a second step by heating in vacuo. The monomer used by us has the following properties: m.p. 27°C; b.p. 56°C/0,4 Torr; n_D^{30} = 1,4541; NMR-signals of /3/ were first reported by Burg (49). NMR data obtained by us are summarized in Table III and IV.

/3/ is polymerizable by several cationic initiators in solution and in bulk at temperatures between -20°C and +150°C. The polymers are colourless waxlike substances; they are readily soluble in water, THF, aromatic hydrocarbons, alcohols and halogenated hydrocarbons.

In the H-NMR spectrum of the polymer only 3 sharp peaks appear at δ = 4,72; 3,67 and 3,65 ppm (see Table III. The peak ratio of M:E = 1:6 agrees with that of the monomer. The 13C-NMR-signals are at δ =95,4; 70,4 and 66,8 ppm (see Table IV). There are no indications of an irregular structure and we therefore conclude, that the polymer at least contains very long blocks of (MEEE)-tetrads and consequently can be described as a sequenced copolymer (eq.viii).

$$\left[-\underset{M}{CH_2O} - \underset{E_3}{(CH_2CH_2O)_3} - \right]_x \quad (viii)$$

After establishing the structure of the polymer we studied in detail the way of formation. Hereto we carried out solution-polymerizations in methylene chloride under argon-atmosphere (50). Monomer concentrations were between 0,15 and 2,5 Mol/l, temperature between -20°C and +20°C. Trifluoromethane sulphonic acid serves as catalyst. After definite times, polymerization was quenched by the additon of some basic aluminium oxide or triethylamine and the

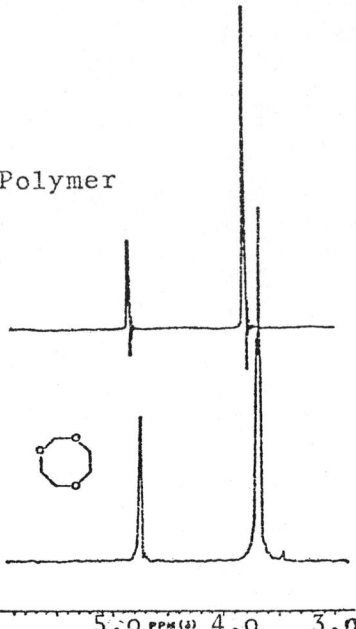

Figure 5. H-NMR spectra of trioxocane /2/ and its polymer

Table III. H-NMR signals of triethylene glycolformal (M_1), the polymer (P) and the oligomers of the general formula/7/

	$-O-CH_2-O-$	$-O-CH_2-CH_2-O-$	
M_1	4,79	3,83	3,69
M_2	4,75	3,72	3,68
M_3	4,75	3,70	3,67
M_4	4,75	3,69	3,67
M_5	4,74	3,69	3,66
M_6	4,74	3,69	3,66
M_7	4,75	3,69	3,66
M_8	4,74	3,68	3,67
P	4,72	3,67	3,65

composition of the reaction mixture was analized by
high pressure gelpermeation chromatography (HP-GPC)
(Waters ALC/GPC 2o1 with Ri detector R 401; Stationary
phase: μ- Styragel; 100 Å + 500 Å mobile phase:
methylene chloride).

It appears that under the applied reaction conditions not only polymers (with molecular weights from 10,000 to 80,000) are formed, but also noticeable amounts of several oligomers (labelled as M_2 to M_i, see Fig.6). If the consumption of monomer M, the formation of oligomer M_2, and the total of all higher oligomers and polymers are plotted as a function of time, time-conversion curves result as shown in Fig.7. One can see, that after about 3o minutes a final state is reached with about 3,5% residual monomer and about 9% M_2.

If pure polymer is treated with trifluoromethane sulphonic acid under the same conditions, eventually exactly the same final state (referring to type and amount of monomer, oligomer and polymer) is reached (see Fig.8). Hence it is surely a matter of a thermodynamic equilibrium polymerization. Within a range of initial monomer concentration between 0,2 to 0,5 Mol/l the equilibrium monomer concentration is constant and amounts at 0°C to (0,0146 ± 0,0016) Mol/l. The equilibrium concentration of the dimer at 0°C is (0,0236 ± 0,0013) Mol/l. The temperature dependence of these concentrations was studied for the polymerization in methylene chloride between -25°C and +30°C with trifluoromethane sulphonic acid as catalyst. A Dainton-plot of the results is shown in Fig.9. We calculated from the slope and the intercept ΔH_{ss} = (-1,9 ± 0,2) kcal/Mol = (-7,95 ± 0,8)kJ/Mol and ΔS^o_{ss} = (+1,5 ± 0,5) cal/Mol·K = (+6,24 ± 2,1) J/Mol·K. The small and positive entropy is noticeable.

Cyclic oligomers of the triethylene glycol formal

From the above mentioned results, it follows that the observed oligomers are not by-products, but are all present in a reversible equilibrium with the monomer and the polymer.

Therefore it is important to know their structure and - if possible - the way of formation.

We succeeded in isolating and identifying the first 8 members of the homologous series of oligomers by preparative GPC. Recently a detailed description has been published by us (51). The substance called M_2

Table IV. ^{13}C-NMR signals of triethylene glycolformal (M_1), the polymer (P) and the oligomers of the general formula /7/

	-O-CH$_2$-O-		-O-CH$_2$-CH$_2$-O-	
M_1	96,2	70,6	70,4	67,8
M_2	95,1	70,8	70,6	66,4
M_3	95,3	70,4	70,3	66,6
M_4	95,3	70,4		66,6
M_5	95,4	70,4		66,7
M_6	95,6	70,5		66,9
M_7	95,6	70,6		66,9
M_8	95,6	70,5		66,9
P	95,4	70,4		66,8

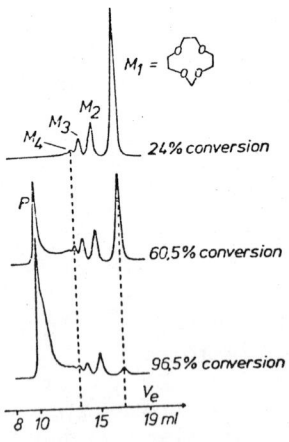

Figure 6. HP–GPC curves of the reaction mixture during the polymerization of triethylene glycol formal /3/

$[M_1]_0$ = 0,5 Mol/l; CH$_2$Cl$_2$; 0°C
$[CF_3SO_3H]$ = 0,1 Mol-%
μ-Styragel 500 Å + 100 Å;
1,0 ml CH$_2$Cl$_2$/min

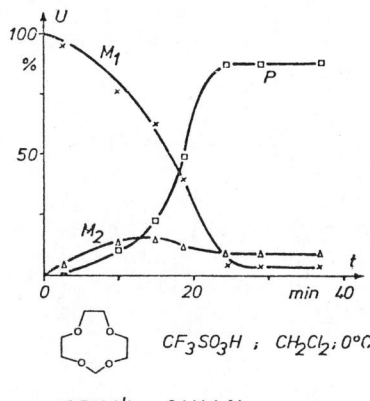

Figure 7. Time–conversion curve for the consumption of monomer (M_1), formation of polymer (P), and dimer (M_2) during the polymerization of triethylene glycolformal /3/ (determined by HP–GPC)

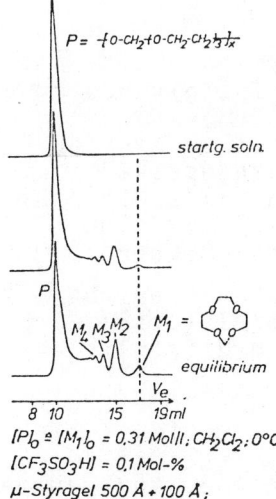

Figure 8. HP–GPC curves of the reaction mixture during depolymerization of a polymer of triethylene glycolformal /3/

in HP-GPC turned out to be the cyclic dimer of compound /3/. The dimer is polymerizable under the same conditions as /3/ and after reaching the equilibrium it leads to the same oligomer distribution as the monomer and the polymer (see Fig.1o).

All oligomers are colourless crystalline compounds. The melting points (see Table V) of the oligomers with even-numbered multiples of the monomers are always higher than the odd-numbered (47). The H-NMR-spectra are nearly identical for all oligomers and lead to the conclusion, that all have analogous structure (compare Tables III and IV). Indications of endgroups are not available either in the NMR- or in the IR-spectra, verifying that it is a matter of cyclic oligomers. The mass spectra of the dimer (M_2) and the trimer (M_3) gave the expected molecular ions. The gel-chromatographic elution volumes for all oligomers are on a common curve which is, however, clearly different from the curve for open chain ethylene glycol oligomers (Fig.11). This proves, that the oligomers occuring at the polymerization of /3/ (catalyzed by trifluoromethane sulphonic acid) have the following general structure /7/.

$$\left[-CH_2O-(CH_2CH_2O)_3- \right]_x \quad 7$$

Whether also the high polymers have ring structure, has hitherto not yet been definitely proved or disproved.

The formation of cyclic oligomers can be explained by two different mechanisms:
a) a stepwise ring extension takes place by insertion at the formal bond without formation of linear intermediates (27);(52)(see Scheme 1)
b) the chain growth proceeds by open chain carboxonium-ions (possibly in equilibrium with ester groups)(53);(54) and the cyclic oligomers arise by back-biting (see Scheme 2).

From our results we cannot decide, which mechanism prevails.

Finally it should be mentioned that during several other polycondensations and ionic polymerizations, the

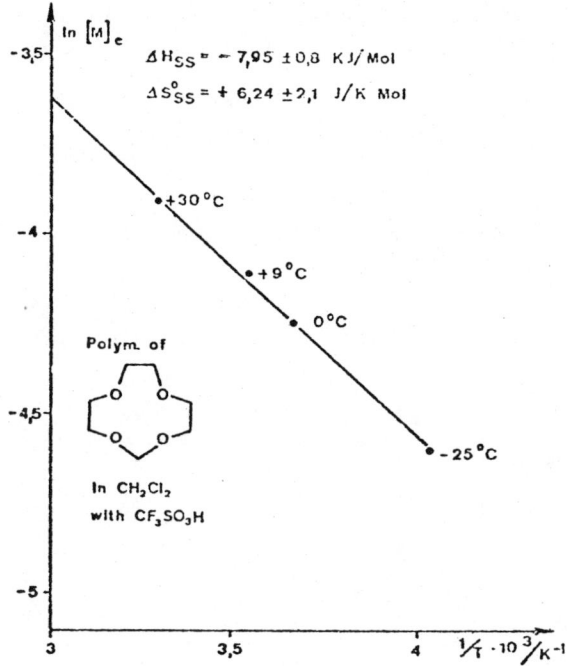

Figure 9. Monomer concentration at equilibrium in the polymerization of triethylene glycol formal /3/

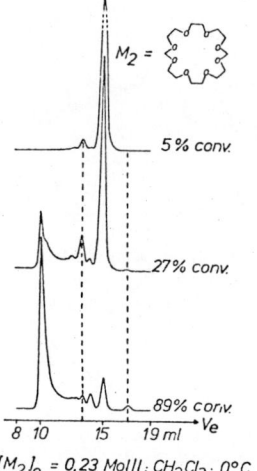

Figure 10. HP–GPC curves of the reaction mixture during the polymerization of the dimer of triethylene glycol formal (M_2)

Table V. Melting points of cyclic oligomers of the general structure /7/

	Degree of polym. x	number of ring atoms	melting point °C
M_1	1	11	27
M_2	2	22	88
M_3	3	33	27
M_4	4	44	56
M_5	5	55	19
M_6	6	66	38
M_7	7	77	23
M_8	8	88	28

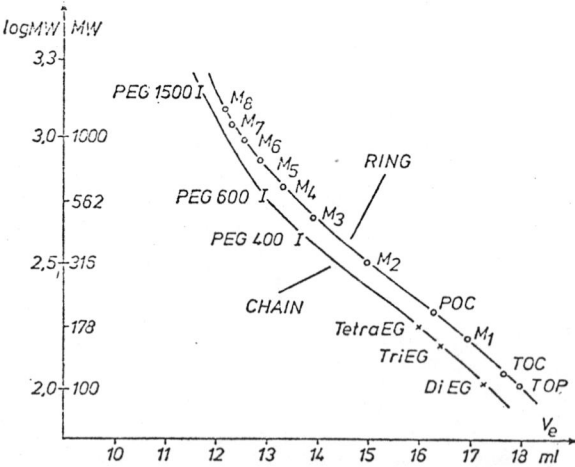

Figure 11. HP–GPC elution curves for homologous series of open chain ethylene glyols and cyclic formals (PEG, polyoxyethylene; EG, ethylene glycols; TOP, /1/; TOC, /2/; POC, /4/; M_1, /3/. M_2, M_3 ... cyclic oligomers of the general structure /7/.

Scheme 1

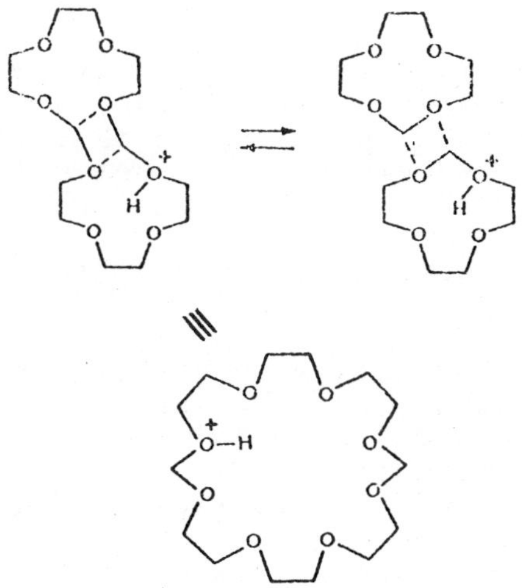

Scheme 2

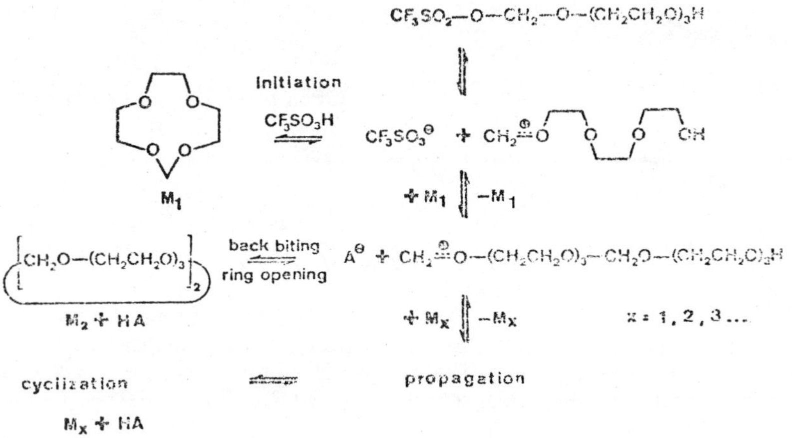

formation of cyclic oligomers has been described, too. E.g. cyclic oligomers of formaldehyde with 3 to 15 units have been found when trioxane is polymerized (49); also at the polymerization of episulphides or dioxolane, cyclic oligomers appear (55); further in the course of the formation of polyesters and polyamides as well as at the metathesis reaction of cycloolefins (57). It is to be expected that oligomers could be shown to be present in other polymerization processes by applying the GPC-technique.

1.3.6.9.12-pentaoxacyclotetradecane (tetraethylene glycol formal) /4/

This compound is not well known; till now it was only mentioned in a patent (58). We managed to prepare it in an analogous way to compound /3/ in good yield (46). After careful purification, colourless crystals with a melting point of 23,5°C are yielded (b.p. 51°C, 10^{-3} torr; H-NMR (CDCl$_3$) δ = 4,70(s;2H); 3,73 (s;8H) 3,63 (s;8H). This cyclic formal is also readily polymerizable by several initiators (as e.g. boron trifluoride etherate, tin tetrachloride, trifluoroacetic acid etc.) (59). At the solution polymerization in methylene chloride ($[M]$ = 2-3 Mole/l) at 0°C with trifluoromethane sulphonic acid (0,5 mole%) waxlike polymers are obtained (molecular weights 20,000 - 30,000), which are easily soluble in water and in organic solvents. In the H-NMR spectra only two peaks appear: δ = 4,73; 3,68 ppm.

Therefore here also, the expected structure with a regular sequence consisting of MEEEE-pentads, is in hand (see eq. IX).

$$\left[-\underset{M}{CH_2O} - \underset{E_4}{(CH_2CH_2O)_4} - \right]_x \quad (ix)$$

According to this structure the polymer is very similar to polyoxyethylene in some chemical and physical properties. It differs, however, from this polymer with respect to its viscosimetric behaviour in water and in organic solutions. The polymer is quickly decomposed by dilute acid solution in contrast the polyoxyethylene (59).

1.3.6.9.12.15-hexaoxacycloheptadecane (pentaethylene glycol formal) /5/

Pentaethylene glycol is prepared by condensation of 1.8-dichloro-3.6-dioxaoctane with 2 moles of ethylene glycol (60). In analogy to the synthesis described above, the cyclic formal is obtained. The H-NMR (in $CDCl_3$) shows the following peaks: δ =4,73 ppm (2H); 3,76 (8H); 3,69 (12H). In IR no OH-groups or carbonyl bands are detectable but broad absorption in ether and acetal region is observed (61).

1.3-dioxacycloundecane (octanediol formal) /6/

Corresponding to the cyclic acetals depicted above, this compound is prepared from 1,8-octandiol and paraformaldehyde (47). It is a colourless liquid (b.p. 196°C; n_D^{25} = 1,4564; d_4^{25} = o,985 g/ml; H-NMR ($CDCl_3$) δ = 4,66 ppm (s; 2H); 3,75 (m;4H); 1,56 (s;12H).

Polymerization in bulk or in solution between 0°C and 30°C with trifluoromethane sulphonic acid (0,2 mole %) leads to solid colourless polymers (H-NMR($CDCl_3$)
 δ = 4,71 ppm (s;2H); 3,57(m;4H); 1,37(m;12H); oligomers are also formed, but under these conditions no residual monomer could be detected. The polymers are soluble in aromatic or halogenated hydrocarbons; they are insoluble in water, alcohol and ether (62).

Acknowledgment

We should like to express our thanks to "Deutsche Forschungsgemeinschaft" for the financial support within the scheme of "Sonderforschungsbereich 41". C.Rentsch thanks the "Schweizerischer Nationalfonds" for granting him a scholarship. We thank "Hüthig and Wepf Verlag Basel" for the permission to publish some if the figures.

Literature Cited
1) Frisch, K.C., Reegen, S.L., (Eds): "Ring-opening Polymerization", pp. 13, 111, 159; M. Dekker, New York, 1969.
2) Hill, F.N., Bailey, F.E., Fitzpatrick, J.T., Ind. Eng. Chem. (1958), 50, 5.
3) Sorenson, W.R., Campbell, T.W., "Preparative Methods in Polymer Chemistry", 2nd ed., p. 367, Interscience Publ., New York, 1968.
4) Eastman, A.M., Advan. Polym. Sci. (1960), 2, 18.
5) Dreyfuss, P., Dreyfuss, M.P., Polym. J. (1976), 8, 81.
6) Kobayashi, S., Danda, H., Saegusa, T., Macromolecules (1974), 7, 415.
7) Seitz, U., Hoene, R., Reichert, K.H.W., Makromol. Chem. (1975), 176, 1689.
8) Plesch, P.H., Westermann, P.H., J. Polym. Sci. C (1968), 16, 3837.
9) Donescu, D., Makromol. Chem. (1974), 175, 2355.
10) Busfield, W.K., Lee, R.M., Makromol. Chem. (1975), 176, 2017.
11) Chen, C.S., J. Polym. Sci., Polym. Chem. Ed. (1975), 13, 1183.
12) Okada, M., Sumitomo, H., Hibino, Y., Polym. J. (1974), 6, 256.
13) Okada, M., Sumitomo, H., Yamata, Y., Makromol. Chem. (1974), 175, 3023.
14) Hall, H.K., Steuck, M.J., J. Polym. Sci., Polym. Chem. Ed. (1973), 11, 1035.
15) Andruzzi, F., Barnes, D.S., Plesch, P.H., Makromol. Chem. (1975), 176, 2053.
16) Bailey, W.J., J. Macromol. Sci., Chem. (1975), A 9, 849.
17) Endo, T., Katsuki, H., Bailey, W.J., Makromol. Chem. (1976), 177, 3231.
18) Beaumont, R.H., Clegg, B., Gee, G., Herbert, J.B.M., Marks, D.J., Roberts, R.C., Sims, D., Polymer (1966), 7, 401.
19) Black, P.E., Worsfold, D.J., Can. J. Chem. (1976), 54, 3325.
20) Dreyfuss, M.P., J. Macromol. Sci., Chem. (1975), A 9, 125, 729.
21) Plesch, P.H., Westermann, P.H., J. Polym. Sci. C (1968), 16, 3837.
22) Yamashita, Y., Okada, M., Suyama, K., Kasahara, H., Makromol. Chem. (1968), 114, 146.
23) Kern, W., Jaacks, V., J. Polym. Sci. (1960), 48, 399.
24) Kern, W., Cherdron, H., Jaacks, V., Angew. Chem. (1961), 73, 177.

25) Chen, C.S.H., J. Polym. Sci., Polym. Chem. Ed. (1976), 14, 129, 143.
26) Enikolopyan, N.S., et al., Polym. Sci. USSR (1976) 17, 742, 759.
27) Plesch, P.H., "IUPAC International Symposium on Macromolecular Chemistry, Budapest, 1969, Plenary and Main Lectures", p. 213, Akadémiai Kiado, Budapest, 1971.
28) Enykolopyan, N.S., J. Macromol. Sci., Chem. (1972), A 6, 1053.
29) Penczek, S., Makromol. Chem. (1974), 175, 1217.
30) Ledwith, A., Sherrington, D.C., Advan. Polym. Sci. (1975), 19, 1.
31) Weissermel,K., Fischer,E., Gutweiler, K., Hermann, H.D., Kunststoffe (1964), 54, 410.
32) Okada, M., Yamashita, Y., Ischii, Y., Makromol. Chem. (1964), 80, 196.
33) Miki, T., Higashimura, T., Okamura, S., J. Polym. Sci. B (1967), 5, 583.
34) Boehlke, K., Jaacks, V., Makromol. Chem. (1971), 145, 219.
35) Mengoli,G., Furlanetto,F., Makromol. Chem. (1975), 176, 143.
36) Gresham , W.F., Bell, C.D., U.S. Pat. 2 475 610 (1949), Du Pont, C.A. (1950), 44, 175b.
37) Fleischer,D., Schulz,R.C., Makromol. Chem. (1975), Suppl. 1, 235; (1976), 177, 3471 (Errata).
38) Duke, A.J., J. Chem. Soc. 1964, 1430.
39) Tran Thi, Q.V., unpublished results, Darmstadt, 1975.
40) Astle, M.J., Zaslowsky, J.A., Lafyatis, P.G., Ind. Eng. Chem. (1954), 46, 787.
41) Fleischer,D., Schulz,R.C., Makromol. Chem. (1972), 162, 103.
42) Weichert, D., J. Polym. Sci., C (1967), 16, 2701.
43) Okada, M., Kozawa, S., Yamashita, Y., Makromol. Chem. (1969), 127, 66.
44) Busfield, W.K., Lee, R.M., Makromol. Chem. (1975), 176, 2017.
45) Hill, J.W., Carothers, W.H., J. Amer. Chem. Soc. (1935), 57, 925.
46) Albrecht, K., unpublished results, Darmstadt, 1975.
47) Albrecht, K., Fleischer, D., Kane, A., Rentsch, C., Tran Thi, Q.V., Yamaguchi, H., Schulz, R.C., Makromol. Chem. (1977), 178, 881.
48) Albrecht, K., Fleischer, D., Rentsch, C.,Yamaguchi, H., Schulz, R.C., Makromol. Chem., in press.
49) Burg, K.H., Hermann, H.D., Rehling, H., Makromol. Chem. (1968), 111, 181.

50) Rentsch, C., Lectured at the Annual Assembly of Schweizerische Chemische Gesellschaft, Geneva, Oct. 1976.
51) Rentsch, C., Schulz, R.C., Makromol. Chem., in press.
52) Cooper, J., Plesch, P.H., JCS., Chem. Commun. 1974, 1018.
53) Plesch, P.H.,Brit. Polym. J. (1973), 5, 1.
54) Matyjaszewski, K., Penczek, S., J. Polym. Sci., Polym. Chem. Ed. (1974), 12, 1905.
55) Semlyen, J.A., Advan. Polym. Sci. (1976), 21, 41.
56) Goethals, E.J., Advan. Polym. Sci. (1977), 23, 103.
57) Höcker, H., Reimann, W., Riebel, K., Szentivanyi, Z., Makromol. Chem. (1976), 177, 1707.
58) U.S. Pat. 3 563 955; C.A. (1967), 66, 116143 r.
59) Kane, A., unpublished results, Mainz, 1976.
60) Krespan, C.G., J. Org. Chem. (1974), 39, 2351.
61) Albrecht, K., unpublished results, Mainz, 1977.
62) Tran Thi, Q.V., unpublished results, Mainz, 1977.

7

Macrocyclic Formals

YUYA YAMASHITA and YUHSUKE KAWAKAMI
Department of Synthetic Chemistry, Faculty of Engineering,
Nagoya University, Nagoya 464, Japan

Recent development of the chemistry of crown ethers pioneered by Pedersen (1) brought new interest to the old work of Carothers (2) on polymerization and ring formation. It is well recognized that cyclic oligomers are often formed during polymerization, and the development of high speed liquid chromatography made it possible to obtain quantitative data on cyclic oligomers. Semlyen (3) studied on thermodynamic equilibrium of cyclic oligomers by using the Jacobson-Stockmayer theory (4). We noticed that considerable amounts of cyclic oligomers were often formed during ring-opening polymerization. This is essentially due to the fact that the elementary reaction in ring-opening polymerization of cyclic compound containing heteroatom is nucleophilic substitution reaction and because the reactivity of the heteroatom in the polymer is not so different from the monomer, they compete with monomer for reaction with the propagating species causing backbiting reactions to form cyclic oligomers.
We have been interested in the cationic polymerization of cyclic formals (5,6). We have synthesized macrocyclic formals which have ether oxygen along with acetal oxygen, and found that the cationic polymerization of these monomers is accompanied with cyclic oligomers, which seems very interesting from kinetic and thermodynamic point of view. Besides, the chelating properties of these monomers and cyclic oligomers are also interesting.

Results and Discussion

Syntheses of Macrocyclic Formals. Quantitative yields of linear prepolymers having molecular weight of several thousands were obtained by refluxing polyethyleneglycols with paraformaldehyde in benzene solution by using p-toluenesulfonic acid as catalyst and by continuous removal of water formed. After distilling off the solvent, thermal decomposition of the prepolymers was carried out by heating without removing the catalyst on an oil bath kept at appropriate temperature. Macrocyclic formals formed by depolymerization were distilled under reduced pressure. In case of triethyleneglycol and tetraethyleneglycol, yields of more than 80% of cyclic formals are obtained at the bath temperature of 250°C. Small amounts of cyclic formals consisting of smaller rings are contaminated in the distillate. This is expressed in the following scheme (Figure 1). Although fairly high yield of cyclic formals is obtained with pentaethyleneglycol, it becomes difficult to obtain pure cyclic formals from hexaethyleneglycol because of the contamination of large amounts of cyclic formals consisting of smaller rings. In the following, we use the abbreviated name such as 11-CF-4 to express eleven membered cyclic formal containing four oxygen atoms.

Properties of Macrocyclic Formals. Macrocyclic formals are hygroscopic liquids and are soluble in ordinary organic solvents and water. Although they are easily hydrolyzed in acidic solution, they are fairly stable in basic solution, and can be used as accelerating agents in nucleophilic synthetic reactions. We compared the rate of an S_N2 reaction between n-butyl bromide and alkali metal acetate accelerated by addition of equimolar amounts of macrocyclic formals.

$$nBuBr + CH_3COOM \longrightarrow nBuOCOCH_3 + MBr$$

Typical results are shown in Figure 2. It is clear that these cyclic formals are less effective on the rate enhancement of S_N2 reaction compared with crown ethers. It was also shown that the equilibrium constant of chelate formation was smaller for these cyclic formals than crown ethers which have same number of oxygen atoms. This might be caused by the less basic acetal oxygen atoms compared with ether oxygen atoms and by the nonplanarity of the chelate complex of cyclic formals.

Cationic Polymerization of 11-CF-4. Cationic polymerization of 11-CF-4 was carried out with boron trifluoride ether complex in dichloromethane. The progress of the reaction was monitored by gas chromatography and liquid chromatography by using n-tetradecane as an internal standard. In Figure 3 is shown the

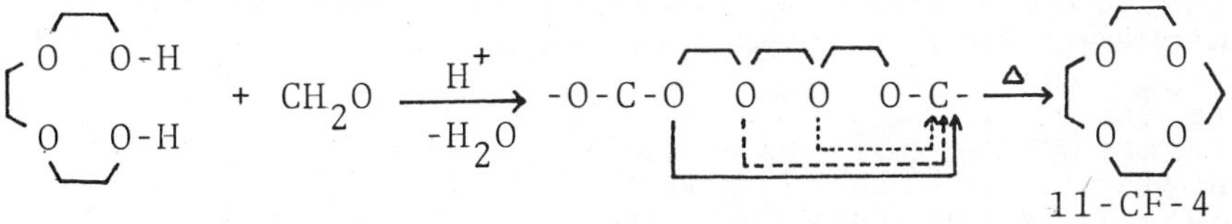

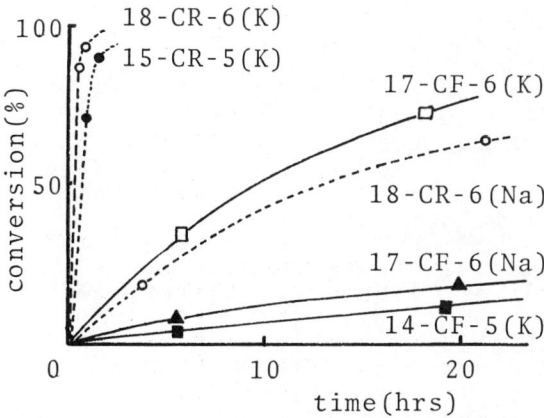

Figure 1. Synthesis of macrocyclic formals

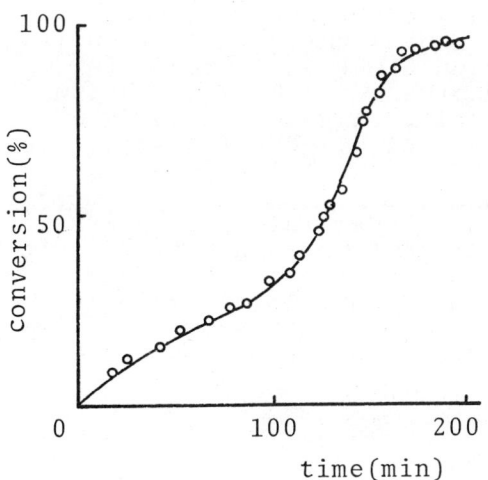

Figure 2. Rate enhancement by addition of macrocyclic formals. nBuBr, 0.01 mol; alkali metal acetate, 0.01 mol; benzene, 9 ml; toluene, 1 ml; chelating agent, 0.01 mol. Temperature, 90°C (reflux).

Figure 3. Time–conversion curve in the polymerization of 11-CF-4. 0°C in dichloromethane; [11-CF-4] = 3.25 × 10^{-1}M, [$BF_3 \cdot Et_2O$] = 1.38 × 10^{-2}M.

time-conversion relationship determined by gas chromatography. It is suggested that there are two stages in monomer consumption. Liquid chromatograms are shown in Figure 4. It is clear that there are two stages in the polymerization. The first stage is a reaction in which oligomers are formed and the second stage is a one in which polymers are formed. The molecular weight of polymers formed in the second stage is about 20×10^3 and does not change with conversion. Only the amount of polymers increase with reaction time. The oligomers formed in the first stage were separated and identified as cyclic. The polymerization mixture was terminated with triethylamine, and after evaporation, the sticky mass was extracted with benzene-hexane mixed solvent. The extracted oligomer mixture was concentrated and recrystallized to yield needle crystal. This was identified as cyclic dimer of 11-CF-4 from mass, ir and nmr spectra. This dimer corresponds to peak A in Figure 4. The oligomers B, C and D in Figure 4 were separated by preparative liquid chromatography and identified in a similar manner as cyclic trimer, tetramer and pentamer of 11-CF-4. The melting point of these cyclic compounds alternate regularly: 24°C for monomer, 85°C for dimer, liquid for trimer, 53°C for tetramer, liquid for pentamer. There are some higher molecular weight oligomers which can be supposed to be cyclic, were not identified. Because the polymers are hygroscopic, it is difficult to determine if the polymers have terminal group or not.

The concentration of each oligomers seem to reach equilibrium shortly after the first stage and does not change during the polymerization. The equilibrium concentration of cyclic dimer is 1.5×10^{-2} mol/l and the equilibrium concentration of cyclic oligomers having higher membered rings decreases with the increase of the number of ring atoms. This trend is consistent with the Jacobson-Stockmayer theory (4), which tells us that the decrease of the equilibrium weight concentration of cyclic oligomers depends on the -1.5 power of the degree of oligomerization. The equilibrium monomer concentration of 11-CF-4 was determined as 1.33×10^{-2} mol/l at 0°C from polymerization of monomer and from depolymerization of polymer. By changing the polymerization temperature from -20°C to 30°C, thermodynamic parameters for the polymerization of 11-CF-4 were determined by plotting the logarithms of the equilibrium monomer concentration against reciprocal temperature to fit the Dainton's equation (7).

$$\ln[M]_e = \frac{\Delta H_{ss}}{RT} - \frac{\Delta S_{ss}^\circ}{R}$$

The obtained value is $\Delta H_{ss} = -2.2 \pm 0.3$ kcal/mol, and $\Delta S_{ss}° = 0.7 \pm 1.1$ e.u. in dichloromethane. The heat of polymerization of cyclic formals changes from 3.6 kcal/mol for five membered ring, 5.3 kcal/mol for eight membered ring and here to 2.2 kcal/mol for eleven membered ring, and seem to become zero for large membered ring as expected. This is shown in Figure 5. The entropy of polymerization of cyclic formals changes from -14 e.u. for five membered ring, -9.3 e.u. for eight membered ring and here to +0.7 e.u. for eleven membered ring. Although the increasing trend is smaller than expected, the driving force for the polymerization of macrocyclic formals seems to depend upon the decreased entropy of large rings for combining the free chain ends.

The effect of initiators and solvents on the two stage polymerization of 11-CF-4 was examined. The effect of changing solvents on time-conversion curve in cationic polymerization with boron trifluoride etherate is shown in Figure 6. The rate of monomer consumption was affected very much by the nature of solvents. The reaction in nitromethane is almost instaneous, and that in dioxane is very slow. However, the product distribution is more or less the same compared with that in dichloromethane and the polymerization proceeds in two stages, forming cyclic oligomers at the first stage and polymers at the second stage. The effects of initiators were also studied. The rate by trifluoromethanesulfonic acid is very fast compared with the rate by boron trifluoride etherate, stannic chloride or tungsten hexachloride. Again the two stage nature is similar in every initiator systems showing the same product distribution in liquid chromatogram. Thus, it can be concluded that the two stage nature of the polymerization of 11-CF-4 forming cyclic oligomers at the first stage and high polymers at the second stage seems to be quite general phenomena.

Mechanism of Cationic Polymerization of 11-CF-4. Several mechanisms are proposed for the cationic polymerization of 1,3-dioxacycloacycloalkanes. 1) The growing species are trialkyloxonium ion and the chain propagation proceeds through an S_N2 mechanism. 2) The growing species are carbocation and the chain propagation proceeds by an S_N1 reaction of the intermediate oxonium ion. 3) The growing species are macrocyclic dialkyloxonium ion and the propagation reaction follows ring-expansion mechanism through four center reaction. 4) The growing species are linear trialkyloxonium ion formed by the reaction of the growing

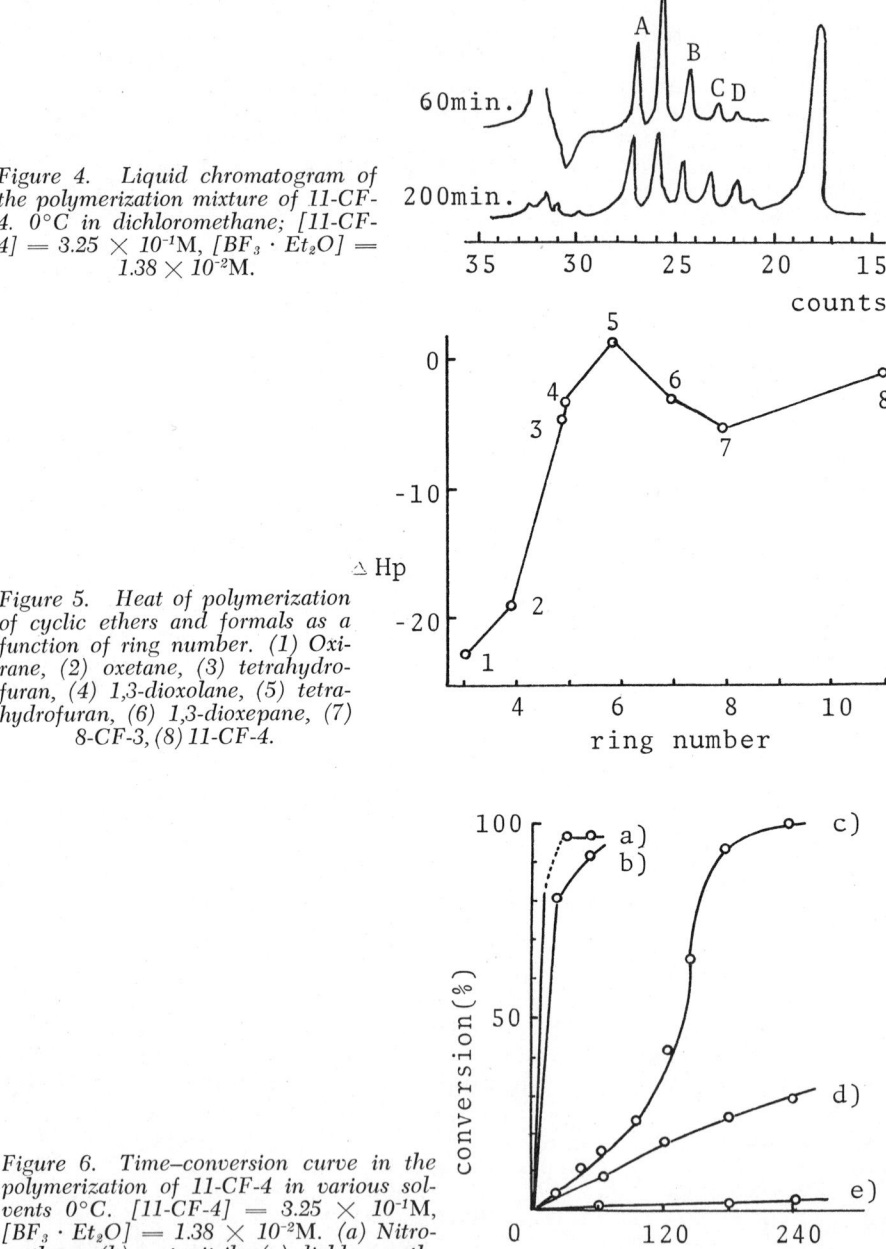

Figure 4. Liquid chromatogram of the polymerization mixture of 11-CF-4. 0°C in dichloromethane; [11-CF-4] = 3.25 × 10^{-1}M, [$BF_3 \cdot Et_2O$] = 1.38 × 10^{-2}M.

Figure 5. Heat of polymerization of cyclic ethers and formals as a function of ring number. (1) Oxirane, (2) oxetane, (3) tetrahydrofuran, (4) 1,3-dioxolane, (5) tetrahydrofuran, (6) 1,3-dioxepane, (7) 8-CF-3, (8) 11-CF-4.

Figure 6. Time–conversion curve in the polymerization of 11-CF-4 in various solvents 0°C. [11-CF-4] = 3.25 × 10^{-1}M, [$BF_3 \cdot Et_2O$] = 1.38 × 10^{-2}M. (a) Nitromethane, (b) acetonitrile, (c) dichloromethane, (d) dimethoxyethane, (e) dioxane.

chain end with an oxygen atom of a polymer chain. Our observation of the formation of cyclic oligomers might seem to favor the ring-expansion mechanism proposed by Plesch (8). However direct observation of the growing species by nmr proved the existence of a carbocation (9), and equilibrium between oxonium ion and carbocation was assumed to explain the polymerization behavior of bicyclic formal (10). Another possibility is that the cyclic oligomers are more reactive than the monomer and might be the reactive intermediate. This postulate is denied by the fact that the cyclic oligomers are not so reactive.

To clarify the nature of the propagating species during the first and the second stage, the following experiments were carried out. First, cationic copolymerization of styrene in dichloromethane with boron trifluoride etherate was studied to examine the nature of the propagating species. The change of the concentration of each component was followed by gas chromatography and liquid chromatography, and shown in Figure 7. Styrene is consumed only at the second stage, where the concentration of the oligomers has reached to equilibrium value, and it is incorporated as a random copolymer with 11-CF-4, and it is not incorporated in cyclic oligomers formed at the first stage. The randomness of the copolymer was proved by fractional reprecipitation from benzene into methanol or into ether and analyzed by liquid chromatography and nmr. There were not any co-oligomers detected in the first stage, although small amount of styrene seems to be incorporated in oligomers formed at the second stage. Thus the active species in the second stage seems to be carbocationic in nature and are responsible for the high polymerization of 11-CF-4 and the copolymerization with styrene.

The mechanism of the oligomer formation was studied by different experiments. The isolated high polymers were depolymerized with boron trifluoride etherate in dichloromethane. From the chromatogram shown in Figure 8, it is clear that the same distribution of cyclic oligomers were found by depolymerization and reached to the same equilibrium concentration. To establish the mechanism of the formation of cyclic oligomers by backbbiting reaction, polymerization of cyclic dimer of 11-CF-4 isolated from the polymerization mixture was studied. The chromatogram in Figure 8 shows that every cyclic oligomers existed in the same distribution as from the polymerization of 11-CF-4 and from the depolymerization. This denies the formation of cyclic oligomers by ring-expansion mechanism, because the formation of 33 and 55 membered rings is difficult to explain starting from 22 membered ring.

Thus the formation of cyclic oligomers is presumed to occur through the back-biting reaction of the oxonium ion existed during the first stage. The nature of the active species present at the first stage is different from carbocation because copolymerization of styrene did not occurred, and more stable species such as oxonium ion back-bites easier than propagates to high polymer.

At present, we can only speculate the nature of the reactive species. In the polymerization of 11-CF-4, oxonium ion responsible for slow propagation and fast back-biting reaction is formed at the first stage and only cyclic oligomers are formed. We noticed that the induction period for high polymer formation in the polymerization of 1,3-dioxolane decreased by decreasing the content of contaminated water (5). In the polymerization of 11-CF-4, the oligomer formation at the first stage was retarded by addition of water or methanol, and also cause to delay the initiation of the second stage. However the rate of the formation of high polymers at the second stage was not retarded by addition of methanol. The carbocation seems to be more stable toward methanol. It is formed only slowly from the oxonium ion and starts to produce high polymers at the second stage.

Two Stage Polymerization of 1,3-Dioxacycloalkanes. In order to establish the two stage character of the polymerization of 1,3-dioxacycloalkanes, the polymerization of various cyclic formals were studied using boron trifluoride ether complex as an initiator at 0°C in dichloromethane. The polymerization were found to proceed in two stages in every case. The selected monomers were 1,3-dioxolane, 1,3-dioxacyclooctane, (dioxepane), 1,3,6-trioxacyclooctane (trioxocane), 8-CF-3, 1,3,6,9,12-pentaoxa-cyclotetradecane 14-CF-5, and 1,3,6,9,12,15-hexaoxa-cycloheptadecane 17-CF-6 besides 11-CF-4. Time-conversion curves are shown in Figure 9, showing the existence of two stages. The rate of the formation of oligomers at the first stage was affected very much by the amount of contaminated water in the system. Contrary to this, the rate of formation of high polymers were scarcely affected by the amount of water. Thus the order of the reaction rate in Figure 9 may not be the true order of the reaction rate because the amount of contaminated water in each monomer is not necessarily the same. The liquid chromatogram of the typical reaction system, the case of 1,3-dioxacyclooctane, is shown in Figure 10, where the formation of oligomers at the first stage reaching equilibrium and the continuous formation of high polymers at the second stage without increasing the molecular weight is observed. Some of

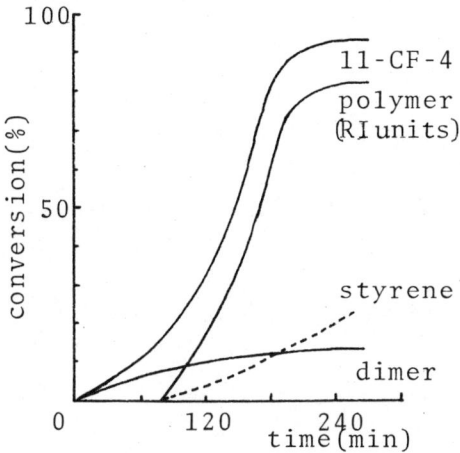

Figure 7. Time–conversion curve for the copolymerization of 11-CF-4 with styrene. 0°C in dichloromethane; $[BF_3 \cdot Et_2O] = 1.16 \times 10^{-2}$M, $[11\text{-}CF\text{-}4] = 3.81 \times 10^{-1}$M, $[St] = 3.70 \times 10^{-1}$M.

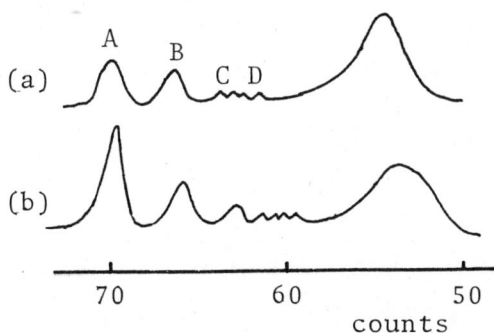

Figure 8. Liquid chromatogram. (a) Depolymerization of the polymer, (b) polymerization of cyclic dimer.

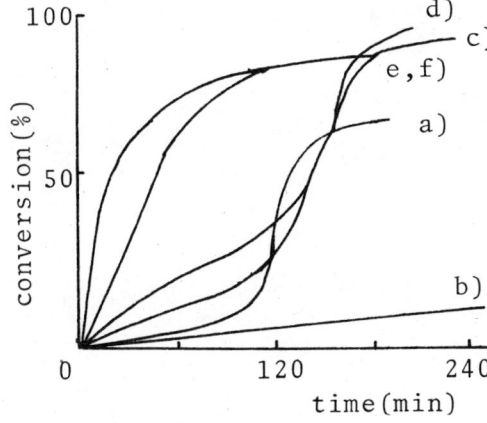

Figure 9. Time–conversion curve for the polymerization of 1,3-dioxacycloalkanes, 0°C in dichloromethane by $BF_3 \cdot Et_2O$. $[BF_3 \cdot Et_2O] = 1.20 \times 10^{-2}$M. (a) [1,3-dioxolane] = 3.77M; (b) [1,3-dioxacyclooctane] = 5.0×10^{-1}M; (c) [8-CF-3] = 5.29×10^{-1}M; (d) [11-CF-4] = 4.40×10^{-1}M; (e) [14-CF-5] = 5.29×10^{-1}M; (f) [17-CF-6] = 6.33×10^{-1}M.

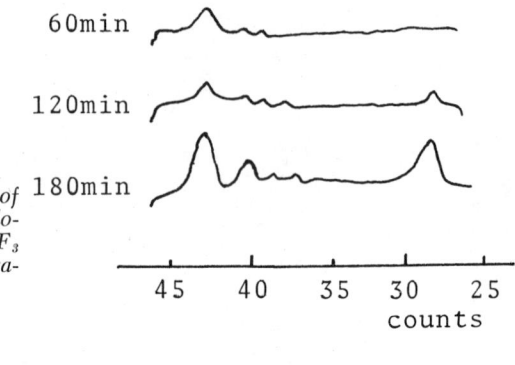

Figure 10. Liquid chromatogram of the polymerization of 1,3-dioxacyclooctane. 0°C in dichloromethane; $[BF_3 \cdot Et_2O] = 1.20 \times 10^{-2}$M, $[1,3\text{-}dioxacyclooctane] = 5.0 \times 10^{-1}$M.

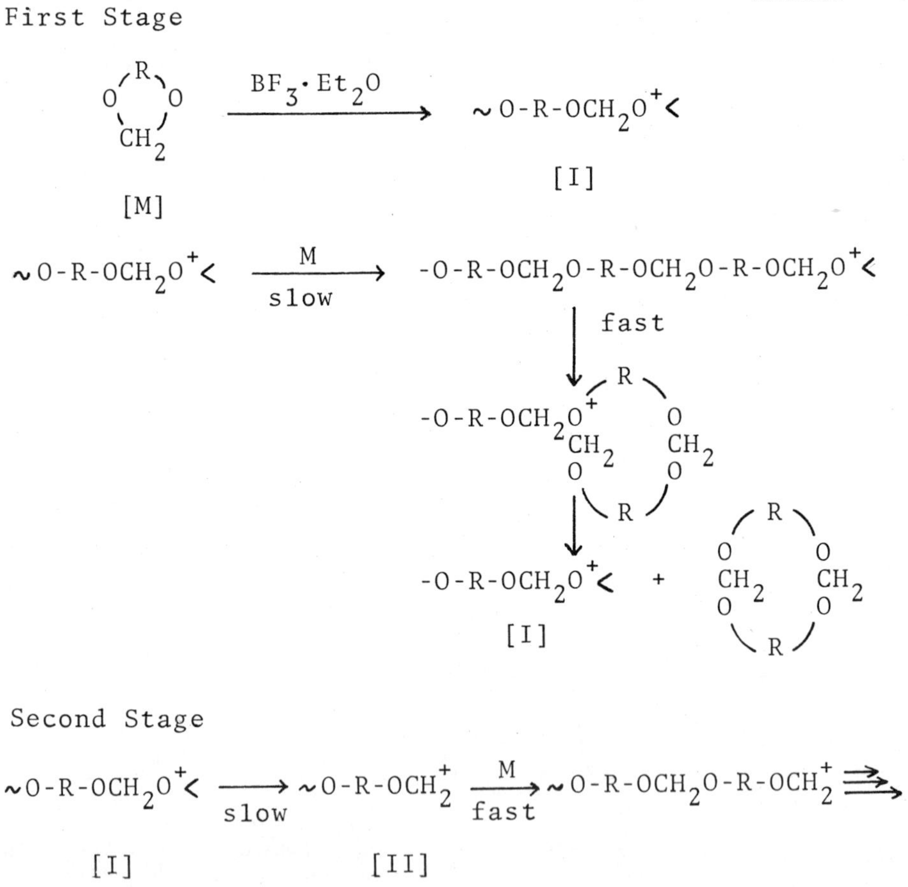

First Stage

$$\underset{[M]}{O\overset{R}{\diagup}O\diagdown CH_2} \xrightarrow{BF_3 \cdot Et_2O} \underset{[I]}{\sim O\text{-}R\text{-}OCH_2O^+\!\!<}$$

$$\sim O\text{-}R\text{-}OCH_2O^+\!\!< \xrightarrow[slow]{M} \text{-}O\text{-}R\text{-}OCH_2O\text{-}R\text{-}OCH_2O\text{-}R\text{-}OCH_2O^+\!\!<$$

$\downarrow$ fast

$$\text{-}O\text{-}R\text{-}OCH_2\overset{+}{O}\underset{CH_2\;\;\;CH_2}{\overset{R}{\diagup\quad\diagdown}}\underset{O\quad\quad O}{}$$

$\downarrow$

$$\text{-}O\text{-}R\text{-}OCH_2O^+\!\!< \quad + \quad \underset{R}{\underset{\diagdown\quad\diagup}{\overset{R}{\overset{\diagup\quad\diagdown}{O\quad O}}\atop CH_2\quad CH_2}}$$

[I]

Second Stage

$$\underset{[I]}{\sim O\text{-}R\text{-}OCH_2O^+\!\!<} \xrightarrow{slow} \underset{[II]}{\sim O\text{-}R\text{-}O\overset{+}{CH_2}} \xrightarrow[fast]{M} \sim O\text{-}R\text{-}OCH_2O\text{-}R\text{-}O\overset{+}{CH_2} \rightrightarrows$$

Figure 11. Mechanism of cationic polymerization of cyclic formals

the oligomers were isolated and identified to be cyclic.

The formation of cyclic oligomers in the cationic polymerization of cyclic formals was reported by several authors (11)(12)(13). The existence of these cyclic oligomers is often claimed to be explained by ring-expansion mechanism (8). However, our observation that cyclic oligomers are formed at different stage from high polymer formation suggests the existence of several species. Tentative schemes are shown in Figure 11. Oxonium ion [I] responsible for the cyclic oligomer formation propagates slowly and backbites fast at the first stage. Carbocation [II] slowly formed at the second stage propagates fast to form high polymers.

Experimental

Monomers: Macrocyclic formals was synthesized from fractionally distilled polyethylene glycol and paraformaldehyde using p-toluenesulfonic acid as catalyst by the similar method in the literature (14). They were purified and dried by distillation over lithium aluminum hydride for four times. The purity was checked by gas chromatography.

Rate measurement of S_N2 reaction: 0.01 Mol of alkali metal acetate was mixed with 0.01 mol of n-butyl bromide in a mixture of 9 ml of benzene and 1 ml of toluene. 0.01 Mol of cyclic formals or crown ethers was added and stirred under reflux at 90°C. The consumption of n-butyl bromide and the formation of n-butyl acetate was followed by gas chromatography of the pulled out samples by using toluene as an internal standard.

Polymerization was carried out under nitrogen by using n-tetradecane as an internal standard for chromatography. The reaction was monitored by gas chromatography and liquid chromatography (Toyo Soda high speed liquid chromatograph model HLC 802 UR) on the pulled out sample from the reaction system with syringe after killing with triethylamine.

The separation and identification of oligomers were carried out as follows. The polymerization mixture was evaporated after killing. The sticky mass was extracted by benzene-hexane mixed solvent and the extract was evaporated and recrystallized to give needle crystals. This was identified by mass, ir and nmr spectra. Parent peak in mass spectra is most important for the identification together with nmr spectra showing corresponding acetal peak area. The melting point of cyclic dimer of 1,3-dioxolane, 1,3,6-trioxocane and 11-CF-4 was 67°C, 55°C and 85°C, respectively.

Literature Cited

(1) C. J. Pedersen, H. K. Frensdorff, Angew. Chem. Internat. Ed., (1972) 11, 16
(2) H. Mark, G. S. Whitby ed. "Collected Papers of W. H. Carothers on High Polymeric Substances", Wiley-Interscience, New York, (1940)
(3) J. A. Semlyen, Advances in Polymer Sci., (1976) 21, 41
(4) H. Jacobson, W. Stockmayer, J. Chem. Phys., (1950), 18, 1600
(5) Y. Yamashita, M. Okada, H. Kasahara, Makromol. Chem., (1968), 117, 242
(6) M. Okada, S. Kozawa, Y. Yamashita, Makromol. Chem., (1969), 127, 66
(7) Y. Yamashita, M. Okada, K. Suyama, H. Kasahara, Makromol. Chem., (1968), 114, 146
(8) Y. Firat, F. R. Jones, P. H. Plesch, P. H. Westerman, Makromol. Chem., (1975), S-1, 203
(9) Y. Yokoyama, M. Okada, H. Sumitomo, Makromol. Chem., (1975), 176, 795
(10) M. Okada, H. Sumitomo, Y. Hibino, Polymer J., (1974), 6, 256
(11) J. M. Andrews, J. A. Semlyen, Polymer, (1972) 13, 142
(12) P. E. Black, D. J. Worsfold, J. Macromol. Sci. Chem. (1975) A 9, 1523
(13) J. W. Hill, W. H. Carothers, J. Am. Chem. Soc., (1935), 57, 925
(14) M. J. Astle, J. A. Zaslowsky, P. G. Lafyatis, Ind. Eng. Chem., (1954), 46, 787

Stereoregularity as a Function of Side Chain Size in Perhaloacetaldehyde Polymerization

D. W. LIPP and O. VOGL

Polymer Science and Engineering, University of Massachusetts, Amherst, MA 01003 and C.N.R.S., Centre de Recherches sur les Macromolécules, Université Louis Pasteur, Strasbourg, France

Aldehyde polymerization, although much less publicized than olefin polymerization, provided some very important contribution to almost all aspects of polymerization reactions. (1,2) It constitutes an alternative to the olefin polymerization, and because of the heteroatom in the main polymer chain, it also provides a link to ring opening polymerization and the preparation of polyesters and polyamides.

Aldehyde polymers have also been important for the understanding of polymer stability, the recognition of the importance of end groups for polymer stability, reaction on polymers by end capping and the limitation of thermal and acidolytic stability of polyacetal chains. (3)

Higher aliphatic aldehydes polymerize readily to isotactic polymers when proper attention is paid to the low ceiling temperature of these polymerizations. The formation of isotactic polymers is apparently favored as the side chain length of these aldehydes increases. All isotactic polyaldehydes, whose crystal structure has been determined, crystallize in a 4_1 helix, and when their melting behavior was studied, a dual melting point was observed. Isotactic polymers of higher aldehydes with aliphatic side chains of chain lengths between C_4 and C_{10} have a melting point indicative for the melting of the paraffin side chain and at higher temperature the melting of the backbone may be observed. It is believed that these isotactic polymers of aldehydes with longer aliphatic side chains crystallize as microphase separated polymer systems : The aliphatic side chains crystallize in the hexagonal paraffin structure and the more polar polyacetal main chains crystallize separately. At shorter side chains (up to C_3), only the main chains and at longer side chains (longer than C_{10}) only the side chains contribute to the crystallization of the isotactic polyaldehydes.(4-6).

Haloaldehyde polymerization provides a new and quite different approach for the study of the polymerization behavior of aldehydes. The properties of perhaloaldehyde polymers are also substantially different from those of polyformaldehyde or of higher aliphatic polyaldehydes. We are discussing here the results of our work on the preparation and polymerization of nine (fluoro, chloro, and bromosubstituted) perhaloacetaldehydes with special emphasis on the stereoregularity of the polmers obtained.(7)

It has earlier been established that chloral can only be polymerized to crystalline, apparently isotactic polymer, no soluble fraction, even of low molecular weight, has ever been observed.(8) Fluoral has been known to exist also in a soluble form.(9) It has become very desirable to establish clearly the space filling size of trihalomethyl sustituent of the perhaloacetaldehyde (which after polymerization becomes the side group of the polyacetal chain) that is necessary to form isotactic polymer and isotactic polymer only.

Aldehydes, both aliphatic and perhaloacetaldehydes polymerize by cationic and anionic mechanisms (Eqn. 1), and all the various intiators were investigated in this work. Fluoral has been reported to have also been polymerized by radical initiators (10). Since our polymerizations could be readily accomplished with well established initiators, it was not found necessary to study other possible ways of initiation.

Polymerization of Aldehydes

$$n \underset{H}{\overset{R}{C}}=O \longrightarrow {\displaystyle \left(\underset{H}{\overset{R}{C}}-O\right)_n}$$

Mechanism : Anionic   Eqn. 1
                 Cationic

Higher aliphatic aldehydes with electrodonating paraffin side group are expected to have a relative higher electron density on the carbonyl oxygen atom with easier alkylation possibilities and, consequently, easier initiation and polymerization by electrophilic means. Perhaloacetaldehydes with the strongly electron withdrahing side group are relatively electron poor on the carbonyl oxygen with easier initiation by nucleophiles and better propagation by the alkoxide anions. This general consideration has actually been observed. Strong electrophiles are necessary for the polymerization of perhaloacetaldehydes, but relatively weak nucleophiles are sufficient for the anionic polymeriation of these aldehydes. The polarization of the carbonyl double bond

$$\underset{H}{\overset{CH_3}{|}}C=O \qquad \underset{H}{\overset{H}{|}}C=O \qquad \underset{H}{\overset{CCl_3}{|}}C=O$$

$$\overset{+}{\underset{H}{\overset{R}{|}}}C-O^-$$

with the partial positive charge on the carbon and the partial negative charge on the oxygen atom is well established. As a consequence, anionic polymerization is usually the preferred polymerization for perhaloaldehydes. (11) (Eqn. 2)

$$R^- \;+\; \underset{H}{\overset{CCl_3}{|}}C=O \;\longrightarrow\; R-\underset{H}{\overset{CCl_3}{|}}C-O^- \;+\; n\,\underset{H}{\overset{CCl_3}{|}}C=O \;\longrightarrow$$

$$\longrightarrow\; R\!\left(\!\underset{H}{\overset{CCl_3}{|}}C-O\!\right)_{\!n}\underset{H}{\overset{CCl_3}{|}}C-O^- \qquad \text{Eqn. 2}$$

Anionically polymerized perhaloacetaldehyde polymers precipitate generally as a gel, are prepared at high polymerization rates and give polymers which cannot be end capped and have apparently occluded end groups with alkoxide character.(12)

Perhaloacetaldehyde polymers are of different physical appearance, powdery, have often -OH end groups and are obtained at a very slow polymerization rate. The suggested mechanism is indicated in Eqn. 3 (13)

Cationic Polymerization of Chloral

$$R^+ + O{=}C(CCl_3)(H) \longrightarrow R{-}O{\overset{+}{=}}C(CCl_3)(H) + n\; O{=}C(CCl_3)(H) \longrightarrow$$

$$\longrightarrow R{-}[{-}O{-}C(CCl_3)(H){-}]_n{-}O{\overset{+}{=}}C(CCl_3)(H) \qquad \text{Eqn. 3}$$

Aldehyde polymerizations are characterized by their low ceiling temperature and by their low heat of polymerization. As a consequence, almost all aldehyde polymerizations are carried out at low temperatures and because of the dependence of the ceiling temperature on the monomer concentration, when the ceiling temperature is very low, the polymerizations are carried out at high monomer centrations or in bulk. It had also been recognized that the ceiling temperature could be utilized for the preparation of solid pieces of polymers as exemplified by the cloral polymerization by using the technique of cryotachensic polymerization. (14)

The study of the perhaloacetaldehyde polymerization provides an ideal example for the investigation of the polymerizability of the aldehydes, the relationship of the polymer stereoregularity as a function of substituent size and shape.

No unusual initiators, such as transition metal complexes or heterogeneous catalysts, are needed for perhaloacetaldehyde polymerizations, the polarization of the carbonyl group of the aldehyde monomer is well defined and does not cause the formation of head to head linkages in the polymer. The shorter carbon oxygen single bond (1.43 Å) which is formed by ring opening of the carbonyl double bond (1.21 Å) has a beneficial effect for the formation of a helical structure for the isotactic polymer and should, consequently, favor the formation of isotactic polymer.

It was expected that the bulkiness as well as the polarizability of the individual atoms and the whole perhalomethyl group of the perhaloacethaldehyde would influence the polymerizability and to which degree of stereospecificity the polymerization of the individual perhaloacetaldehyde could be carried out. The type and combination of the halogen atoms in the trihalomethyl group would also be a determining factor for the rate of polymerization of the perhaloacetaldehyde and for the location of the ceiling temperature of polymerization.

The diameter of the trihalomethyl side group can be calculated, and is for the CF_3 group : 3.3 Å, for the CCl_3 group : 4.3 Å, and the CBr_3 group : 4.8 Å. These values do not give an indication for the polarizability of the individual halogen atoms but it is undoubtedly greater as the atomic number of the halogen atoms increases. The methyl group has a diameter of 2.2 Å and the tertiary butyl group of about 4.4 Å (without being polarizable).

Chloral was known to give only insoluble, presumably isotactic polymer, fluoral could be polymerized to insoluble and soluble presumable atactic polymer. The question was now to determine the results of the polymerization experiments with perhaloacetaldehydes with increasing substituent size as shown in Eqn. 4.

$$\begin{array}{c} CX_3 \\ | \\ C=O \\ | \\ H \end{array} \longrightarrow \begin{array}{c} CX_3 \\ | \\ -(C-O)_n- \\ | \\ H \end{array}$$

CX_3 : CCl_3 : Crystalline, intractable polymer only

CF_3 : Crystalline and amorphous polymer

CCl_2F : $CClF_2$: ? ?

Eqn. 4

PREPARATION OF MONOMERS AND POLYMERS

Fluoral was prepared from the commercially available fluoral hydrate by dehydration with sulfuric acid and ultimately with P_2O_5. It was polymerized with pyridine.

Fluorochloroacetaldehydes : Difluorochloroacetaldehyde (DFCA) was prepared by dehydration of the mixed hemiacetal-hydrate of DFCA with sulfuric acid and again with P_2O_5. The hydrate was obtained from the reverse $LiAlH_4$ reduction of $CClF_2COOCH_3$ which was commercially available.(16)

Fluorodichloroacetaldehyde (FDCA) was also prepared by $LiAlH_4$ reduction at -78°C, but of methyl fluorodichloroacetate, which in turn was made from methyl trichloroacetate and SbF_3. (Eqn.5) (17)

The aldehydes were purified by low temperature distillation and final bulb to bulb distillation from P_2O_5 and contained impurities of not more than 100 ppm as judged by gas chromatogrphy. Physical characteristics of the aldehydes are shown in the Table.

Aldehyde Syntheses :

$$CClF_2-COOCH_3 \xrightarrow{LiAlH_4} \xrightarrow{H_2O} CClF_2-CH(OH)_2 \xrightarrow{} CClF_2CHO$$

$$CCl_2F-COOCH_3 \xrightarrow[\text{rev.add.}]{-78°C} CCl_2FCH(OH)_2 \xrightarrow{H_2SO_4} CCl_2FCHO$$

Eqn. 5.

DFCA had been polymerized at room temperature and at -78°C and gave polymer in about 50 % yield. The polymer obtained at -78°C was completely soluble. High yields of polymers were also obtained with $AlEt_3$, LTB and $SbCl_5$ as the initiators. The polymer prepared with $SbCl_5$ as the initiator is also completely acetone soluble. Most other DFCA polymers obtained with various initiators contained soluble fractions of the polymer, even the polymer from DFCA and LTB had a 11 % soluble fraction.

Soluble poly-DFCA can be acetylated with acetic anhydride and shows a broad NMR peak between 5.5 and 6 ppm indicative of the acetal proton which is bound to a carbon atom which has as the fourth valency a strongly electron withdrawing group attached.

The fine structure of the band is very complicated ; it is clearly influenced by the splitting of the acetal proton by the 2 fluorine atoms of the difluorochloromethyl group. The proton resonance is further broadened by the fact that the polymer has an atactic structure. Line broadening because of the rigidity of the backbone chain, which prevents proper averaging of the proton signal may also be operative.

FDCA was also polymerized with a number of initiators at room temperature and at $-78°C$. Typical anionic initiators, LTB at $25°C$ or Ph_3P at $-5°C$ and $SbCl_5$ at $-78°C$ gave high yields of polymers of FDCA. All FDCA polymers were insoluble although they were prepared under a variety of reaction conditions, and with anionic and cationic initiators. (Table 1).

Fluorobromoacetaldehydes : Difluorobromoacetaldehyde (DFBA) and fluorodibromoacetaldehyde (FDBA) were prepared by two different routes (18). In the first approach 1.1-difluoro-2,2-dibromoethylene was oxidized with oxygen at $0°C$ and gave in good yield a mixture of difluorobromoacetyl bromide and fluorodibromoacetyl fluoride. This reaction undoubtedly goes via the 1,1-difluoro-2,2-dibromoethylene oxide, which could not be isolated.

The mixture of difluorobromoacetyl bromide and fluorodibromoacetyl floride was treated with methanol to give the methyl esters. Methyl difluorobromoacetate was readily separated from methyl fluorodibromoacetate by distillation.(Eqn. 6)

Sunthesis of CBr_2FCHO and $CBrF_2CHO$

$$CBr_2=CF_2 \xrightarrow[0°C]{oxygen} CBr_2 \underset{O}{\overset{}{\triangle}} CF_2 \longrightarrow \begin{array}{c} CBr_2FCOF \\ CBrF_2COBr \end{array}$$

$$\begin{array}{c} CBr_2FCOF \\ CBrF_2COBr \end{array} \xrightarrow{CH_3OH} \begin{array}{c} CBrFCOOCH_3 \\ CBrF_2COOCH_3 \end{array}$$

Eqn. 6.

The individual methyl esters were reduced with $LiAlH_4$ at $0°C$ and gave the corresponding aldehyde hydrates, which were dehydrated in the usual way. This route is the preferred method for the preparation of FDBA. (Eqn. 7)

$$CBr_2FCOOCH_3 \xrightarrow[0°C]{LiAlH_4} CBr_2FCH(OH)_2$$

$$CBr_2FCH(OH)_2 \xrightarrow{H_2SO_4} CBr_2FCHO$$

An alternate better route for the synthesis of DFBA starts from trifluorochloroethylene. Bromination

Table 1
Bulk Polymerization of Perhaloacetaldehydes.

Aldehyde	Initiator Type	Polymerization Time, in hrs.	Bath Temp., in °C	Polymer Yield, in %
CF_3CHO	tertiary amines, acids	several	room temp.	good
CF_2ClCHO	LTB	1	-78	91[b]
	Ph_3P	1	-78	46[c]
	$SbCl_5$	1	-78	56[c]
CF_2BrCHO	LTB	24	25	50[b]
	pyridine	24	25	40[b]
	H_2SO_4	24	25	90[b]
$CFCl_2CHO$	LTB	2	25	85
	Ph_3P	48	-5	70
	$SbCl_5$	1	-78	80
$CFBr_2CHO$	LTB	24	-78	16
	pyridine	24	25	6
	H_2SO_4	24	25	80
CCl_3CHO	LTB	1	0	85
	pyridine	1	0	85
	$SbCl_5$	1	0	60
CCl_2BrCHO	LTB	3	-30	80
	pyridine	3	-30	72
	$SbCl_5$	70	-10	58
$CClBr_2CHO$	pyridine	72	-45	52
	$SbCl_5$	72	-45	24
CBr_3CHO[d]	LTB	72	-78	16
	pyridine	72	-78	46

a) Initiator concentration : 0.3 to 2.0 mole %.

b) Contains acetone soluble polymer portion.

c) Completely acetone soluble.

d) Some toluene used as diluent.

at 0°C in Freon 113 gave in 85 % yield 1,1,2-trifluoro-
-2-chloro-1,2-dibromo ethylene which was treated with
fuming sulfuric acid and HgO to give in 75 % yield
difluorobromoacetyl fluoride. Treatment with methanol
at 0°C gave in 75 % yield methyl difluorobromoacetate
which was reduced with $LiAlH_4$ at $-78°C$. Decomposition
of the reaction product with water and dehydration of
the hydrate with sulfuric acid gave in 65 % yield DFBA.
(Eqn. 8)

Alternative Synthesis of CBr_2CHO

$$CF_2=CClF \xrightarrow[\text{Freon 113}]{0°C} CBrF_2-CClFBr$$

$$CBrF_2-CClFBr + \text{fuming } H_2SO_4 + HgO \longrightarrow CBrF_2COF$$

$$CBrF_2COF + CH_3OH \xrightarrow{0°C} CBrF_2COOCH_3 \xrightarrow[-78°C]{LAH} \xrightarrow{H_2SO_4} CBrF_2CHO$$

The monomers were purified by distillation from P_2O_5
and polymerization grade monomers which contained less
than 200 ppm of impurities were obtained.

Polymerizations of fluorobromoacetaldehydes were
carried out for one day with an initiator concentration
of one mole percent. With $LiAlH_4$ as the initiator,
DFBA gave polymer in up to 50 % yield at -78°C and
+25°C. The polymer prepared at room temperature had a
10 % soluble fraction, but polymer prepared with Ph_3P
at room temperature, and obtained at 25 % yield, is
completely soluble. With pyridine, sulfuric acid and
$SbCl_5$ as the initiator, DFBA polymerized to give polymers, all of which had varying amounts of an insoluble
fraction.

FDBA did not polymerize readily at room temperature, obviously because of the low ceiling temperature
of polymerization. Several nucleophiles, which are
normally good initiators for perhaloacetaldehyde polymerization gave little or no polymer. Even at low temperatures (-78°C). LTB gave polymer in only 16 % and
pyridine in 6 % yield. Cationic initiators, such as
sulfuric or triflic acid were effective initiators even
at room temperature and gave polymers of FDBA in as
much as 80 % yield.

<u>Chlorobromoacetaldehydes</u> : Dichlorobromoacetal-
dehyde (DCBA) and chlorodibromoacetaldehyde (CDBA)

were also prepared by two different synthetic methods. (19,20). The reaction product of chloral and Ph_3P, which is dichlorovinyloxy triphenylphosphonium chloride, was brominated to give 1,1-dichloro-1,2-dibromoethoxy triphenylphosphonium chloride. This compound is very sensitive to water and can be hydrolyzed in aqueous dioxane to the hydrate of DCBA. This material was now dehydrated with sulfuric acid to the free aldehyde. An alternate route for the preparation of DCBA is the bromination of dichloroacetaldehyde diethylacetal followed by the reaction of the bromination mixture with sulfuric acid to DCBA. Although this reaction also gives DCBA in good yield, impurities are retained in this preparation of DCBA which could not be removed and which interfered with the polymerization.

CDBA was prepared by bromination of chloroacetaldehyde diethylacetal. The decomposition of the bromination product with sulfuric acid is very easy and no impurities which were detrimental to the polymerization of CDBA were found. An alternate route for the preparation of CDBA was also studied which used the reaction product of bromal and Ph_3P as the starting material. Chlorination of 1,1-dibromovinyloxytriphenyl phosphonium bromide gave a complicated mixture, from which, after hydrolysis, no CDBA could be isolated.

DCBA was most conveniently polymerized at -30°C. Initiator and monomer were mixed at room temperature and the polymerization was carried out for periods of 3 hours to 1 week. These times are not really the times necessary to achieve the conversions indicated in the Table but times which were convenient to terminate the reaction. More accurate polymerization times can be estimated from the curves of the rate studies of some polymerizations with typical initiators. With pyridine as the initiator a 72 % yield was obtained in 3 hours and with LTB an 80 % yield was relized. $SbCl_5$ gave a 58 % and triflic acid a 40 % yield poly-DCBA.

Polymerization of CDBA needed as low a temperature as -45°C. With pyridine as the initiator a 52 % yield and with $SbCl_5$ a 24 % yield of poly- CDBA was obtained, when the polymerization was stopped after 3 days.

Bromal, which for a long time was considered incapable of polymerization, was purified by heating it with SbF_3 at elevated temperatures. This treatment eliminated 2 impurities which are the apparent inhibitors of the polymerization. Bromal polymerized at -78°C with pyridine or LTB as the initiators in isolated yields of up to 46 %. Bromal and chlorobromoacetaldehydes gave only insoluble polymers.

RATES OF PERHALOACETALDEHYDE POLYMERIZATION

The rate of polymerization of some perhaloacetaldehyde polymerizations was studied by observing the disappearance of the aldehyde proton signal, and consequently the disappearance of monomer. Because of convenience, the chlorobromoacetaldehydes were more extensively studied. The rate of polymerization of DCBA, CDBA and bromal at -78°C. at an initiator concentration of 2 mole percent pyridine is shown in Figure 1. DCBA and CDBA (curves B and C) similarly to chloral are polymerized in 20 minutes and reach an equilibrium at 85 % monomer conversion. Bromal polymerizes much slower ; only a 50 % conversion is obtained in one hour and the ultimate conversion.

In order to compare the rates of polymerization of perhaloacetaldehyde polymerization with different initiators, the polymerization of DCBA was studied at -10°C. and at initiator concentrations of 2 mole percent. The rate of polymerization of DCBA with pyridine is very fast. The rate of DCBA polymerization with sulfuric acid is much slower but is faster than that of chloral polymerization which is shown in curve A. The sulfuric acid initiated polymerization shows induction periods not normally encountered in anionic polymerizations of perhaloacetaldehydes.

The values of these rate studies should not be taken as obsolute but as comprative values because the polymers precipitate during the polymerization, and have all different morhpologies, depending upon the initiator used. The nature of the precipitating polymer determines the rate of polymerization and ultimate conversion of slow polymerizations.

COPOLYMERIZATIONS OF PERHALOACETALDEHYDES

Perhaloacetaldehydes copolymerize with each other, most prominently with chloral or with isocyanates. A typical copolymerization scheme with DCBA is given in Eqn. 9.

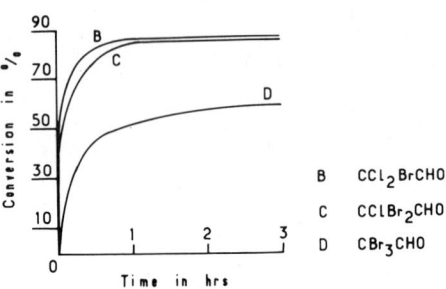

Figure 1. Rate of perhaloacetaldehyde polymerization: NMR study. Initiator, pyridine; initiator concentration, 2 mol %; polymerization bath temperature, $-78°C$. B, DCBA; C, CDBA; D, bromal.

B CCl_2BrCHO
C $CClBr_2CHO$
D CBr_3CHO

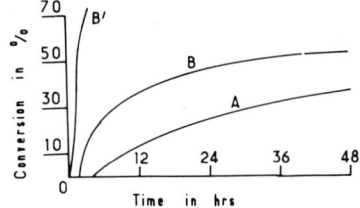

Figure 2. Rates of perhaloacetaldehyde polymerization: NMR study. Bulk polymerization; initiator concentration, 2 mol %; polymerization bath temperature, $-10°C$.

A CCl_3CHO (H_2SO_4)   B CCl_2BrCHO (H_2SO_4)
B' CCl_2BrCHO (pyridine)

Copolymerization of CCl_2BrCHO

$$n \underset{H}{\overset{CCl_2Br}{C}}=O \xrightarrow{R^-} \begin{cases} \xrightarrow{CCl_3CHO} & +(\underset{H}{\overset{CCl_2Br}{C}}-O)_n(\underset{H}{\overset{CCl_3}{C}}-O)_m \\ \xrightarrow{C_6H_5NCO} & +(\underset{H}{\overset{CCl_2Br}{C}}-O)_n\underset{\overset{\|}{O}}{C}-\underset{}{\overset{C_6H_5}{N}}- \end{cases}$$

Copolymerization between individual perhaloacetaldehydes are sensitive to reaction conditions, reaction temperature and initiator type. Perhaloacetaldehydes polymerize quite readily with chloral ; fluorosubsitituted perhaloacetaldehydes are more preferentially incorporated into the copolymers. Chloro- and bromo- substituted perhaloacetaldehydes were not very reactive in the copolymerization and chloral rich polymers were obtained. Bromal can be incorporated into copolymers with chloral only with great diffuculty and from a feed mixture containing 25 more % bromal only 1.5 mole % bromal was incorporated into the copolymer.

Copolymerizations of aldehydes with isocyanates are well known and for perhaloacetaldehydes, aromatic isocyanates are the best comonomers. The polymerization of phenylisocynates with chloral has been most extensively studied. (21,22). The copolymerization of BDCA is described in Eqn. 9 and the results of copolymerization of various perhaloacetaldehydes with phenylisocyanate are shown in Table 2. The copolymers are generally prepared with anionic initiators, for example with pyridine or Ph_3P, but LTB was also used as effective initiator, particularly for chloral copolymerizations. As in homopolymerizations, low temperature conditions must be employed for an effective copolymerization and yields of 30 % to nearly quantitative yields have been obtained. Most copolymers of a perhaloacetaldehyde and phenylisocyanate are insoluble when the copolymer contains only small amounts of phenylisocyanate, but polymers with more than 20 mole % of phenylisocyanate are normally soluble. A truly alternating copolymer of a perhaloacetaldehyde and phenylisocyanate has only been prepared with bromal as the monomer. Many attempts to prepare an alternating copolymer with chloral failed. This result seems to indicate, that the degree of polarization has an influence on the ease of copolymerization of isocyanates with perhaloacetaldehydes.

CEILING TEMPERATURE OF PERHALOACETALDEHYDE POLYMERIZATION

Initiation of perhaloacetaldehyde polymerization must be done above the polymerization threshold temperature (at one molar monomer solutions, this is the ceiling temperature of polymerization) in order to provide complete mixing of initiator and monmer prior to polymerization. To form a homogeneous mixture is particularly important, when the initiation equilibrium is very much on the side of the addition of the initiator to one mole of monomer (Effective initiation), because the polymer which forms rapidly, precipitates and occludes unused initiator. Growing polymer ends are also occluded and the polymerization comes to a standstill ; both effects cause the polymers to be formed in low yield and/or low molecular weight.

Initiation of perhaloacetaldehyde polymerization above the threshold temperature of polymerization allows complete initiation, and maximum yields of polymers may be obtained. (cryotachensic polymerization).

We have been able to determine the ceiling temperature of polymerization of the perhaloacetaldehyde polymerization by determining the threshold temperature of polymerization at various monomer concentrations and extrapolating the Arrhenius plot of ln(M) vs 1/T to one molar concentrations of monomers. The determination of the threshold temperature was carried out by an optical method, which determined the point of rapid change of the opacity in the polymerization medium and is a very accurate method for the determination of the onset of polymerization in systems where the polymer even at very low molecular weight precipitates from the mixture.

In Table 3 the ceiling temperature of polymerization of all the polyaldehydes are listed together with spectral characteristics of the perhaloacetaldehydes. It may be seen that the T_c of fluorosubstituted acetaldehydes are higher that those of chloro- and bromosubstituted perhaloacetaldehydes. The "mixed" perhalaloacetaldehydes have T_c's somewhere in between those of the three halo-substituted perhaloacetaldehydes.

Table 2
Copolymerization of Selected Perhaloacetaldehydes with Phenyl-isocyanate.

Perhalo-acetaldehyde	PhNCO, in Mole %	Initiator Type	Polymerization Temp. in °C	Polymerization Time in Days	Yield in %
CCl_3CHO	5 to 20	many	0	0.05	85
CCl_2BrCHO	5 to 50	pyridine	−78	3	65 − 85
$CClBr_2CHO$	10 to 40	pyridine	−78	3	50 − 90
CBr_3CHO	5 to 50	pyridine	−78	3	30 − 75
CCl_2FCHO	10	Ph_3P	−5	2	50
$CBrF_2CHO$	30	Ph_3P	−25	1	50

Table 3
Physical Characterization of Perhaloacetaldehydes : CX_3CHO.

R=CX_3	PMR (neat) in ppm	CMR in ppm CX_3	CMR C=O	Infra-Red C=O Bands in cm^{-1} Gas	Infra-Red C=O Bands in cm^{-1} Hexane Sol.	Infra-Red C=O Bands in cm^{-1} Neat	Dens. in g/ccm
CF_3	9.35	−	−	1785	1780	--	1.47
CF_2Cl	9.25	5.52[a]		1770	1775	--	1.45
$CFCl_2$	9.15	0.32[a]		1760	1772	1770	1.43
CCl_3	8.95	93.7	175.3	1777	1768	1760	1.45
CF_2Br	9.10	8.53[a]		1770	1762	1755	1.80
$CFBr_2$	8.85	0.45[a]		1760	1755	1751	2.25
CCl_2Br	8.87	79.4	176.7	1774	1763	1754	1.87
$CClBr_2$	8.70	63.5	176.3	1768	1758	1750	2.27
CBr_3	8.45	45.5	176.9	1765	1754	1742	2.73

a) F^{19}NMR, external standard : CF_3COOH

Temperatures

	b.p.	T_c, °C	T_c, K
CF_2Cl	18	63	336
$CFCl_2$	56	41	314
CCl_3	98	11	284
CF_2Br	43	48	321
$CFBr_2$	116	−7	266
CCl_2Br	127	−15	258
$CClBr_2$	148	−40	233
CBr_3	174	−75	198
CF_3	−18	85	358

It appears that there is a linear relationship of the contribution of temperature increment of each of the C-X bonds of the trihalomethyl groups to the value of the ceiling temperature of polymerization for each of the perhaloacetaldehydes. Some of the measurements of the threshold temperatures of the more volatile perhaloacetaldehydes are still in the process of being refined. Our calculations and final values will be presented at a later time.

Another attempt was made in this work to correlate the spectral characteristics of the perhalotaldehydes with their polymerizability as it is expressed in the ceiling temperature of polymerization. As indicated earlier, the polymerization is influenced by the electron distribution of the carbonyl double bond, its polarization and the spacefilling size and polarizability of the side group. We have therefore determined the PMR spectrum for the position of the proton signal, the CMR spectrum for the signals of the C-atoms of the CX_3 group and the carbonyl group ; the data were accumulated for the perhaloacetaldehydes in hexane solution, neat and in the gas phase. The data were taken at the same instrument, as a consequence, an accurate relative relationship of these values was obtained.

Introduction of fluorine atoms into the CX_3 group caused a significant down field shift and introduction of bromide atoms an upfield shift of the aldehyde protons. The carbonyl carbon frequency is relatively little influenced by the substitution changes, but, as expected the CX_3 carbon is very much effected. Even chloro substitution caused a downfield shift to 93.7 ppm. but the carbon atom of the tribromothyl group has its resonance upfield 45.5 ppm.

The infrared stretching frequency of the carbonyl groups of the various perhaloacetaldehydes reflect also the different degrees of inductive effects of the individual perhalomethyl groups. The difference between the extremes is 25 wave numbers, depending upon the type of measurement, and state of the material. The difference between the values as gas or in bulk reflects the degree of association and can be seen in the position of the carbonyl absorption.

The trifluorosubstituent causes the carbonyl frequency to be shifted to shorter wave numbers ; bromosubstitution gives values at longer wave numbers, which indicated that a greater electron availability in the double bond and also explains why bromal polymerized easier with cationic initiators.

The qualitative conclusion of our work at this time is that all halogen substitution causes the change in the spectral behavior and polarization of the acetaldehyde with fluorine substitution. The aldehyde proton is shifted downfield, the carbon resonance of the CX_3 group is shifted downfield and the wave number of the stretching frequency of the carbonyl double bond is shifted to shorter wave numbers ; these changes resulted in an increase in the ceiling temperature of polymerization. Bromine substitution as particularly exemplified with bromal, caused an upfield shift of the aldehyde proton, an upfield shift of the CX_3 carbon and a shift of the carbonyl stretching frequency to higher wave numbers which, together with the increased size of the side group results in the lowering of the ceiling temperature of polymerization for these perhaloacetaldehydes.

ABSTRACTS

Nine fluoro-, chloro-, or bromosubstituted perhaloacetaldehydes were synthesized and/or purified to polymerization grade monomers. They could be homopolymerized to substituted polyoxymethylenes which were crystalline and presumably isotactic. Similarly to polyfluoral, polydifluorochloroacetaldehyde and polydifluorobromoacetaldehyde exist also in the form of a soluble, presumably atactic polymer of reasonable molecular weight. The rate of polymerization of the perhaloacetaldehydes measured under comparable conditions depended on the type of substituent ; the polymerization was fastest for fluorine substitutions and lowest for bromine substituted perhaloacetaldehydes. In copolymerization with chloral, fluorosubstituted perhaloacetaldehydes are more readily incorporated into the copolymer than bromosubstituted aldehydes. The ceiling temperature for the perhaloacetaldehydes also reflected the case of polymerization and is highest for fluorosubstituted and slowest for bromosubstituted perhaloacetaldehyde polymerization. Attemps were made to correlate the ceiling temperature to the monomer structure with special emphasis to the key spectral properties of monomeric perhaloacetaldehydes

Acknowledgements : This work was in part supported by the Materials Research Laboratory of the University of Massachusetts, and by the National Science Foundation. Some of the experimental work was done by R.W. Campbell and is published in detail elsewhere.

REFERENCES

1. J. Furukawa and T. Saegusa, Polymerization of Aldehydes and Oxides, Wiley-Interscience, New-York, 1963.
2. O. Vogl, Polyaldehydes, Marcel Dekker Inc., New-York 1967.
3. O. Vogl, Makromol. Chem., 175, 1281 (1974).
4. O. Vogl, J. Polymer Sci., 46, 261 (1960).
5. I. Negulescu and O. Vogl, J. Polymer Sci., Polymer Letters, B13, 17 (1975)
6. J. Wood, I. Negulescu and O. Vogl, J. Macromol. Sci., Chem., in print.
7. D.W. Lipp, R.W. Campbell and O. Vogl, Preprints, ACS Division of Polymer Chemistry, 18(1), 40 (1977).
8. A. Novak and E. Whalley, Trans. Faraday Soc., 55, 1490 (1959).
9. S. Temple and R.L. Thornton, J. Polymer Sci., A-1, 10, 7°9 (1972).
10. W.K. Busfield and I.J. McEwen, Europ. Polymer J., 8, 789 (1972).
11. O. Vogl, H.C. Miller and W.H. Sharkey, Macromolecules, 5, 658 (1972).
12. O. Vogl, J. Macromol. Sci., Revs. Macromol. Chem., C12(1), 109 (1975).

Mechanism of the Cationic Polymerization of Lactams*

M. ROTHE and G. BERTALAN**
Lehrstuhl Organische Chemie II, University of Ulm, 7900 Ulm, Germany

Lactams are strongly resonance stabilized monomers showing low carbonyl reactivity als well as exceptional thermal stability. Small amounts of initiators, however, are sufficiently effective to start ring-opening polymerizations through transacylation reactions but they are generally active only at elevated temperatures above 200°.

In all ring-opening polymerizations one lactam molecule acts as the acylating agent, i.e. as an electrophile, and the other one as the substrate which undergoes acylation, i.e. as a nucleophile. The initiators serve for activation of the inactive amide group which subsequently reacts with free lactam through successive transamidations leading to polyamides with different endgroups. Transamidations are well known to form part of carbonyl reactions which are catalyzed by acids and bases. These additional electrophiles or nucleophiles provide either an increase in the electrophilicity of the carbonyl carbon of the acylating lactam molecule (acidic initiation) or an increase in the nucleophilic character of the lactam substrate (basic initiation).

Therefore, all initiators used for lactam polymerization so far may be divided into two types: 1. strong bases which are capable of removing the amide proton to form a lactam anion and thus can start an anionic polymerization (1), and 2. compounds with active hydrogen which can protonate the amide bond so that cationic polymerizations become possible (2). An excellent

---

* The Cationic Lactam Polymerization, Part VIII; Part VII: Rothe, M., Bertalan, G., and Mazánek, J., Chimia (1974), 28, 527.
** A. v. Humboldt Research Fellow from the Department of Organic Chemical Technology, Technical University of Budapest, Hungary.

review on the mechanism of lactam polymerization has been published by J. Sebenda (3).

In the following a survey of the present state of cationic lactam polymerization is given. During the more recent years a detailed insight into the complicated reaction mechanisms has been gained by means of investigations about the structure of the endgroups in the oligomer and polymer range and by kinetic measurements. This has led to a clear picture of the reaction course of the various types of cationic lactam polymerization (4).

According to a suggestion we made some time ago (2), this term is now being used for the ring-opening polymerizations with the following initiators: 1. strong, anhydrous acids (4, 5), 2. Lewis acids (2), 3. salts of primary and secondary amines (5, 6, 7), 4. carboxylic acids (8, 9), and 5. even water as well as amino acids and salts of amines with carboxylic acids splitting off water at elevated temperatures (1o, 11, 12).

The Initiation Reaction

These acidic initiators will coordinate with a lactam molecule in a rapid preequilibrium to give a lactam cation which is the reactive species in the polymerization. This type of initiation may also take place with weakly acidic compounds which cannot actually transfer a proton to the lactam, but which are able to form a hydrogen bond with it.

In all cases initiation and propagation are due to the high acylating properties of the lactam cation formed which in turn reacts with the strongest nucleophilic species present in the polymerization medium. The high reactivity of the lactam cation (or the Lewis acid addition product) may be attributed to the decreased electron density at the carbonyl carbon atom thus making it more attractive to nucleophilic attack by the lactam amide bond.

Protonation of amides is known to occur on the oxygen (13) because of the resonance stabilization of the cation formed, but the N-protonated amidium form may be present in very low concentration in a tautomeric equilibrium (2)(Equation 1).

$$O=C-NH \underset{}{\overset{H^{\oplus}}{\rightleftharpoons}} HO\cdots \overset{\oplus}{C}\cdots NH \rightleftharpoons O=C-\overset{\oplus}{N}H_2 \quad (1)$$

Although no direct evidence on the reactive acylating species is available the acylation reactions are preferable described by means of the N-conjugate acid which should be more reactive due to the lack of resonance stabilization. Furthermore, it possesses a better leaving group than the O-protonated form in which the amino fragment is eliminated as an anionic species. Finally, the equations using the N-protonated form for the initiation and growth reaction are more readily comparable to those of the anionic polymerization as we shall see below.

In each of the initiation steps the lactam cation reacts with the strongest nucleophile present, as mentioned above. In the polymerization initiated by strong anhydrous Brønsted acids the free lactam is acylated with the formation of an aminoacyllactam, in the amine salt initiated polymerization the corresponding amine is converted to the amino acid amide, and in the hydrolytic polymerization the acylation of water (or OH$^-$) yields the unsubstituted amino acid (2) (Equation 2).

$$H_2\overset{\oplus}{N}\text{-CO} + \begin{cases} + HN\text{-CO} \rightleftharpoons H_3\overset{\oplus}{N}\ CO\text{-}N\text{-}CO \\ + H_2N\text{-}R \rightleftharpoons H_3\overset{\oplus}{N}\ CO\text{-}NH\text{-}R \\ + H_2O \rightleftharpoons H_3\overset{\oplus}{N}\ COOH \\ + {}^{\ominus}OCO\text{-}R \rightleftharpoons H_2N\ CO\text{-}OCO\text{-}R \end{cases} \quad (2)$$

Accordingly, with weak carboxylic acids an acylation of the carboxylate anion may be assumed leading to a mixed anhydride of the carboxylic acid and the amino acid (14). At the same time, an acylation of the monomer is assumed to take place yielding an ammonium group, as described above (15).

As is well known, N-alkyl lactams can be polymerized only in exceptional cases, and that by a cationic mechanism (16). So far, only the polymerization with strong acids has been studied more thoroughly. In this case the free lactam cannot be acylated to give an aminoacyllactam owing to the N-substituent. Instead, the weakly nucleophilic anion of the initiating acid, e.g. Cl$^-$, is acylated by the lactam cation to give the amino acid chloride (17) (Equation 3).

$$R\text{-}\overset{\oplus}{N}H\text{-}CO + Cl^{\ominus} \rightleftharpoons R\text{-}NH\underset{\smile}{COCl} \qquad (3)$$

Accordingly, amino groups are formed in all cationic polymerizations during the initiation step, the ammonium ion being the predominant form.

The Structure of the Oligomers and Polymers

The initiation reactions mentioned above were studied in great detail in our group (2, 18, 19) and by others (2o) using chromatographical and spectroscopical techniques to elucidate the structure of the products formed in the initial stages of the polymerizations.

Under suitable conditions, e.g. short reaction times, relatively low temperatures and high initiator concentrations, mixtures of oligomers were obtained. These low molecular weight products sufficiently differ from each other in their physical properties and, thus, can be separated into monodisperse homologues by chromatographical means. Their structure was determined unequivocally by IR spectroscopy and, in particular, by identification of the first members of the homologous series with authentic samples which had been prepared by a stepwise synthesis.

The chain propagation was studied in a similar manner. For this purpose we examined the behavior of the oligomers under polymerization conditions. Pure monodisperse oligomers were heated to moderate temperatures either in the presence of equivalent amounts of the monomer or without any additional compound. If this reaction is interrupted after a short time, too, the polymerization is restricted to a few growth steps without producing side reactions to a considerable degree. Model reactions using monofunctional reactants with the same structure as the intermediates of the polymerization provide information about the course and the mechanism of each step of the reaction. Finally, details about the change in the concentration of the endgroups formed during the polymerization may be obtained from potentiometric titrations. In this way the strongly acidic as well as the weakly acidic and the basic groups of the resulting oligomers and polymers can be detected (21).

Some examples are given in the following. Reactive endgroups are easily detected by IR-spectroscopy (17, 18, 19), e.g. N-acyllactam groups ($\nu_{C=O}$ = 1695/cm) or acid chloride groups ($\nu_{C=O}$ = 18o6/cm). Anhydride groups, however, have not been found (15). Therefore, an alternative mechanism will be discussed below.

Chromatographic and electrophoretic methods allow the separation and identification of complete series of oligomers of identical structure. This was proved for the first 6 members of the oligo-ε-aminocaproyl-caprolactams (in the polymerization initiated by strong acids)(18, 19), the first 3 oligo-ε-aminocaproyl-butylamides and benzylamides (using the amine hydrochloride as initiator)(2o, 22), and the first 4 oligo-ε-aminocaproic acid derivatives with semicyclic amidine endgroups (in all cationic polymerizations)(2o).

Finally, the concentration of all functional groups during the polymerization was determined by potentiometric titrations both before and after chemical modifications of the polymers. The equivalence points were assigned to the endgroups with the aid of monodisperse oligomers and model compounds having the corresponding structures (4)(Figure 1).

Thus, strongly acidic groups (lactam salts, acid chlorides, acylamidine salts) and weakly acidic groups (ammonium and carboxyl groups) are determined consecutively by titration with tetraalkylammonium hydroxide. Amidines are strong bases; hence, their salts are not included. By treatment with mercuric acetate, however, they are transformed to the free bases which subsequently can be titrated with acids (23, 24). When the polymers are treated with an excess of alkali new carboxylic groups are formed by hydrolysis of acyllactam and acylamidine groups (25). Aminolysis with hydroxylamine results in the formation of hydroxamic acids which are determined after complexation with Fe(III) salts (19).

Unfortunately, the curves shown in Figure 1 which result from the titration of model compounds cannot be obtained with the polymers. Only the sum of the strong acids as well as the sum of the weak acids can be titrated. An excess of amide groups has been found to cause a levelling effect on the determination of the various strongly acidic groups (Figure 2), which therefore cannot be determined separately so far. This also holds for amine salts and carboxylic groups.

The Propagation Reaction

Similarly, the propagation reactions were investigated by determination of the endgroups (15, 21, 23, 24, 25), by model reactions (19) and, in particular, by extensive kinetic studies (21, 26, 27, 28).

From these results it follows that the growth

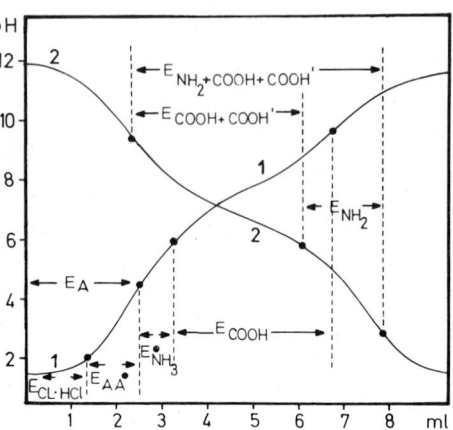

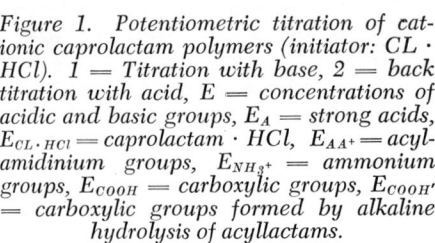

Figure 1. Potentiometric titration of cationic caprolactam polymers (initiator: $CL \cdot HCl$). 1 = Titration with base, 2 = back titration with acid, E = concentrations of acidic and basic groups, E_A = strong acids, $E_{CL \cdot HCl}$ = caprolactam $\cdot$ HCl, E_{AA^+} = acylamidinium groups, $E_{NH_3^+}$ = ammonium groups, E_{COOH} = carboxylic groups, $E_{COOH'}$ = carboxylic groups formed by alkaline hydrolysis of acyllactams.

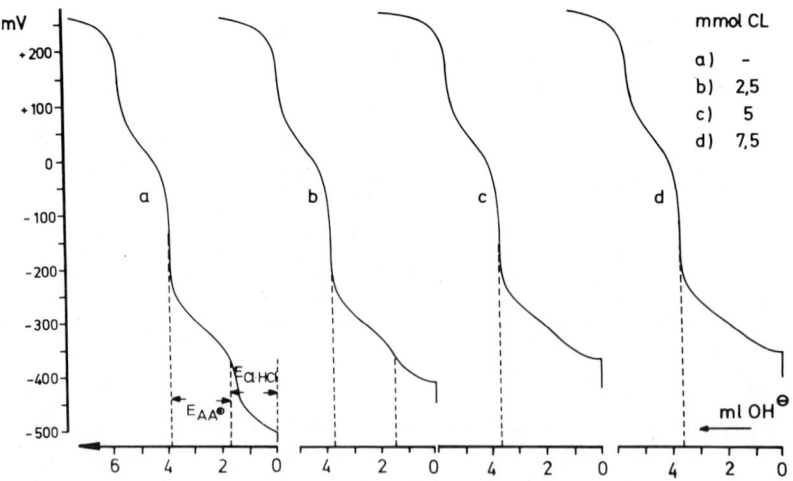

Figure 2. Potentiometric titration of $CL \cdot HCl$ + acylamidine $\cdot$ HCl in the presence of CL

reactions occur by transamidations between lactam rings and the ammonium groups formed in the initiation step, both free lactam and acyllactam endgroups are involved. During the propagation reaction a proton transfer occurs first leading from the amine salt to the lactam (or acyllactam) to give lactam (or acyllactam) cations (Equation 4) which in turn acylate the free amine formed with the regeneration of an ammonium (or amidium) group (Equation 5).

$$\ldots\text{-}\overset{\oplus}{N}H_3 + OC\text{-}NH \rightleftharpoons \ldots\text{-}NH_2 + OC\text{-}\overset{\oplus}{N}H_2 \quad (4a)$$

$$+ OC\text{-}N\text{-}CO\text{-} \rightleftharpoons + OC\text{-}\overset{\oplus}{N}H\text{-}CO\text{-}\ldots \quad (4b)$$

$$+ O\overset{\oplus}{C}\text{-}NH_2 \rightleftharpoons \ldots\text{-}NH\text{-}CO\quad\overset{\oplus}{N}H_3 \quad (5a)$$

$$\ldots\text{-}NH_2 + O\overset{\oplus}{C}\text{-}NH\text{-}CO\text{-} \diagup \begin{matrix} \ldots\text{-}NH\text{-}CO\quad\overset{\oplus}{N}H_2\text{-}CO\text{-}\ldots \\ \\ \ldots\text{-}NH\text{-}CO\text{-} + O\overset{\oplus}{C}\text{-}NH_2 \end{matrix} \quad (5b)$$

The rate of the propagation reaction following this mechanism is particularly high when the aminolysis occurs at the carbonyl group of an activated acid derivative (such as acyllactam or acid chloride) formed in the initiation step; it is slower when the amide group of the monomer is involved. The type of reaction described in Equation 5b corresponds to a bimolecular condensation of two molecules of aminoacyllactam or amino acid chloride (21, 26).

The reaction course of the polymerization initiated by carboxylic acids may be discussed in analogous terms. Here, the aminolysis of the anhydride formed in the initiation reaction by another amino acid anhydride molecule should result in chain propagation with the regeneration of an anhydride group after each step. As anhydride groups could not be detected during the polymerization up to now, an alternative mechanism may operate. Possibly, the tetrahedral intermediate formed from the O-protonated lactam cation and the carboxylate anion decomposes directly via a four-centre transition state to give the higher oligomer with a carboxylic endgroup.

The acylations follow an addition-elimination mechanism as all carbonyl reactions do. At first, a tetrahedral intermediate is formed which leads to chain propagation in the following elimination step. At the same time, water may be split off with the formation of amidines (24, 29)(Equation 6).

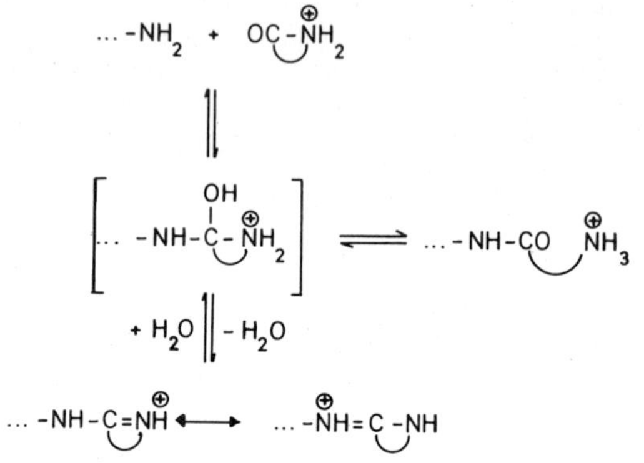

$$\ldots-NH_2 + OC-\overset{\oplus}{N}H_2$$
$$\Updownarrow$$
$$\left[\ldots-NH-\underset{OH}{\overset{|}{C}}-\overset{\oplus}{N}H_2\right] \rightleftharpoons \ldots-NH-CO\quad\overset{\oplus}{N}H_3 \qquad (6)$$
$$+H_2O \Updownarrow -H_2O$$
$$\ldots-NH-\overset{\oplus}{C}=NH \longleftrightarrow \ldots-\overset{\oplus}{N}H=C-NH$$

Similarly, acylamidinium ions result from dehydration of the tetrahedral intermediates formed during the reaction of acyllactams with ammonium groups. Such groups arise inside the polymer molecules; two structures being possible, as shown in Equation 7.

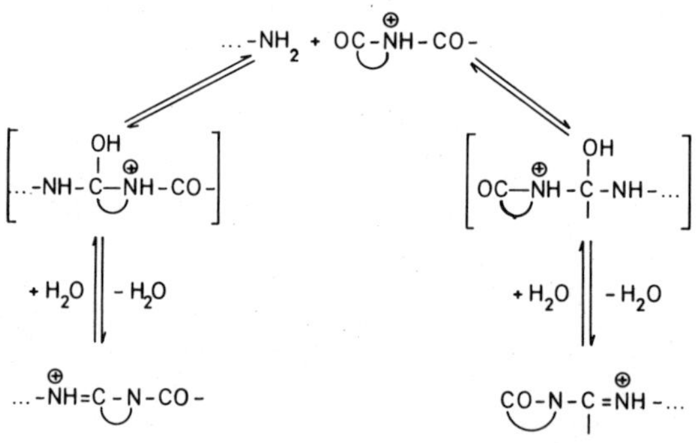

$$\ldots-NH_2 + OC-\overset{\oplus}{N}H-CO-$$

$$\left[\ldots-NH-\underset{OH}{\overset{|}{C}}-\overset{\oplus}{N}H-CO-\right] \qquad \left[OC-\overset{\oplus}{N}H-\underset{|}{\overset{OH}{\overset{|}{C}}}-NH-\ldots\right] \qquad (7)$$

$$+H_2O \Updownarrow -H_2O \qquad\qquad +H_2O \Updownarrow -H_2O$$

$$\ldots-\overset{\oplus}{N}H=C-N-CO- \qquad\qquad CO-N-\underset{|}{C}=\overset{\oplus}{N}H-\ldots$$

The water released in these reactions subsequently hydrolyzes acyllactams, acylamidine salts and lactam salts to yield carboxylic groups.

The following scheme (Table I) shows the endgroups formed during the cationic lactam polymerization. In all types of this polymerization ammonium and amidinium groups form the N-terminal chain end whereas acyllactam, carboxylic and alkylamide residues are present at the C-terminal end. Semicyclic acylamidines are formed

Initiator	Endgroups	
	N-terminal	C-terminal
H_2O	$H_2N-...$	$...-COOH$
	$NH-\overset{\oplus}{C}=NH-...$	$...-COO^{\ominus}$
$R-\overset{\oplus}{N}H_3$	$H_3\overset{\oplus}{N}-...$	$...-CO-NHR$
	$NH-\overset{\oplus}{C}=NH-...$	$...-COOH$
$H^{\oplus}$	$H_3\overset{\oplus}{N}-...$	$...-CO-N-CO$
	$NH-\overset{\oplus}{C}=NH-...$	$...-\underset{...-\overset{\oplus}{N}H}{\overset{\|}{C}}-N-CO$
		$...-COOH$

Table I. Endgroups formed in cationic lactam polymerizations.

inside the polymer molecules. The structure with a side-chain lactam ring represents a potential C-terminal group as it is hydrolyzed to a side-chain carboxylic group. The reactivity of all the endgroups formed decisively determines the further course of the polymerization.

The various types of cationic polymerization of lactams are thus attributed to different endgroups which are formed in the initiation step and then may lead to different consecutive reactions owing to their differing reactivities. As a consequence, the mechanism of the reaction may be principally changed. In this connection the formation of amidines has the main influence (Figure 3). Their concentration increases extraordinarily with increasing acidity and concentration of the initiator and, particularly, with increasing temperature. In the course of the acid and the amine salt initiated polymerization nearly all amine salt groups are converted to amidine salts shortly after initiation (24). Amidines are also observed during the hydrolytic polymerization though to a considerably lower degree (3o).

These strongly basic groups bind the initiating acid very firmly. Amidinium salts initiate the polymerization of lactams much less effectively than ammonium salts. Therefore, their formation leads to a high decrease of the polymerization rate (2o, 23, 24, 25) which is typical for all cationic lactam polymerizations.

The effect of Lewis acids may be interpreted analogously. Lactams react with molar amounts of $POCl_3$ or

P$_2$O$_5$ to give cyclic acylamidines. In this way N-(azacyclohepten-(1)-yl-(2))-caprolactam as well as the corresponding butyrolactam derivative were obtained in 7o-8o% yield on a preparative scale (31). They are obviously formed by elimination of water from the tetrahedral intermediate which results from the reaction of the Lewis acid adduct and free lactam during the initiation step. In the presence of cocatalysts, however, which are capable of transforming the Lewis acid into a proton acid, polymerization proceeds following the usual mechanism, e.g. with boron trifluoride and water.

Simultaneously with the formation of amidines and from the very beginning of the polymerization initiated by strong acids water is eliminated in amounts nearly equivalent to the initiator used. Subsequently, carboxylic groups are formed in high concentrations by hydrolysis of the active carboxylic acid derivatives mentioned above (Figure 4). These groups can now act as initiators of the polymerization.

Moreover, hydrolytic polymerization can occur as well after long reaction times. This follows from the increase of the N-terminal groups beyond the initial initiator concentration (broken line in Figure 6). This can only be explained by the formation of an additional initiator, viz. water.

In that way the complicated inflection kinetics of the cationic polymerization (26) with the decrease and the reincrease of the reaction rate may be interpreted.

According to the titration studies of the strong acid-catalyzed polymers, only low amounts, if any, of acylamidinium salts can be present. As was shown by model compounds, the concentration of these groups do not only decrease by hydrolysis, but also by thermal decomposition yielding amidines and acid chlorides as well as amides and imide chlorides (32)(Table II).

Table II. Acylamidine salts used as model compounds.

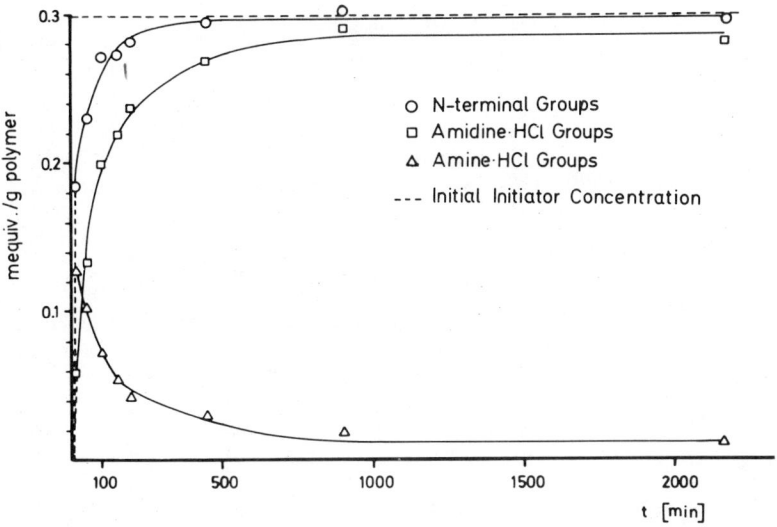

Figure 3. Concentration of N-terminal groups. Initiator, $3.54 \cdot 10^{-2}$ mol CL $\cdot$ HCl/mol CL; $T = 216°C$.

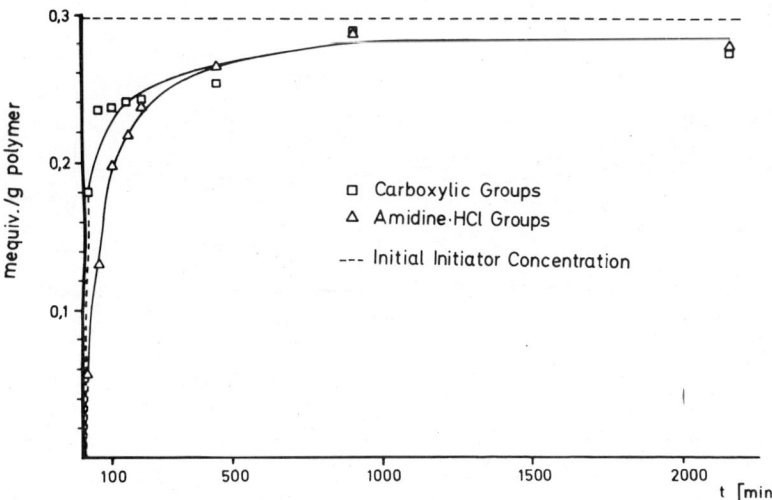

Figure 4. Concentration of carboxylic and amidine $\cdot$ HCl groups. Initiator, $3.54 \cdot 10^{-2}$ mol CL $\cdot$ HCl/mol CL; $T = 216°C$.

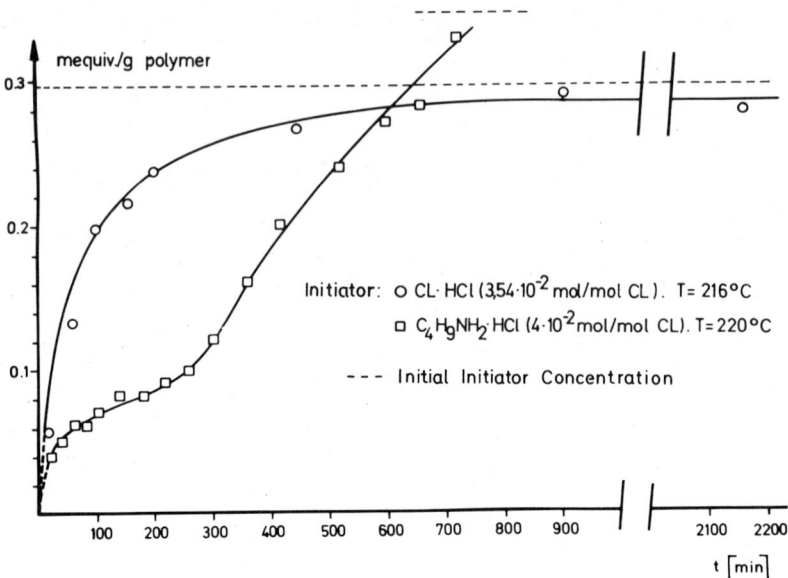

Figure 5. Concentration of amidine · HCl groups

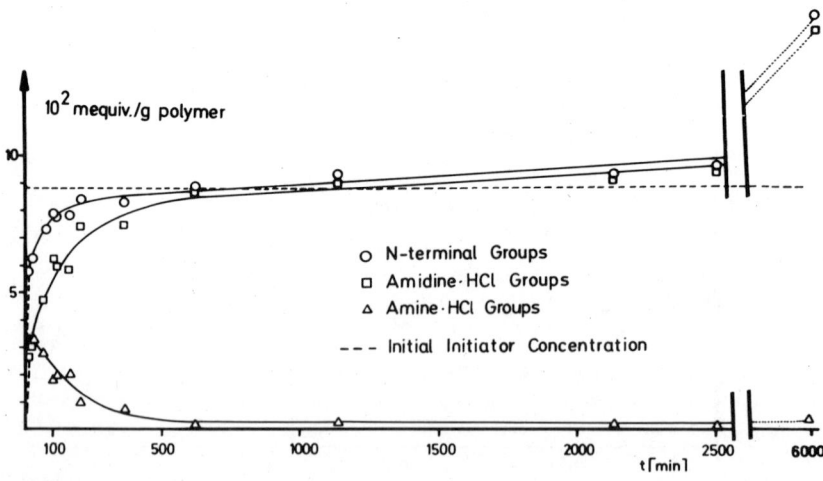

Figure 6. Concentration of N-terminal groups. Initiator, 10^{-2} mol CL HCl/mol CL; $T = 216°C$.

The hydrolysis of the acylamidine groups represents a second possibility of formation of amidines which occurs via polycondensation of aminoacyllactams. This reaction, however, only occurs during the polymerization using Brønsted acids but not during the polymerization initiated by amine salts because acyllactams are not formed at all in the latter case (4).

Figure 5 shows the change in the concentration of the amidinium groups during both these cationic types of polymerization. The considerably larger content of amidines in the acid initiated polymerization can therefore only be derived from acyllactams. This result shows the higher reaction rate of these residues with ammonium groups and a larger contribution of this reaction in the propagation.

The mechanism of lactam polymerization previously referred to as particularly clear has thus turned out to be very complicated. However, from the recent investigations a simple scheme follows consisting of only two equations. In this scheme all reactions are included which proceed during initiation, propagation, and exchange reactions as well as depolymerization and cyclization to higher ring oligomers (Equation 8).

$$-CO-NH- + -CO-\overset{\oplus}{N}H_2- \rightleftharpoons \left[-CO-N-\overset{HO}{\underset{|}{C}}-\overset{\oplus}{N}H_2- \right] \begin{array}{c} \nearrow -CO-N-CO- + -\overset{\oplus}{N}H_3 \\ \searrow -CO-N-C=\overset{\oplus}{N}H- + H_2O \end{array}$$

(8)

$$...-\overset{\oplus}{N}H_3 + -CO-NH- \rightleftharpoons \left[...-NH-\overset{HO}{\underset{|}{C}}-\overset{\oplus}{N}H_2- \right] \begin{array}{c} \nearrow ...-NH-CO- + -\overset{\oplus}{N}H_3 \\ \searrow ...-NH-C=\overset{\oplus}{N}H- + H_2O \end{array}$$

Furthermore, a close analogy to the mechanism of the anionic polymerization results (Table III). In both cases the formation of the initiator, the initiation reaction, the rapid proton-exchange, and the growth reaction run parallel, the lactam cation acting as the electrophilic, acylating species whereas the lactam anion functions as the nucleophilic substrate in all steps. Moreover, in both kinds of the polymerization

the structure (i.e. endgroups) and some of the side
reactions correspond with each other. Thus, a satisfactory picture of the chemistry of lactam polymerization has now been developed.

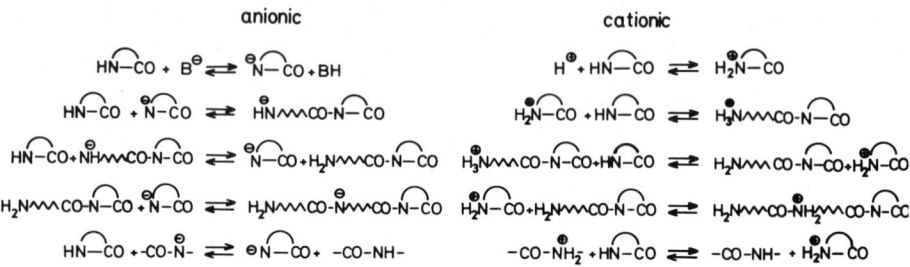

Table III. Comparison of the mechanisms of anionic and
cationic lactam polymerizations.

We gratefully acknowledge generous financial support
by the Deutsche Forschungsgemeinschaft, the Fonds der
Chemischen Industrie, and the BASF AG, Ludwigshafen,
Germany.
Figures 5 and 6 were taken from Chimia (1974), $\underline{28}$, 527
by courtesy of BAG Brunner Verlag AG, Zürich.

Abstract

The mechanism of the various types of cationic
lactam polymerization is discussed in detail. This term
is used for the ring-opening polymerization initiated
by strong and weak Brønsted acids and salts of primary
and secondary amines as well as for the hydrolytic
polymerization. In all cases initiation and propagation
reactions are due to the high acylating properties of
the lactam cation formed which reacts with the strongest nucleophilic compound present in the polymerization
medium. In addition, large amounts of amidine (and
acylamidine) groups are formed from the tetrahedral intermediates originated during the acylation reactions.
These strongly basic groups bind the initiating acid
very firmly. Therefore, their formation leads to a high
decrease of the polymerization rate.

Literature Cited.

(1) Wichterle, O., Sebenda, J., and Králicek, J., Adv. High Polymers (1961), 2, 578.
(2) Rothe, M., Reinisch, G., Jaeger, W., and Schopov,I. Makromol. Chem. (1962), 54, 183.
(3) Sebenda, J., J. Macromol. Sci.-Chem. A (1972), 6, 1145.
(4) Rothe, M., Bertalan, G., and Mazánek, J., Chimia (1974), 28, 527
(5) Van der Want, G. M., and Kruissink, Ch. A., J. Polymer Sci. (1959), 35, 119.
(6) Wiloth, F., Makromol. Chem. (1958), 27, 37.
(7) Yumoto, H., and Ogata, N., Makromol. Chem. (1957), 25, 71.
(8) Majury, T. G., J. Polymer Sci. (1958), 31, 383.
(9) Puffr, R., and Sebenda, J., J. Polymer Sci. C (1973), 42, 21.
(1o) Hermans, P. H., Heikens, D., and van Velden, P. F., J. Polymer Sci. (1958), 3o, 81.
(11) Heikens, D., Hermans, P. H., and van der Want, G. M., J. Polymer Sci. (196o), 44, 437.
(12) Wiloth, F., Z. Phys. Chem. (1958), 11, 78.
(13) Homer, R. B., and Johnson, C. D., in: Zabicky, J. (Edit.) "The Chemistry of Amides", 188, Wiley/Interscience, New York 197o.
(14) Wyness, K. G., Makromol. Chem. (196o), 38, 189.
(15) Lánská, B., and Sebenda, J., Coll. Czech. Chem. Commun. (1975), 4o, 1524.
(16) Masar, B., and Sebenda, J., Coll. Czech. Chem. Commun. (1974), 39, 11o.
(17) Masar, B., and Sebenda, J., Coll. Czech. Chem. Commun. (1974), 39, 2581.
(18) Rothe, M., Boenisch, H., and Kern, W., Makromol. Chem. (1963), 67, 9o.
(19) Rothe, M., Boenisch, H., and Essig, D., Makromol. Chem. (1966), 91, 24.
(2o) Csürös, Z., Rusznák, I., Bertalan, G., Trézl, L., and Körösi, J., Makromol. Chem. (197o), 137, 9.
(21) Doubravszky, S., and Geleji, F., Makromol. Chem. (1967), 11o, 246.
(22) Rothe, M., Angew. Chem. (1968), 8o, 245.
(23) Csürös, Z., Rusznák, I., Bertalan, G., Anna, P., and Körösi, J., Makromol. Chem. (1972), 16o, 27.
(24) Bertalan, G., and Rothe, M., Makromol. Chem. (1973), 172, 249.
(25) Rothe, M., and Mazánek, J., Makromol. Chem. (1971), 145, 197.
(26) Doubravszky, S., and Geleji, F., Makromol. Chem. (1967), 1o5, 261.

(27) Doubravszky, S., and Geleji, F., Makromol. Chem. (1968), 113, 270.
(28) Doubravszky, S., and Geleji, F., Makromol. Chem. (1971), 143, 259.
(29) Schlack, P., Pure Appl. Chem. (1967), 15, 507.
(30) Csürös, Z., Rusznák, I., Bertalan, G., and Körösi, J., Makromol. Chem. (1970), 137, 17.
(31) Bredereck, H., and Bredereck, K., Chem. Ber. (1961), 94, 2278.
(32) Rothe, M., and Kerschbaumer, F., unpublished data.

10

Ring-Opening Copolymerization of Some Cyclic Compounds Containing Oxygen and Nitrogen Atoms

H. L. HSIEH

Phillips Petroleum Co. Research and Development, Bartlesville, OK 74004

The copolymerization of cyclic polar monomers can be used for the preparation of various classes of linear, hetero-chain copolymers. Cyclic compounds of the same chemical type, differing from one another only in the number of units in the ring or the presence of various substituents, can be copolymerized to form products, some of which find a wide variety of application. The copolymerization of various oxides to form linear polyethers has been extensively studied.[1] A number of investigations have been made of the copolymerization of lactones[2,3] and lactams[4,5] to form polyesters and polyamides respectively.

Cyclic compounds of different chemical type can also be polymerized to produce copolymers with hetero-bonds in the macromolecular chain derived from both copolymerizing monomers. Lactones can polymerize with cyclic ethers such as epoxides, tetrahydrofuran, oxetans and trioxane as well as imines. The copolymerization of lactones with epoxides, for example, should lead to the formation of copolymers containing ether and ester links in the chain.

$$x(CH_2)_n \overset{C=O}{\underset{O}{|}} + yCH_2-\overset{R}{\underset{O}{CH}} \longrightarrow \left[-O-(CH_2)_n-\overset{O}{\underset{||}{C}}-\right]_x \left[-OCHCH_2-\atop R\right]_y$$

It was also reported that lactones undergo copolymerization with cyclic phosphites upon heating or in the presence of a basic catalyst.[7]

$$\begin{matrix}CH_2-C=O\\ |\quad\quad|\\ CH_2-O\end{matrix} + C_6H_5OP\overset{O-CH_2}{\underset{O-CH_2}{\diagdown}} \xrightarrow{150°C\ or\ above} \left[\begin{matrix}O\\||\\-P-OCH_2CH_2COOCH_2CH_2-\\|\\OC_6H_5\end{matrix}\right]_x$$

The alternating copolymerization of epoxides and dibasic acid anhydride resulted in formation of polyesters.[7,8,9,10]

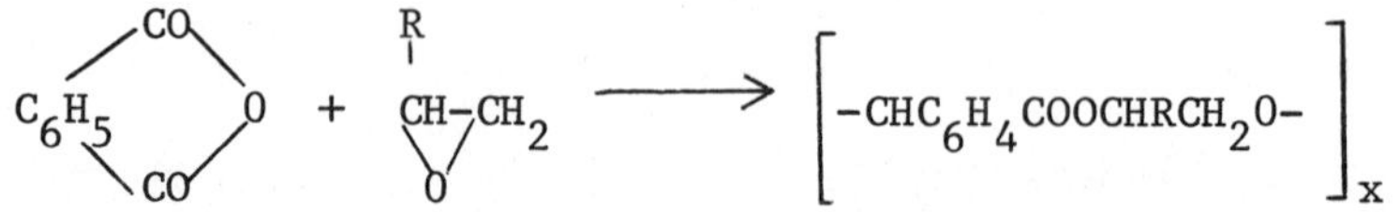

$$C_6H_5\text{-phthalic anhydride} + \underset{O}{\overset{R}{CH-CH_2}} \longrightarrow [-CHC_6H_4COOCHRCH_2O-]_x$$

Alternating terpolymers of epoxides, dibasic acid anhydrides and tetrahydrofuran or oxetane were successfully prepared by using trialkylaluminum as catalyst.[10] This unique family of polymers has repeating ether-ester-ester linkages along the chain.

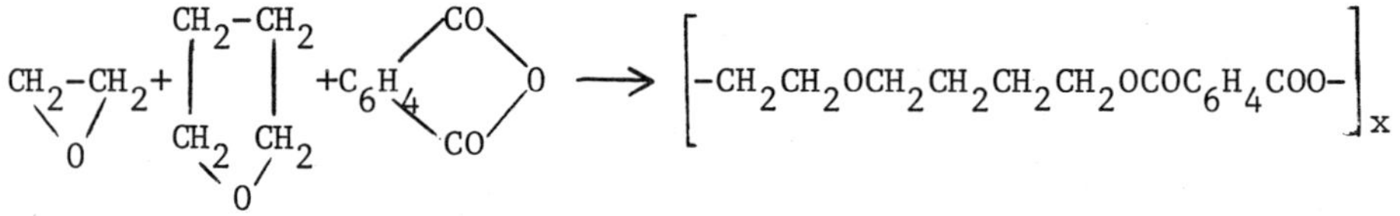

$$CH_2\text{-}CH_2\text{(oxide)} + THF + C_6H_4(CO)_2O \longrightarrow [-CH_2CH_2OCH_2CH_2CH_2CH_2OCOC_6H_4COO-]_x$$

Another interesting reaction is the copolymerization of aziridines with cyclic imides, which leads to the formation of crystalline polyamides.[11]

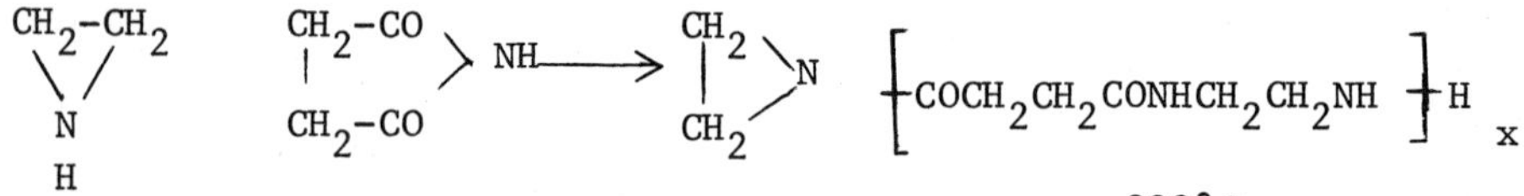

$$\underset{\text{Aziridine}}{\underset{H}{\overset{CH_2-CH_2}{N}}} + \underset{\text{Succinimide}}{\overset{CH_2-CO}{\underset{CH_2-CO}{|}}}\!\!\!>NH \longrightarrow \overset{CH_2}{\underset{CH_2}{|}}\!\!>N\text{-}[\text{-}COCH_2CH_2CONHCH_2CH_2NH\text{-}]_x\text{H}$$

m.p. 300°C

High molecular polyurethanes have been prepared by the ring-opening copolymerization of aziridines with cyclic carbonates.[12]

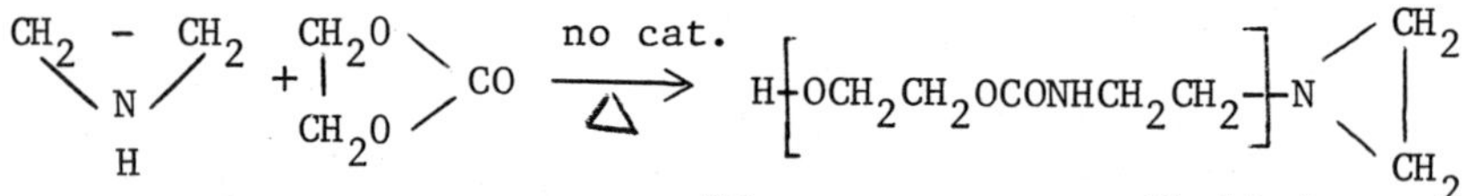

$$\underset{H}{\overset{CH_2-CH_2}{N}} + \underset{CH_2O}{\overset{CH_2O}{|}}\!\!>CO \xrightarrow[\Delta]{\text{no cat.}} H\text{-}[OCH_2CH_2OCONHCH_2CH_2\text{-}]\text{-}N\!\!<\!\!\overset{CH_2}{\underset{CH_2}{|}}$$

HO group   alkylimino group

There are many other examples of this type of copolymerization which involves the ring-opening of two or more heterocyclic monomers. For this report, I will discuss the formation of polyamidoesters by means of this kind of reaction.

Experimental

epsilon-Caprolactone was distilled, and epsilon-caprolactam was melted and purged with nitrogen, before use. Phthalic

anhydride and N-phenylaziridine were used as received. The
initiator, R_4AlLi, was obtained from Foote Mineral and its
chemical formula for the R group is not known although the
molecular weight is 253. It is soluble in hydrocarbon solvent to
give a viscous solution. Toluene was dried by countercurrent
scrubbing with nitrogen.

All polymerizations were done in beverage bottles. Solid
monomers were weighed into the bottle first and then the bottle
was flushed with nitrogen. Toluene was added and the bottle was
flushed with nitrogen again before capping. Caprolactone was then
added by hypodermic syringe. Initiator was generally added at
room temperature. Polymers, in most of the runs, were insoluble
in toluene and came out of solution. They were stirred in acidi-
fied isopropyl alcohol and dried in the vacuum oven.

Results and Discussion

A. **N-Substituted Aziridine and Dibasic Acid Anhydride.** Just as
alkylene oxide under appropriate condition can alternatingly
copolymerize with acid anhydride to yield polyester, aziridine
compounds can also copolymerize similarly with acid anhydride to
form polyamidoester.

$$n \; \underset{CH_2-CH_2}{\triangle^O} + n \; \text{(phthalic anhydride)} \longrightarrow \{CH_2CH_2OC\text{-}C_6H_4\text{-}CO\}_n$$

$$n \; \underset{CH_2-CH_2}{\triangle^{N\phi}} + n \; \text{(phthalic anhydride)} \longrightarrow \{CH_2CH_2\}NOC\text{-}C_6H_4\text{-}CO\}_n$$

Triisobutylaluminum, a very effective initiator for alkylene
oxide-dibasic acid anhydride copolymerization,[10] was used to
initiate the copolymerization of N-phenylaziridine and phthalic
anhydride (Table I).

Both the conversion and the elementary analysis indicated the
two monomers are present in equal mole ratio. Since phthalic
anhydride cannot be homopolymerized,[10] it is concluded that the
product is an alternating copolymer. Surprisingly, when the same
experiment was carried out without initiator, the result was the
same. Obviously, these two monomers copolymerize readily by
simply heating. The low softening point of this polymer, however,

TABLE I

N-PHENYLAZIRIDINE AND PHTHALIC ANHYDRIDE COPOLYMERIZATION

N-Phenylaziridine	0.06 mole (7.1 g)
Phthalic anhydride	0.06 mole (9.0 g)
Toluene	100 ml
Triisobutylaluminum	4 mmole
Temperature, °C	70
Time, hours	16

Experimental Data

Total Monomers Charged, G	Polymer Recovered, G	Polymer Soft Point, °C	Polymer % N	Polymer % O
16.1	16.5	60	4.6[a] (5.2)[b]	18.6[a] (18.0)[b]

a – Found
b – Calculated based on 1 to 1 mole ratio

limits its usefulness. Endic anhydride, chloroendic anhydride and succinic anhydride also copolymerize with N-phenylaziridine to form low-melting solids, but in much lower yields.

B. epsilon-Caprolactone and epsilon-Caprolactam. Another interesting method for preparing polyamidoester is the copolymerization of a lactone such as caprolactone with a lactam such as caprolactam.

$$n \; \overset{\overset{C=O}{|}}{(CH_2)_5\underset{O}{|}} \; + \; n \; \overset{\overset{C=O}{|}}{(CH_2)_5\underset{NH}{|}} \; \longrightarrow \; [O(CH_2)_5 CONH(CH_2)_5 CO]_n$$

In the first experiments, five organometallic compounds were screened as initiators. It is known that caprolactone polymerizes readily in the presence of triisobutylaluminum, butyllithium, potassium tert-amyloxide, and lithium tetraalkylaluminate. However, the mixture of caprolactone and caprolactam in toluene formed polymer only in the presence of the last compound (Table II).

The fact that polymer in over 50% conversion was formed indicated both monomers participated in the reaction, and that the product seemed homogeneous and insoluble in toluene (caprolactone homopolymer is toluene-soluble) prompted further experimentation with R_4AlLi. The results are shown in Table III.

TABLE II

CAPROLACTONE AND CAPROLACTAM COPOLYMERIZATION
WITH ORGANOMETALLIC COMPOUNDS

Caprolactone	10 g
Caprolactam	10 g
Toluene	200 ml
Organometallic compound	2 mmoles
Temperature, °C	70
Time, hours	16

Experimental Data

Organometallic Compound	% Conversion
$(i\text{-Bu})_3Al$	0
Et_2AlCl	trace
$n\text{-BuLi}$	0
tert-AmylOK	0
R_4AlLi	60[a]

a – Polymer precipitated

TABLE III

LITHIUM TETRAALKYLALUMINATE AS CATALYST FOR
CAPROLACTONE AND CAPROLACTAM COPOLYMERIZATION[a]

Caprolactone, Grams	Caprolactam, Grams	Polymer, Grams	% N	% Caprolactam[b]	Solubility in Toluene
100	0	100[c]	0	0	Yes
70	30	37	3.7	30	Yes
50	50	72[d]	6.7	54	No
30	70	80	9.9	79	No
0	100	0	---	--	--

a – In 1 liter toluene with 5.1 grams (20 mmoles) R_4AlLi initiator. Polymerization was carried out at 70°C for 16 hours.
b – Based on % N in polymer.
c – Waxy solid; melting point 60°C.
d – Melting point 180°C. Nylon 6 melts at 220°-230°C.

As was anticipated, caprolactone was readily polymerized by R_4AlLi to yield homopolymer which is low-melting and soluble in toluene, tetrahydrofuran and chloroform. Caprolactam, on the other hand, did not homopolymerize. The composition of the copolymers varied with the charge ratio of the monomers, indicating it is not an exclusively alternating process. Polymers containing high lactam content are high-melting and completely insoluble in the common solvents. This leads to the conclusion that the products are not a mixture of homopolymers.

To further elucidate the structure of these copolymers, phase transition behavior of three copolymers of caprolactone and caprolactam (see Table IV) were determined.

Three methods were used to determine the phase transition behavior of these polymers.
  a. Capillary dilatometry, using mercury as the displacement fluid, from -38°C to +65°C.
  b. Dynamic measurements (Vibron), at 110 Hz, from -80°C to the upper melting point (120°-240°C).
  c. Differential scanning calorimetry from 40°C to 250°C.

Also included are results from a physical blend of polycaprolactone and polycaprolactam made in a Brabender Plastograph at 255°C.

From the data on the physical blend it appears that the two homopolymers are incompatible in both the amorphous and crystalline states. Only the expected transitions of the two homopolymers were observed. Apparently the three experimental polymers are random copolymers with some homopolymer block on or admixed. The polymer near 50/50 in composition showed a very broad transition around 0°C in both the dilatometric experiment and in the Vibron. The only other transition was a melting point (170°C (V), 192°C [DSC]). The other two experimental polymers had similar broad transitions near 0°C. In addition, these polymers displayed dispersion regions which appear to be associated with the transitions

TABLE IV

SUMMARY OF STUDIES OF PHASE TRANSITION BEHAVIOR OF COPOLYMERS

Caprolactam, %	Composition,[b] %		
	Copolymer[c]	Polycaprolactone	Polycaprolactam
30	60	40	0
54	100	0	0
79	40–50	0	50–60

a – Based on % N in polymer.
b – From the solubility data, it seems most likely the homopolymers are present in the form of block.
c – About 50-50 composition.

of the major components. The results would support the identification of the experimental polymers as copolymers of ca. 50-50 composition plus excess homopolymer (possibly block).

In conclusion, N-substituted polyamidoesters can be readily formed by heating N-substituted aziridines with dibasic acid anhydrides. The low softing point of these polymers limits their usefulness. Copolymerization of caprolactam and caprolactone leads to very interesting products. They are generally high melting and by adjusting monomer charge ratio either polyester or polyamide block copolymer can be produced.

ABSTRACT

N-Phenylaziridine and phthalic anhydride copolymerize in alternating order to give polyamidoesters upon initiation with triisobutylaluminum or by heating. The low softing point of this polymer limits its usefulness.

epsilon-Caprolactone and epsilon-Caprolactam copolymerize in the presence of R_4AlLi. The composition of the copolymers varied with the feed ratios of the monomers. Polymers containing over 50 per cent lactam are high-melting and completely insoluble in common solvents. From the results of studying the phase transition behavior of these polymers it was concluded that they are about of 50/50 composition plus excess homopolymer possibly in block form.

LITERATURE CITED

1. Furukawa, J. and Saegusa, T. "Polymerization of Aldehydes and Oxides", John Wiley & Sons, New York, 1963.
2. Tada, K., Numata, Y., Saegusa, T., and Furukawa, J., Makromol. Chem. 77, 220 (1964).
3. Yamashita, Y., Tsuda, T., Ishikawa, J., and Himidy, T., J. Chem. Soc. Japan, Ind. Chem. Sect., 66, 1493 (1963).
4. Glickman, S.M. and Miller, E. S., U.S. Patent 3,016,367 (1962).
5. Hedrick, R. M., Motters, E. H., and Butler, T. M., U.S. Patent 3,120,503 (1964).
6. McConnel, R.L. and Coover, H. W., U.S. Patent 3,062,788 (1962).
7. Fish, W., Hoffman, W., and Koskikallio, J., Chem. and Ind., 756 (1956).
8. Fisher, R. F., J. Polymer Sci., 44, 155 (1960).
9. Tsuruta, T., Matsumura, K., and Inoue, S., Makromol. Chem. 75, 211 (1964).
10. Hsieh, H. L., J. Macromol. Sci-Chem., A7 (7), 1525 (1973).
11. Kagiua, T., Narisawa, S., Manobe, K., and Kobata, M., J. Polymer Sci., A1, 2081 (1966).
12. Drecksel, E. K., U.S. Patent 2,824,857 (1958).

11

Ring-Opening Polymerizations: Mechanism of Polymerization of ε-Caprolactone

R. H. YOUNG, M. MATZNER, and L. A. PILATO
Union Carbide Corp., Bound Brook, NJ 08805

The first synthesis of ϵ-caprolactone was reported by Carothers (1). He also investigated its polymerization under the influence of heat and catalysts. Since then the polymerizations of this as well as that of other lactones were studied by many researchers. Throughout the 1950's to the 1970's the polymer formation and its properties were the subject of several investigations in our laboratories (2-5). Union Carbide is presently the commercial producer of the monomer and of a series of polymers which range in molecular weights from 500 to 40,000. The starting ϵ-caprolactone is produced by the peracetic acid oxidation of cyclohexanone as shown in Equation (I).

$$\text{cyclohexanone} + HOOCCH_3 \longrightarrow \text{caprolactone} + HOOCCH_3 \qquad (I)$$

$$\underset{1}{} \qquad \underset{2}{} \qquad \underset{3}{} \qquad \underset{4}{}$$

In spite of the number of investigations that were devoted to the polymerization of lactone 3, the exact mechanism whereby the polymer is formed is still not entirely clear. It is the purpose of this paper to present the various factors that influence the reaction and to describe its complexity when it is performed in the melt in the presence of either anionic or coordination catalysts.

Mechanisms:

In principle, the polymerization of a lactone should follow mechanism(s) similar to the catalyzed reactions of simple esters. The transformations that are observed are a function of the catalyst and can be subdivided into (a) cationic, (b) anionic, and (c) coordination type. A simplified description for the three mechanisms is shown with ϵ-caprolactone as an example.

(a) <u>Cationic</u>

It was suggested (4,6,7) that the cationic catalyzed polymerization proceeds via the steps shown in Equation (II). First an equilibrium of the cationic species, 5, with the intermediate, 6, is established. This is followed by ring-opening to 7 which then propagates until a high polymer is obtained.

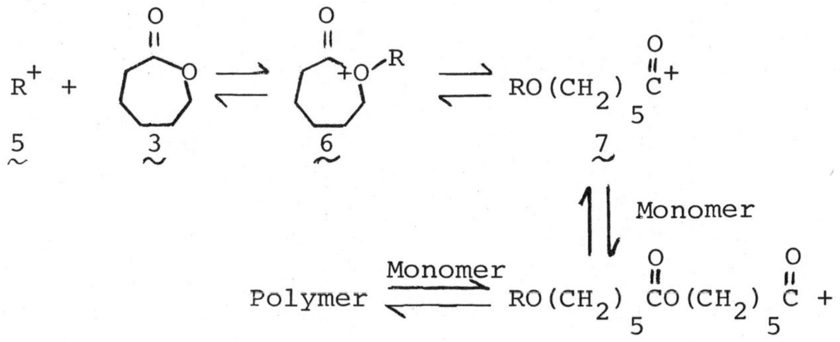

$$R^+ + \underset{3}{\text{(lactone)}} \rightleftharpoons \underset{6}{\text{(oxocarbenium)}} \rightleftharpoons \underset{7}{RO(CH_2)_5 \overset{O}{\overset{\|}{C}}{}^+}$$

$$\underset{7}{\downarrow} \text{Monomer}$$

$$\text{Polymer} \underset{\longleftarrow}{\overset{\text{Monomer}}{\rightleftharpoons}} RO(CH_2)_5 \overset{O}{\overset{\|}{C}} O(CH_2)_5 \overset{O}{\overset{\|}{C}}{}^+ \qquad (II)$$

(b) Anionic

In this case, attack of the base upon the carbonyl group of the cyclic ester is the characteristic feature in the initiation process($\underline{4}$) (Equation (III)).

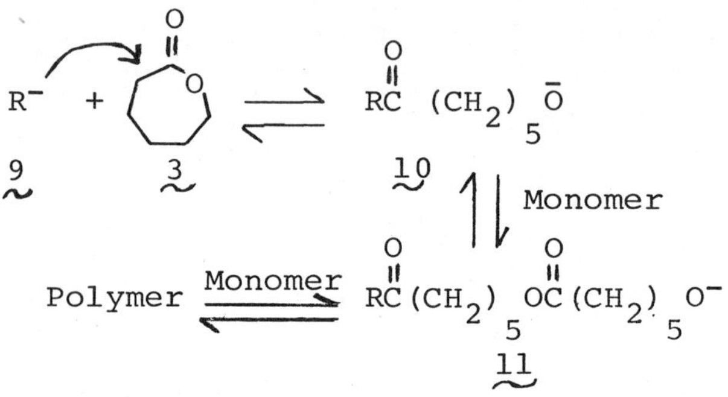

$$R^- + \underset{\underset{\underline{3}}{}}{\text{(cyclic ester)}} \rightleftharpoons \underset{\underline{10}}{RC(=O)(CH_2)_5 O^-} \overset{\text{Monomer}}{\rightleftharpoons} \underset{\underline{11}}{RC(=O)(CH_2)_5 OC(=O)(CH_2)_5 O^-} \overset{\text{Monomer}}{\rightleftharpoons} \text{Polymer} \quad (III)$$

$\underline{9}$

Once produced anion $\underline{10}$ is then the propagating intermediate until the final polymer is formed.

(c) Coordination Type

The coordination catalyzed polymerization is defined for the purpose of this paper as one which involves a concerted insertion with concurrent cleavage of a covalent polymer-catalyst bond($\underline{4}$). It is illustrated in Equation (IV).

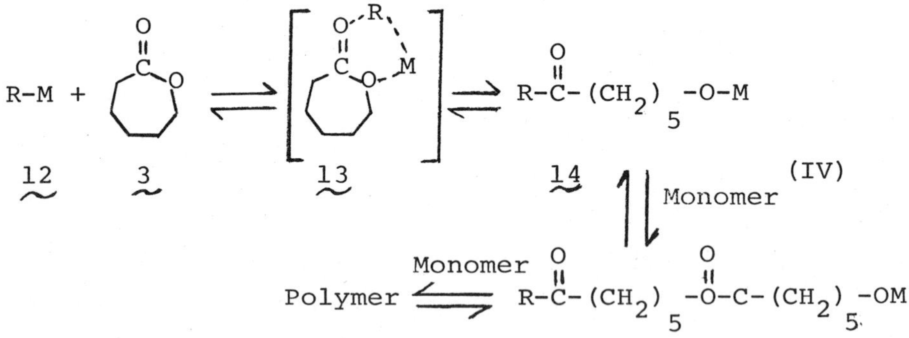

$$R-M + \underset{\underset{\underline{3}}{}}{\text{(cyclic ester)}} \rightleftharpoons \left[\underset{\underline{13}}{\text{complex with } O-R, M}\right] \rightleftharpoons \underset{\underline{14}}{R-C(=O)-(CH_2)_5-O-M} \overset{\text{Monomer}}{\rightleftharpoons} R-C(=O)-(CH_2)_5-O-C(=O)-(CH_2)_5-OM \overset{\text{Monomer}}{\rightleftharpoons} \text{Polymer} \quad (IV)$$

$\underline{12}$

Note that the cationic and anionic mechanisms as depicted above are "limiting" cases. Depending upon the reagents and experimental conditions the "whole spectrum" of mechanisms is observed (Figure 1).

As shown the coordination mechanism is basically the "intermediate" case between the two other modes of reaction.

Both the polymerizability and the mechanism of polymerization are obviously dependent on the ring size of the lactone:

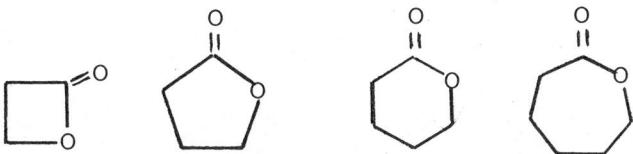

An interesting generalization in this regard was made by Hale(8). He finds that (1) in the case of the five- and six-membered ring-containing monomers, the ease of polymerization varies with the class to which the monomers belong (i.e., lactone, urea, imide, anhydride, lactam, etc.); (2) monomers containing four-, seven-, and eight-membered rings appear to polymerize in all cases; and (3) alkyl or aryl substitution of the ring has a deleterious effect on the polymerization.

Experimental Approach

A correlation between the intrinsic viscosity of poly-ϵ-caprolactone and its weight-average molecular weight has been reported previously(3). A commonly known relationship between the melt viscosity at elevated temperatures and the weight-average molecular weight has been shown(9) to also hold for poly-ϵ-caprolactone (Equation VI).

$$[\eta] = 9.9 \times 10^{-5} \bar{M}_w^{0.82} \quad \text{(in benzene)} \quad \text{(V)}$$

$$\mu = \alpha \bar{M}_w^{3.4} \quad \text{(VI)}$$

Thus, an excellent monitor for ϵ-caprolactone polymerizations is following the viscosity as a function of time. This procedure was adapted and a typical "reaction profile" (at 204°C, neat) is shown in Figure 2.

There are essentially three stages of the reaction. For the time period of t_o to t_1 a rapid rise in viscosity is observed, designated as portion a of the curve. At time t_1 the viscosity reaches its maximum value, v_1. Following this, a decrease of the viscosity takes place, although the change is not as rapid (portion b of the curve). At time t_2, the viscosity levels off to a practically constant value, v_2 (portion c of the curve).

The shape of the curve which reflects changes related to $\bar{M}_w$ is significant and has a direct bearing on the reaction mechanism. It is different from the results reported for solution polymerizations carried out at lower temperatures(10).

Kinetic Considerations

The overall mechanism of an anionic-coordination catalyzed polymerization of ϵ-caprolactone is dependent upon a number of factors. The most important of these are the type of catalyst and whether a coinitiator is used. A very large number of both have been reported(2). If R_1M represents the catalyst-initiator, R_2OH an active hydrogen containing coinitiator, and CL the ϵ-caprolactone monomer, the following steps have to be considered in order to arrive at a meaningful kinetic expression:

(a) Preequilibrium:

$$R_1-M + R_2OH \underset{K_{-1}}{\overset{K_1}{\rightleftarrows}} R_1-H + R_2-OM \qquad (1)$$

(b) Initiation:

$$R_1-M + CL \underset{K_{-2}}{\overset{K_2}{\rightleftarrows}} R_1-\overset{O}{\overset{\|}{C}}-(CH_2)_5-OM \qquad (2)$$

$$R_2-OM + CL \underset{K_{-3}}{\overset{K_3}{\rightleftarrows}} R_2-O-\overset{O}{\overset{\|}{C}}-(CH_2)_5-OM \qquad (3)$$

$$R_2-OH + CL \underset{K_{-4}}{\overset{K_4}{\rightleftarrows}} R_2-O-\overset{O}{\overset{\|}{C}}-(CH_2)_5-OH \qquad (4)$$

We have shown that reaction (4) is very slow in comparison to reactions (2) and (3). When ϵ-caprolactone is heated with an alcohol in the absence of any catalyst under our normal experimental conditions no viscosity change was observed.

(c) Propagation:

$$R_1\text{-C(=O)-(CH}_2)_5\text{-OM} + \text{CL} \xrightleftharpoons[K_{-5}]{K_5} R_1\text{-C(=O)-(CH}_2)_5\text{-O-C(=O)-(CH}_2)_5\text{-OM} \rightleftarrows \text{etc.} \quad (5)$$

$$R_2\text{OC(=O)-(CH}_2)_5\text{-OM} + \text{CL} \xrightleftharpoons[K_{-6}]{K_6} R_2\text{O-C(=O)-(CH}_2)_5\text{-O-C(=O)-(CH}_2)_5\text{-OM} \rightleftarrows \text{etc.} \quad (6)$$

(d) Termination:

$$R\text{-[O-C(=O)-(CH}_2)_5\text{]}_n\text{-OM} + R_2\text{OH} \xrightleftharpoons[K_{-7}]{K_7} R\text{-[O-C(=O)-(CH}_2)_5\text{]}_n\text{-OH} + R_2\text{OM} \quad (7)$$

(e) Chain Transfer:

$$R\text{-[O-C(=O)-(CH}_2)_5\text{]}_n\text{-OM} + R\text{-[C(=O)-(CH}_2)_5\text{]}_m\text{-OH} \xrightleftharpoons[K_{-8}]{K_8}$$

$$R\text{-[O-C(CH}_2)_5\text{]}_n\text{-OH} + R\text{-[O-C(=O)-(CH}_2)_5\text{]}_m\text{-OM} \quad (8)$$

(f) Ester Interchange
(both intra- and inter-molecular):

$$\sim\sim(1)\sim\sim\text{C(=O)-O}\sim\sim(2)\sim\sim + \text{HO}\sim\sim(3)\sim\sim \xrightleftharpoons[K_{-9}]{K_9}$$

$$\sim\sim(1)\sim\sim\text{C(=O)-O}\sim\sim(3)\sim\sim + \text{HO}\sim\sim(2)\sim\sim \quad (9)$$

The relative rates of the above processes will determine the kinetics of the polymerization and, consequently, the molecular weight of the polymer and its molecular weight distribution. Needless to say, this is a complex reaction and the data that follows must be considered in that context.

RESULTS AND DISCUSSION

Various authors(2-4, 11) have suggested that the polymerization of ϵ-caprolactone proceeds via a "living" mechanism. This should yield polymers with very narrow molecular weight distributions referred to as a "Poisson distribution"(3,12). In this case, the kinetics of the polymerization would be expected to be rather simple and straight forward.

Figure 3 illustrates three possible viscosity profiles for the polymerization of ϵ-caprolactone at elevated temperatures.

A Poisson molecular weight distribution occurs only if the following requirements are fulfilled: 1) the rate of initiation is much faster than the rate of polymerization; 2) propagation occurs by addition of the monomer to the polymer chain end; and 3) there is no termination, chain transfer or any other secondary reaction.

If the polymerization of ϵ-caprolactone were to proceed in this manner ($\bar{M}_w/\bar{M}_n=1$) it would be followed by ester interchange reactions until the establishment of the most probable distribution. This would result in a continuing increase in $\bar{M}_w$ and hence of the melt viscosity (Figure 3). The polymerization of ϵ-caprolactone does in fact fulfill the second requirement above. However, fulfillment of conditions one and three are questionable making two alternative polymerization profiles possible.

In the first alternative the "normal" distribution of $\bar{M}_w/\bar{M}_n=2$ is established during the reaction. In that case, no further change in $\bar{M}_w$ is expected irregardless of the fact that ester-interchange may continue to occur. As a result the melt viscosity of the polymer after having reached a plateau would remain essentially constant.

Another alternative consists in the polymerization reaching a molecular weight distribution of >2. The subsequent ester interchange reactions should then result in a decrease in $\bar{M}_w$ and melt viscosity until the normal distribution is reached.

A typical profile for melt viscosity as a function of time for a polymerization reaction was shown in Figure 2. The shape of the curve reflects the kinetic processes which are occuring during the polymerization. All of the reactions were carried out neat, at $\sim$200°C. The viscosity/time profiles that were observed with both anionic and coordination type catalysts were essentially the same. The data do not fit a normal "living" mechanism, with no side reactions. There is

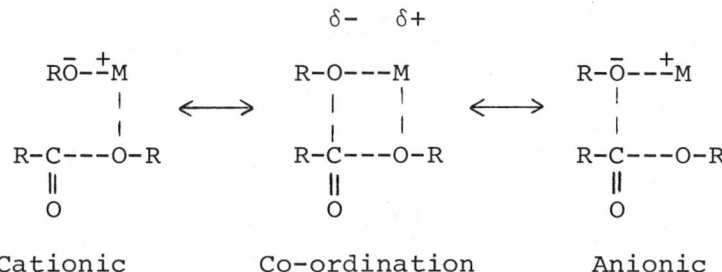

Figure 1. Mechanisms of initiation and polymerization

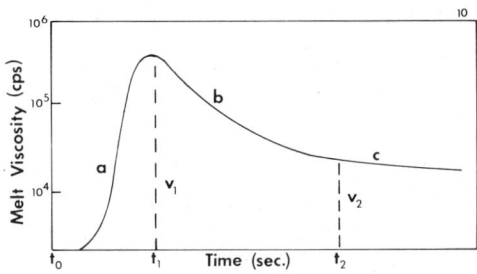

Figure 2. Melt viscosity vs. time for the polymerization of ε-caprolactone

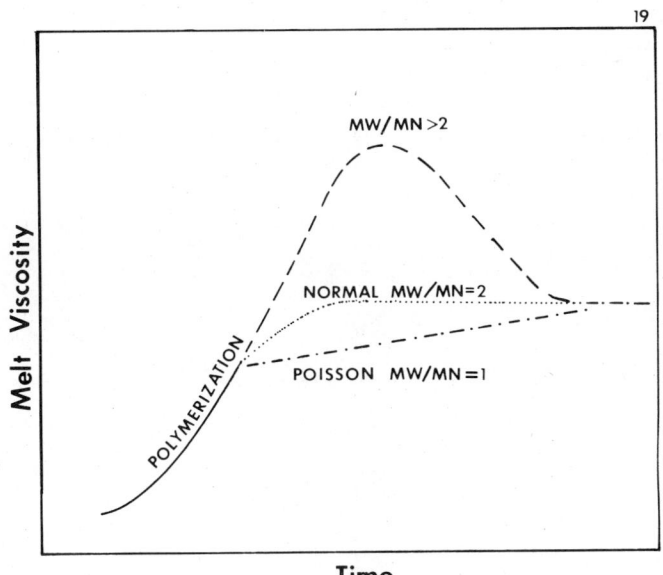

Figure 3. Suggested melt viscosity profiles for polymerization of ε-caprolactone

no doubt that in addition to the usual initiation and propagation steps, several other processes are occurring. These include ester interchange reactions, side-reactions involving the co-initiator, chain transfer phenomena, formation of cyclics in depolymerization reactions, etc.

As stated above, the viscosity profile had the same general shape both in the presence and absence of a co-initiator (Figure 4) and catalyst (Figure 5). The concentration of the catalyst has a significant effect on both the rate of polymerization and on the rate of subsequent reactions resulting in the decrease in melt viscosity (Figure 5).

In order to rationalize these results, molecular weight distribution measurements were performed as a function of reaction time and the results are recorded in Figure 6. The variation in molecular weight distribution with time indicates that $\frac{\bar{M}_w}{\bar{M}_n}$ is initially >2 and that $\bar{M}_w$ decreases with time. This is consistent with the pattern predicted from viscosity measurements (Figures 2, 4 and 5). In addition to ester interchange reactions, there is a second post-polymerization equilibrium reaction that occurs.
It was possible to show reformation of cyclic monomer as well as the formation of other cyclic oligomers.

Gas liquid phase chromatography was carried out during the course of the polymerization. It was observed (Figure 7) that there is a decrease in monomer concentration to an equilibrium level (0.2%). Simultaneously the appearance of both cyclic dimer and cyclic trimer oligomers formed in a depolymerization reaction was noted.

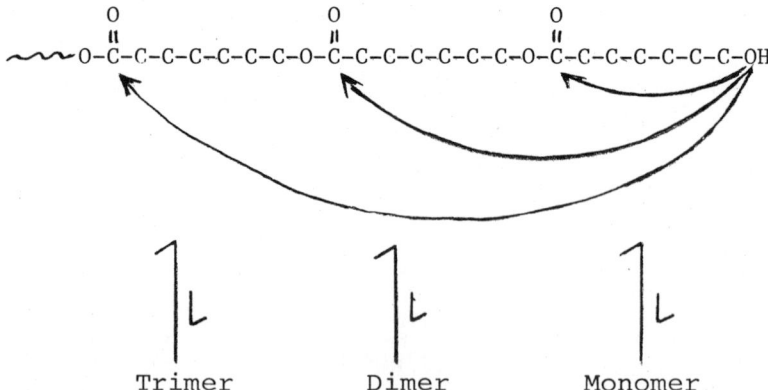

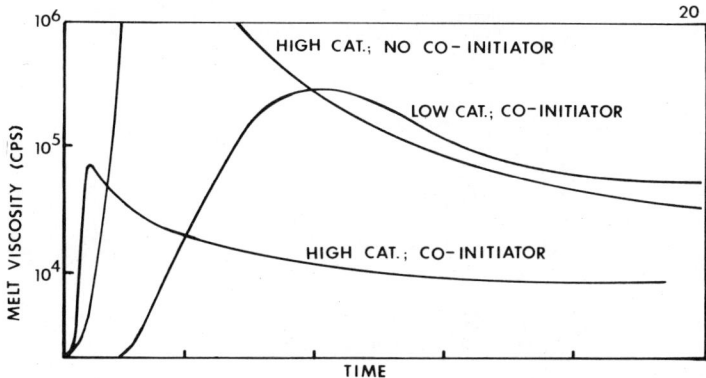

Figure 4. Viscosity–time relationship for the polymerization of ε-caprolactone in the presence of an anionic catalyst with a hydroxyl containing co-initiator

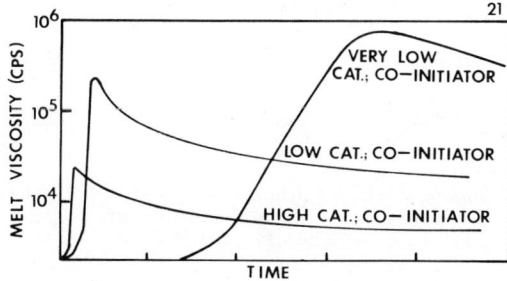

Figure 5. Polymerization ε-caprolactone at constant co-initiator concentration (alcohol) at different levels of typical coordination catalysts

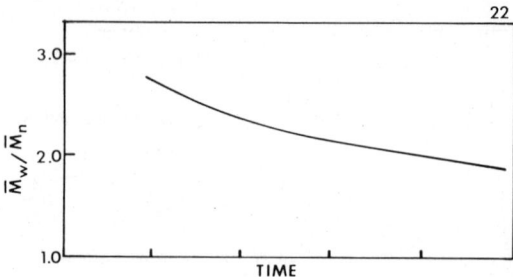

Figure 6. Change in molecular weight distribution for the polymerization of ε-caprolactone

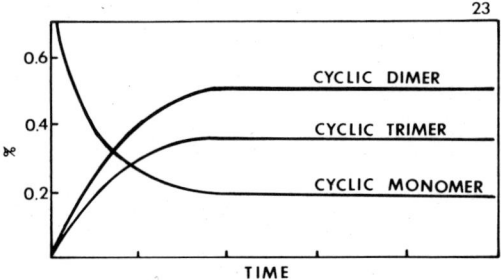

Figure 7. Concentration of cyclic oligomers as a function of time for high-temperature polymerizations of ε-caprolactone

It is interesting that these depolymerization processes have very similar appearance profiles to those of polymer viscosity and $\bar{M}_w/\bar{M}_n$. However, they do not occur at sufficiently high levels to affect either $\bar{M}_w$ or the overall melt viscosity of the reaction mixture.

In step reaction (condensation) polymerizations, the molecular weight distribution is related to the number of propagating polymeric branches. The molecular weight distribution becomes narrower with increasing functionality, as shown in the equation below:

$$\bar{M}_w/\bar{M}_n = 1 + \frac{1}{f}$$

where f is the number of polymeric branches.

The effect of the number of reactive sites of the co-initiator upon the molecular weight distribution was established. This was relatively easy to perform by simply using mono- and dihydroxy compounds as co-initiators. As predicted, at the maximum $\bar{M}_w$ (v_1, Figure 2) the value of $\bar{M}_w/\bar{M}_n$ was higher when the monohydroxy initiator was used.

It is clear from our data that the high temperature neat reaction is extremely complex. It may be due in part to the severe polymerization conditions which were used. These results are not in agreement with those reported by Teyssie([10]). Under milder reaction conditions he has observed the formation of a polymer with a molecular weight distribution close to one. He describes the polymerization as a "perfectly 'living' process". The difference in reaction conditions could have caused the observed differences. Further work will be required to make these discrepancies fully understood.

CONCLUSIONS

The ring opening polymerization of ε-caprolactone at high temperatures follows a pattern which is radically different from the one observed in solution under mild low temperature conditions. It is postulated that the phenomenon is best explained by assuming that several secondary processes occur simultaneously with the primary initiation and polymerization reactions.

Literature Cited

(1) VanNatta, F. J., Hill, J. W. and Carothers, W. H., J. Amer. Chem. Soc. (1934), $\underline{56}$, 455.

(2) Lundberg, R. D. and Cox, E. F. "Ring-Opening Polymerization", Ed. Frisch, K. C. and Reagen, S. L.; Marcel Dekker, N. Y. (1969), pp. 247-302.

(3) Lundberg, R. D., Koleske, J. V. Wischman, K. B., J. of Poly. Sci. (1969), $\underline{7}$, 2915.

(4) Brode, G. L. and Koleske, J. V., J. Macromol. Sci. (1972), A6(6), 1109.

(5) Cox, E. F. and Hostettler, F. (to Union Carbide Corp.) U.S. Patents 3,021,309(1962).

(6) Ludwig, E. B. and Bebenbaya, B. G., J. Macrmol. Sci.(1974) A8(4), 819.

(7) Sekiguchi, H. and Clarisse, C., Die Makromol. Chem. (1976), 177, 591.

(8) Hall, H. K. and Schneider, A. K., J. Amer. Chem. Soc. (1958), $\underline{80}$, 6409.

(9) Jones, T. R., Union Carbide, unpublished results.

(10) Ouhadi, T., Stevens, C. and Teyssie, P., Die Makromol. Chem. Suppl. (1975), $\underline{1}$, 191.

(11) Teyssie, P., Provate Communication.

(12) Billmeyer, F. W., Jr., "Textbook of Polymer Science", 2nd Ed., Wiley, Interscience, New York (1971).

New Prospects in Homogeneous Ring-Opening Polymerization of Heterocyclic Monomers

PH. TEYSSIÉ, J. P. BIOUL, A. HAMITOU, J. HEUSCHEN, L. HOCKS, R. JÉRÔME, and T. OUHADI

Laboratory of Macromolecular Chemistry and Organic Catalysis, University of Liège, Sart-Tilman, 4000 Liège, Belgium

The importance of ring-opening polymerization has been recognized from the very early days of the science of macromolecules (1,2), and this increasingly broad field has been illustrated by many interesting fundamental studies on various types of monomers (including lactones, lactams, oxiranes, oxetanes, tetrahydrofurans, dioxolanes, thiiranes, thietanes, aziridines, leuch's anhydrides, cyclosiloxanes and others).
On the other hand, several of the products obtained exhibit remarkably useful properties, which have promoted extensive physical evaluation and industrial application, e.g. nylon 6, regular polyesters (polycapro- and -pivalolactones), and polyethers (elastomeric homo- and co-polymers of propylene oxide with f.i. epichlorohydrin).

In most cases, these polymerizations have been achieved using different types of initiators, belonging respectively to acid-base, ionic and coordination catalysis. The highly active coordination type initiators have been extremely helpful in controlling the chain-growth processes and their stereospecificity : in particular, they are the only ones able to promote the ring-opening polymerization of methyloxirane (propylene oxide, PO) to high molecular weight, eventually stereoregular, polyethers.

However, most of these catalytic systems do not lend themselves to a straightforward analysis of their structural and kinetic behaviour, owing to the often ill-defined composition and structure of the active site precursors; it is the purpose of this contribution to show how the design of a coordination ring-opening catalyst, having a well-defined composition which can be systematically modified, can lead to some interesting advances in the field.

I. Bimetallic μ-oxoalkoxides : a new family of ring-opening coordination catalysts

Both the incentives and the experimental bases of the present work came from an analysis of the previously described catalytic systems. These were essentially obtained by controlled hydrolysis of ferric organic salts (3,4,5), or of zinc (6,7)and aluminium (8,9) alkyls. Other complex salts catalysts have been described, including the zinc xanthates and thiocarbamates studied by Lal (10) and the hexacyanometallate complexes investigated by Herold (11).

From the indications obtained in these studies, two important requirements for the efficient production of high molecular weight, stereoregular PPO become apparent :
(a) the method of preparation of the active catalysts generally imply that they include M...X...M µ-bridged groupings (mainly M...O...M ones formed by the hydrolysis reactions);
(b) the possibility of obtaining stereoregular crystalline polymers from monomers (like PO) which do not induce an important "chain end control" of the tacticity implies, in turn, the following "catalyst site control" as proposed by Vandenberg (12) and Tsuruta (13) :

$$\sim\!\!\!\sim\!\!\!\sim\!\!O\cdots \overset{\displaystyle C}{C}-O$$
$$\quad\quad | \quad\quad\quad\quad\quad |$$
$$\quad\quad Al \quad\quad\quad\quad Al$$

This type of control, as well as the linear 3-centers structure implied by the rear attack of the unsubstituted carbon, definitely necessitate the presence of several metal atoms in a polynuclear site, as indicated experimentally by several studies (5,15,16,17).

On the basis of these conclusions, it appeared worthwhile to synthesize purposely, in a reproducible procedure, well-defined compounds containing several metal atoms linked together by µ-oxo bridges, and carrying an OR group which would foreshadow the growing polymer chain (like the M-R structure in the Ziegler-type catalysts for olefins polymerization).

A. Synthetic methods

As already reported elsewhere (18,19) a reevaluation of potential direct methods yielding these -M-O-M-OR groupings led to the development (20) of a straightforward 2-steps condensation process between metal acetates and alkoxides, according to the following general scheme :

$$2\,M^1(OR)_n + Y_m M^2(OCOCH_3)_2 \xrightarrow[C_{10}H_{18}]{200°C} (RO)_{2n-2}M^1_2 M^2(Y)_m O_2 + 2ROCOCH_3 \nearrow$$

The composition of these complexes has been confirmed by elemental and functional analysis, and an alternative synthetic route has been devised, involving the carefully controlled hydrolysis of a Meerwein's double alkoxide (21) and yielding products displaying similar

$$2M^1(OR)_n \cdot M^2(OR)_2 \xrightarrow[ROH\ (exc.)]{2H_2O} (RO)_{2n-2}M^1_2M^2O_2 + 4\ ROH\uparrow$$

compositions and properties.

This approach proved to be particularly general and versatile, allowing the synthesis of a broad family of compounds including most metals of the periodic table, and different types of OR groups introduced by quantitative displacement of the small $Oi.C_3H_7$ group by another alcohol).

B. General properties

From the experimental kinetic and structural data gathered up to now (18,19), these μ-oxo-alkoxides are believed to have the following structure : $\left[(RO)_n M^1-O-M^2(Y)_m-O-M^1(OR)_n\right]_{\bar{n}}$; $\bar{n}$, as determined by cryoscopic measurements, indicates the mean degree of association of the compounds and ranks from 1 to 8 in benzene or cyclohexane solutions. It is indeed obvious that these compounds will tend to fulfil their vacant coordination positions on metals M^1 and M^2 by using the electron pairs available on the OR groups of the same or other oxoalkoxide molecules, resulting in a reversible coordinative association (characteristic of all metal alkoxides). This $\bar{n}$ value depends as expected on the nature of the solvent, of the metal atoms, and strikingly of the R group. It is important to note that some specific solvents or ligands like alcohols lead to a complete dissociation of these aggregates. This aggregation might also explain the incredibly high solubility of the compounds in saturated hydrocarbons (practical miscibility), as being due to a compact oxide structure surrounded by a lipophilic layer of alkoxide groups. It also accounts for the electronic delocalization put in evidence by spectroscopic and magnetic measurements.

C. Catalytic Properties

In perfect agreement with the structural hypotheses discussed above, these compounds rank among the best catalysts known for the ring-opening polymerization of several heterocyclic monomers : a practical indication of their activity is given in Table I

Monomer	Catalyst	(Zn) $M \times 10^{-3}$	(Mon) M	Solvent	T°C	$t_{1/2}$ (min)
Methyl-oxirane	$Al_2ZnO_2(On.Bu)_4$	16.6	1.0	heptane	30	20
Methyl-thiirane	$Al_2ZnO_2(On.Bu)_4$	16.6	1.0	heptane	30	13
ε-capro-lactone	$Al_2ZnO_2(Oi.Pr)_4$	5.0	1.0	toluene	0	6

Table I. Activity of bimetallic oxoalkoxides

Although many oxoalkoxide derivatives are active, the aluminium-zinc ones are the most interesting, the more as they are less harmful (if left in the product) than those containing transition metals. They also promote the equilibrium polymerization of isocyanates to polyamides-1 of very high molecular weight (10^6).
As expected, all these polymerizations have been shown (by structural analysis of oligomeric fractions) to proceed by an insertion process into the M^1-OR, and later M^1-OP, bonds.

Accordingly, these soluble compounds represent an interesting model inbetween homogeneous and heterogeneous catalysis; owing to their high activity, their well-defined composition, and the possibility to modify systematically their structure, they offer an attractive tool to study the ring-opening polymerization processes, in particular the eventual topochemical influence of the aggregate on the active site behaviour, and the possibility to generate new types of products. The following sections will summarize our recent advances in these prospects.
It might be also worthwhile to point out an additional point of interest of these derivatives, i.e. their capability to bind and activate molecular oxygen at room temperature (22), when M^2 is a transition metal like Fe(II), Cr(II) or Mo(II).

II. Oxiranes polymerization

As already indicated above, these catalytic sytems rank among the most efficient ones for the conversion of typical oxiranes (like PO) into high molecular-weight, partially stereoregular, polyethers : in particular, they have been considered for an eventual production of PO rubbers (23). Kinetic and structural data point towards a coordinative-anionic mechanism, proceeding through β-stereoselective opening and insertion into the Al-OR bond of the catalyst. Furthermore, the catalytic aggregate on which the reaction takes place is apparently not dissociated, but may undergo a more or less thorough rearrangement depending on the nature of its R groups and of the monomer. The overall behaviour fits with a "flip-flop" mechanism between 2 (or more) metal atoms, in agreement with the proposals of Vandenberg and Tsuruta.

A rather extensive description of these reactions has been already published (18), and more detailed accounts are in preparation; accordingly, this chapter will concentrate on a few new trends which have emerged in our exploratory research and seem to deserve further investigation.

A. Specific controls of the catalytic behaviour in homopolymerization

The kinetic and structural studies mentioned above (18) have put in evidence the existence of 2 competing parallel reactions, proceeding by the same insertion mechanism already mentioned, and both yielding (after the hydrolysis of the catalyst) linear

OH-terminated polyethers. However, the first one proceeds by a random ring-opening and produces atactic oligomers ($\overline{DP}$: 2 to 40) having an intriguing non-monotonous molecular weight distribution, while the other one, which involves only a minor fraction of the active sites (less than 4 %), is stereospecific and gives rise to partially isotactic high molecular weight (10^6) polyether.

It has been proposed (24) that these two types of propagation take place on active sites which have the same chemical structure, but very different steric environments. This view is supported by several experimental results : the relative importance of these 2 processes may be modified up to the pratical exclusion of one or the other by changing the type of aggregate (in particular the value of $\overline{n}$); the degree of stereospecificity depends also (from 5 to 75 %) on the same parameter; and this control of the catalytic behaviour is sensitive not only to the size ($\overline{n}$) but also to the shape of the aggregate (as indicated by the different results obtained with two closely related compounds having different coordination sphere geometries, e.g. the blue $Al_2CoO_2(On.Bu)_4$ and the red-violet $Al_2CoO_2(Oi.Pr)_4$). In other words, it might be suggested that these soluble systems exert some degree of topochemical control on the kinetic and stereochemical course of the polymerization reactions.

On the other hand, and maybe for similar reasons the stereospecificity of the processes is highly dependent on the structure of the monomer for a given catalyst, as illustrated by the fact that $Al_2ZnO_2(On.Bu)_4$ polymerizes, with rather similer rates, allylglycidylether to an essentially amorphous polymer and phenylglycidylether to a highly isotactic crystalline material.

The understanding of such a sensitive and subtle balance in the catalytic behaviour, as related to the size and shape of the aggregates, is certainly a provocative and worthwhile challenge.

B. "On purpose" Modifications of the relative reactivity ratios in copolymerizations

As already reported in ref. (18), the extent of incorporation of a given monomer in a growing copolymer chain depends on other factors than its simple intrinsic reactivity versus a given bimetallic oxoalkoxides catalyst. E.g., it has been possible, by playing with the nature of the solvent, to favour the preferential insertion of either PO or of epichlorohydrin in random copolymerization experiments of these two monomers. In another similar approach, equimolar mixtures of PO and methylthiirane (PS) have yielded products containing essentially PO units or PS units, depending on the use of a $Al_2FeO_2(OR)_4$ or of a $Al_2ZnO_2(OR)_4$ catalyst.

A tentative interpretation involves, in sharp contrast with other types of catalyses, a powerful thermodynamic control of these apparent reactivity ratios, due to the corresponding relative formation constants of the different competing complexes formed between the catalyst and the monomers; obviously, the overall rate of the copolymerization process should still be kinetically determined by the relative intrinsic reactivities of these monomers as appearing in the homopolymerization rates. Anyhow, we have a new and very powerful tool for studying mechanistic behaviours, and for controlling the composition of new products, in particular azeotropic situations in binary copolymerization reactions.

III. Lactones polymerization

A. Homopolymerization reactions.

For this type of monomer also, in particular for ε-caprolactone which is an interesting petrochemical product, the bimetallic oxoalkoxides are at least as active as the best catalytic systems already known (see table 1).

The overall course of these reactions is very similar to that one described for oxiranes; again, kinetic and structural data indicate a typical anionic-coordinated mechanism (18, 25, 26). The molecular weight increases proportionaly to the conversion : the perfectly "living" character of these polymerization reactions has been ascertained by the linear relationship between $\overline{DP}$ at 100% conversion and (M)/(C) ratios, as well as by the resumption of the polymerization on addition of fresh monomer to a polymerized reaction mixture (with a proportional increase of $\overline{DP}$). High molecular weights (up to 200.000) as well as narrow distributions ($\overline{M_w}/\overline{M_n} \geq 1.05$) can be controlled by avoiding side reactions. A structural analysis of the first products of the chain propagation indicates clearly that this reaction proceeds through insertion of the lactone units in the Al-OR bonds, with a specific cleavage of the acyl-oxygen bond, resulting in the permanent binding of the growing chain to the catalyst through an alkoxide link (rather than a carboxylate one) :

$$\geq Al-OR + n \text{ caprolactone} \rightarrow \geq Al\{O(CH_2)_5CO\}_nOR \xrightarrow{H^+} H\{O(CH_2)_5CO\}_nOR$$

An interesting and mechanistically important point is that the number of active sites (potentially 4 per trinuclear catalytic molecule) depends in fact on the type of aggregation of the oxoalkoxides : in other words, as indicated also by N.M.R. measurements (19), there are 2 different types of OR groups depending on their bridging in the aggregates, and only one is active in the polymerization process which results (before hydrolysis) in a catalytic star-shaped entity. These views have been confirmed by the fact that dissociated catalysts (under the influence of the solvent or added alcohols) generate 4 growing chains per $Al_2MO_2(OR)_4$ molecule. The reaction itself seems to proceed by the usual flip-flop mechanism, involving either one Al atom with a vacant cis-coordination position, or two more saturated atoms, again depending on the value of $\overline{n}$.

B. Block copolymerization

The perfectly living character of these lactone polymerizations was obviously a tempting tool to undertake the synthesis of block copolymers, and different successful approaches have been developed, which are summarized hereafter.

Polylactone A - Polylactone B copolymers have been studied first as models, to reach a good control of these reactions. As expected, consecutive (and quantitative) polymerization of 2 different lactone monomers was easy to perform. However, high yields of block copolymers ($\geq 90\%$) are attained only under complete dissociation of the aggregates, ensuring the use of all potentially active OR groups which otherwise might start homopolymer chains (due to a rearrangement or a further dissociation of these aggregates) on addition of the second monomer.

As a typical example, different poly(caprolactone-b-propiolactone) samples have been prepared, which exhibit interesting mixed crystalline morphologies, depending on the thermal history of the samples (27).

Polylactone - PolyH copolymers, where H is any heterocyclic monomer susceptible to undergo ring-opening polymerization by oxoalkoxide catalysts, have also been easily obtained in high yields under the same dissociative conditions. Typical examples include poly-(caprolactone-b-oxiranes), poly-(caprolactone-b-thiiranes), and poly-(caprolactone-b-isocyanates). Since these A.B chains are always OH-terminated (cfr polymerization mechanism), any efficient coupling technique may lead to the corresponding A-B-A structures (eventually thermoplastic elastomers).

Polylactone - PolyX copolymers, where polyX is any preformed polymer chain carrying a suitable functionnal group able to react with the oxoalkoxide catalyst, are the most interesting products obtained.

On the basis of the structural and kinetic behaviour of the oxoalkoxide entities (see above), a specific straightforward procedure has been developed (again under dissociative conditions): it involves the metathetic quantitative displacement of one OR group by the terminal hydroxyl function of a preformed polymer, followed by caprolactone polymerization by the polymer-supported catalyst so obtained (29) :

$$\supset Al-OR + HO\sim\sim\sim^{PX} \rightarrow \supset Al-O\sim\sim\sim^{PX} + ROH \xrightarrow{n\ CL}$$

$$\supset Al-[O-(CH_2)_5-CO]_n\ O\sim\sim\sim^{PX} \xrightarrow{H^+} HO - PCL - PX$$

grafted, A-B diblocks, A-B-A and B-A-B triblocks may be synthesized by using these techniques.

Since the OH-terminated preformed PX block is often prepared by anionic polymerization with a good control of $\overline{M}_n$ and molecular weight distribution (as it is also the case for the polylactone block), it is possible by this technique to tailor a whole family of new materials with a broad range of carefully matched molecular characteristics; e.g. such copolymers where PX is polystyrene, polybutadiene or polyether have been obtained in 90% yields for individual block molecular weights ranging from 1.000 to 150.000.

IV. Application properties for new materials obtained by ring-opening polymerization.

It has already been reminded that this type of polymerization is a fruitful source of interesting materials for practical applications and it seemed worthwhile, in the frame of this contribution, to illustrate that point by some examples generated from the study of the oxoalkoxide catalysts. In this prospect, a particularly significant achievement is constituted by the block copolymers of polystyrene and polycaprolactone (29) (prepared as described in section III.B.), and the present section will be devoted to a brief demonstration, on these materials, of some potentialities of this type of approach.

A. Physical properties of poly-(caprolactone-b-styrene).

As expected (at least for high M.W. blocks), these products exist under the form of heterophasic materials. Electron micrographs show that depending on the sample composition, one may have either a very fine dispersion of PCL domains (eventually amorphous) in a rigid PSt matrix (PCL content lower than 35 %); or of PSt domains in a crystalline PCL matrix (PCL content above 40 %). The size of these domains may be systematically modified from about 150 to 500 Å by controlling the molecular characteristics of the blocks. Obviously, the crystallisation behaviour (kinetics, morphology) is also very sensitive to these composition and molecular size parameters.

The mechanical consequences of this heterophasic structure are well illustrated by the torsion modulus curves of fig.1 (recorded with a Gehman's type apparatus). Not only the usual influence of the Tg and melting point of PCL blocks (around -65°C and + 50°C) and of the PSt blocks Tg (100°c) are clearly and independently apparent, but they also suggest that the whole material has a viscoelastic behaviour up to ca. 140°C in contrast with pure PCL. If one considers the variation of the same modulus in function of the PCL content of the copolymers, above the PCL crystalline melting point e.g. at 70°C, a definite transition is apparent (fig.2) around the same composition where electron microscopy observations show a phase inversion : this result suggests that indeed an heterophasic morphology (amorphous-amorphous) persists in the product.

Another illustrative behaviour may be found in the stress-strain curves recorded on these materials (fig.3). An important cold drawing is observed, implying the formation of a well-

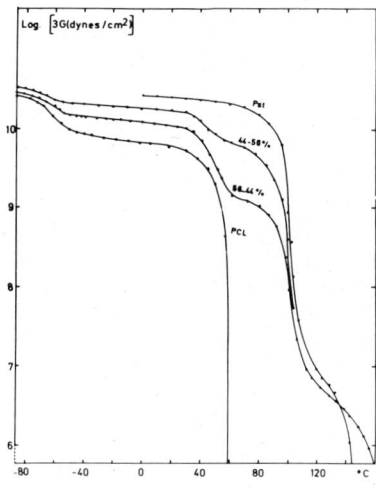

Figure 1. Torsion modulus of poly-(caprolactone-b-styrene) and corresponding homopolymers in function of temperature (10 sec)

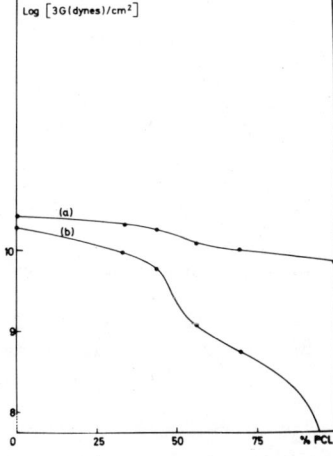

Figure 2. Torsion modulus of poly(CL-b-St) samples in function of the caprolactone content. (a) At $0°C$, (b) at $70°C$.

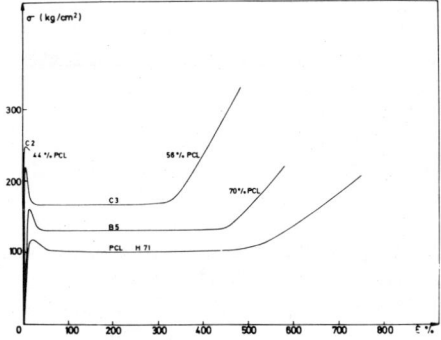

Figure 3. Stress–strain curves for poly-(CL-b-St) samples and pure PCL (molded films). Rate, $600\%/min$ at $23°C$.

organized fibrillar structure (evidenced by other techniques); it is followed by another zone of elastic behaviour (still displaying a good modulus for the samples with a relatively high PSt content). It must be emphasized that the stretched material has a very high resistance to break even at very low temperatures, and peculiar fracture behaviour.

In other words, this type of material offers an attractive additive combination of the properties of both components (high Young's modulus and tensile strenght, viscoelastic behaviour up to 140°C, good stability), plus some unexpected and valuable beneficial points (like the fracture behaviour at low temperature). In conclusion, these results represent a new confirmation of the power of the "properties additivity concept" in block copolymers.

B. Application of the Poly-(caprolactone-b-styrene) to morphology controls in polymer blends.

It has already been known for some time that a PA-PB block copolymer was able to bridge the compatibility gap between two homopolymers PA and PB, and to help in establishing a fine phases dispersion in their mixtures.
Thanks to the apparent compatibility of PCL with other PX polymers, we have tried to extend this concept to the dispersion of mixtures of PA and PX homopolymers with a PA-PCL block material.

The validity of this concept is demonstrated by the optical photomicrographs (fig.4) of films casted from 80PVC/20PSt solution mixtures containing respectively 0, 5 and 10 % of a 56/50 poly-(caprolactone-b-styrene) sample (B_3).
The corresponding torsion modulus diagrams (Gehman's curves, fig.5) show the existence of 3 phases in the mixtures : a PSt phase (transition at 100°C) including both the homo-PSt and PSt blocks, a PVC phase (transition at 80°C) containing the pure PVC, and a third one consisting probably of PVC "plasticized" by PCL blocks (40-50°C).

In other words, the block copolymer realizes an "anchorage" at the interface of the two homopolymers; increasing its concentration promotes a development of the interface, i.e. a decrease in the size of the domains, as confirmed by the microscopic observations. Another important consequence is the increase of the mixtures modulus between 80 and 100°C when increasing the block copolymer content. As expected, the other physico-mechanical properties are also accordingly improved.

Obviously, these concepts and techniques can be applied to a variety of block copolymers, using bimetallic oxoalkoxides or even other catalytic systems : another interesting example is the synthesis of a nylon.6-polybutadiene-nylon.6 triblock copolymer using anionic techniques. The resulting material exhibits again a fine heterophasic structure where small PBD domains (mean diameter around 400 Å) are homogeneously dispersed in the crystalline nylon matrix (30).

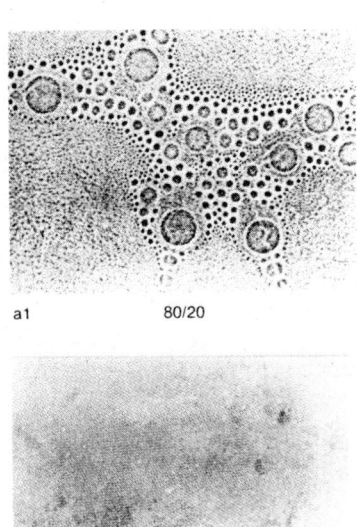
a1   80/20

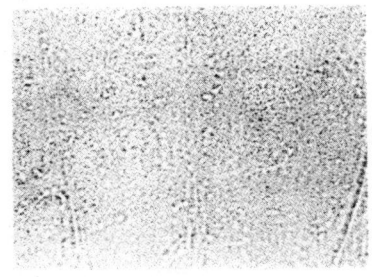
a2   80/20 + 5 % B3

a3   80/20 + 10 % B3

Figure 4. Optical photomicrographs (G = 180) of films obtained from solution mixtures of 80 PVC/ 20PSt, in the presence of 0 (a1), 5 (a2), and 10% (a3) of 50/50 poly-(CL-b-St)

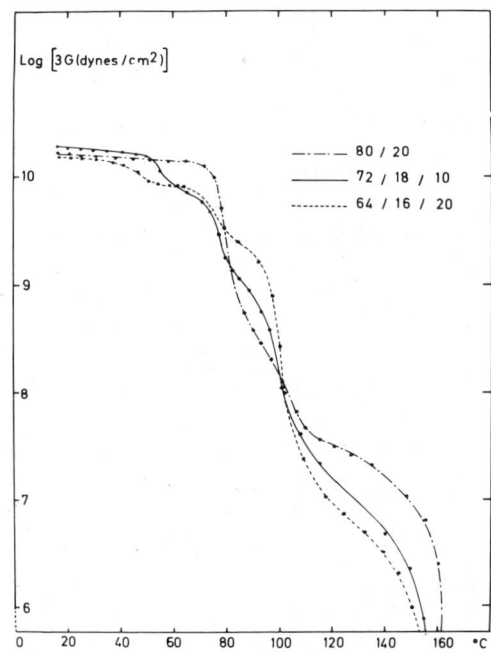

Figure 5. Torsion modulus in function of temperature for blends of PVC, PSt, and P(CL-b-St)

Conclusions

The good control and knowledge of new catalysts like these bimetallic oxoalkoxides is a powerful tool, leading to a better understanding and mastering of the key factors, sometimes unexpected, which govern the ring-opening polymerization processes.

This approach also allows to fully develop the very rich potentialities of this type of polymerization for the synthesis of new materials, mainly block copolymers, enjoying original and useful sets of physical properties.

Acknowledgments The authors want to recognize the pioneering work of M. Osgan (I.F.P.), and the contributions of N. Kohler, J.J. Louis, C. Stevens and P. Condé. They are also grateful to I.F.P. (France), Union Carbide C° (U.S.A.), I.R.S.I.A.(Belgium) and C.R.I.F. (Belgium) for their material and scientific support.

o

o   o

Literature Cited

(1) Wurtz A., Annalen (1859) 110; Ber. (1877) 10
(2) Staudinger H., Ann. Chem., (1933) 41, 505
(3) Pruitt, Jackson and Bagguet, U.S. Pat. (1955), 2.706.181
(4) Colclough R.C. and Gee G., J. Polymer Sci., (1959) 34 171
(5) Osgan M., J. Polymer Sci. (1968), A1,6, 1249
(6) Furukawa J., Tsuruta T. and Saegusa T. Kogyo Kagaku Zasshi, (1959), 62, 1269
(7) Sakata R. and Tsuruta T., Makromol. Chem. (1960) 40, 64
(8) Vandenberg E.J., J. Polymer Sci. (1969), A1,7, 525
(9) Coclough R.C. and Wilkinson K.,J. Polymer Sci.(1964), 4, 311
(10) Lal J., Polymer Letters (1967), 5, 793
(11) Belner R.J., Herold R.J. and Milgrom J., U.S. Pat. (1969), 3.427.256 and 3.427.334.5
(12) Vandenberg E.J., J. Polymer Sci. (1960) 47, 489
(13) Tsuruta T., Int. Sci. Technol., (1967), 71, 66
(14) Hirano T., Kogyo Kagaku Zasshi (1963), 66, 1158
(15) Gurgiolo A.E., Rev. Macromol. Chem. (1966), 1, 39
(16) Gee G. and Higginson W., Polymer (1962), 3, 231
(17) Vandenberg E.J., J. Polymer Sci. (1969) A1,7, 525
(18) Teyssié Ph., Ouhadi T. and Bioul J.P., Int. Rev. of Sci., Phys. Chem. Ser. 2, 8, 192. Butterworths, London 1975
(19) Ouhadi T., Bioul J.P., Stevens C.,Warin R., Hocks L. and Teyssié Ph. Inorg. Chim. Acta (1976) 19, 203
(20) Osgan M. and Teyssié Ph., Polymer Letters (1967) B5, 789

(21) Osgan M., Pasero J.J. and Teyssié Ph., Polymer Letters (1970), B8, 319
(22) Teyssié Ph. et al., in "Catalysis, heterogeneous and Homogeneous", Ed. B. Delmon and J. Jannes, p. 289, Elsevier, Amsterdam 1975.
(23) Osgan M., Teyssié Ph. and Wauquier J.P., A.C.S. Symp. 156 (1968) Div. Petr. Chem., Prepr. A, 89
(24) Bioul J.P., Ph. D. Thesis, University of Liège (1973)
(25) Ouhadi T., Hamitou A., Jérôme R. and Teyssié Ph., Macromolecules (1976) 9, 927
(26) Hamitou A., Ouhadi T., Jérôme R. and Teyssié Ph., J. Polymer Sci., (1977), A1, in press
(27) Huynh Ba Gia, Licence Thesis, University of Liège (1975)
(28) Heuschen J., Jérôme R. and Teyssié Ph., Fr. Pat. (1977) dep. nr B 7501
(29) Heuschen J., Ph. D. Thesis, University of Liège (1977)
(30) Petit D., Ph. D. Thesis, University of Liège (1975)

13

Optically Active Poly[oxy(1-alkyl)ethylene]

TEIJI TSURUTA

Department of Synthetic Chemistry, Faculty of Engineering, University of Tokyo, Bunkyo-ku, Tokyo, Japan 113

The optical rotation of poly(R-oxypropylene) has different signs at the sodium D line (589 nm) in different solvents. The specific rotation, $[\alpha]_D$, of poly(R-oxypropylene) is positive in cyclohexane and in chloroform, but negative in benzene and trifluoroethanol. Since Price and Osgan (1) first reported this phenomenon, several attempts have been made to interpret this in terms of a collision complex formation (2) with the solvent or a change of conformation of polymer molecules in response to the nature of solvent. According to our knowledge, however, any decisive conclusion has not yet been drawn concerning the optically active behaviors of poly(R-oxypropylene). To solve this problem, a series of studies on optically active poly[oxy(1-alkyl) ethylene] has been carried out. It was found from these studies that the influence of solvent on the ORD spectra decreased with the increase in bulkiness of the alkyl substituent of the oxyethylene unit (3),(4).

The bulkiness of the alkyl substituent also exerts an enormous influence upon the nature of stereoselective polymerization of alkyloxiranes. When the R,S-copolymerization of t-butyloxirane was carried out starting with a monomer mixture consisting of R/S=76/24 using t-BuOK as initiator, R-monomer was found to be incorporated into polymer chain preferentially over S-monomer (5). This is explained in terms of the growing chain control mechanism, in which the chiral structure of the growing polymer chain is responsible for the stereoselection.

A unique and significant effect of the bulky substituent has recently been found also in a cationic oligomerization of (R)-t-butyloxirane with boron trifluoride etherate as initiator, where a cyclic tetramer was formed in an excellent yield (6).

In the present review article, the author intends to discuss on the effect of the bulky substituent on the physical properties of poly(alkyloxirane) as well as the mechanism of selective polymerization and oligomerization reactions.

1. Conformation and Optical Rotatory Behavior of Poly[Oxy(1-alkyl)ethylene]

For the discussion of conformation of poly(R-oxypropylene) molecules, we have to consider stereochemistry with respect to C-C and C-O bonds along a polymer chain, $\{O-CH(CH_3)-CH_2\}_n$. In order to get information on the rotational isomers around the C-C bond, a deuterated poly(R-oxypropylene) was prepared starting from *trans*-deuterated methyloxirane monomer (7),(8). The vicinal coupling constants, $|J_{AC}|$, between the methylene and the methine protons were found to be 5.3, 4.9 and 5.2 (Hz) in cyclohexane, chloroform and benzene, respectively. By assuming the standard values $^3J_{60°}$=2.6 (Hz) and $^3J_{180°}$=9.3 (Hz) reported in a variety of 1,2-dioxygen-substituted propane derivatives, it was possible to estimate the population of the three rotational isomers in cyclohexane, chloroform and benzene, respectively.

From the results stated above, the distributions of the three rotational isomers were estimated to be almost the same for poly(R-oxypropylene) in the three solvents, the conformational differences being not so large as to explain the sign difference in the $[\alpha]_D$ of the oxirane polymer.

Studies on the dipole moment of an *isotactic* poly(oxypropylene) suggested the distribution of rotational isomers around C-O bonds to be scarcely changeable in benzene and in cyclohexane as shown in Table I.

We therefore carried out a series of studies on the circular dichroism spectrum of poly(R-oxypropylene) in a number of solvents in the vacuum ultraviolet region under the cooperation with W.C. Johnson, Oregon State University. In the circular dichroism(CD) spectra of poly(R-oxypropylene), two CD bands were observed for cyclohexane, acetonitrile, and trifluoroethanol(TFE) solutions. The CD spectrum was extended to 140 nm and three bands were measured in a 1,1,1,3,3,3-hexafluoro-2-propanol(HFIP) solution.

A Kronig—Kramers transform of the two CD bands observed in cyclohexane accounts for the observed positive ORD spectrum. In contrast, a third large and negative ORD band centered at 155.5 nm is responsible for the negative ORD spectrum observed in HFIP. In the latter solution as well as in benzene, the ORD spectrum was found to fit the Drude one term equation with λ_0=150 nm.

In the light of the results obtained, it is most probable to conclude that the interaction between polymer main chain and solvent molecules should be the major cause for the different sign of ORD in the two groups of solvent in the visible region.

Since poly[oxy(1-alkyl)ethylene] is expected to possess lower degrees of solvent interaction as the alkyl-substituent becomes bulkier, poly(R-isopropyloxirane) (3) and poly(R-t-butyloxirane) (4) were synthesized and their optical rotatory behaviors were examined in a number of solvent.

As shown in Fig. 1, the influence of solvent on the ORD spectra decreases with the increase in bulkiness of the alkyl substituent of the oxyethylene unit. The bulky t-butyl substituent seems to make the main chain relatively "rigid" and reduces the accessibility of the main chain in the preferred conformation (perhaps a local helix) to the solvent molecule.

In order to get independent information concerning the degree of rigidity of poly[oxy(1-alkyl)ethylene] molecule, partially relaxed FT ^{13}C NMR spectra were examined. The ^{13}C spin-lattice relaxation times for the three polyoxiranes are listed in Table II, in which n denotes the number of hydrogen atom bound to the relevant carbon atom. The bulkier the alkyl substituent, the smaller nT_1 values were obtained, which was regarded as a consequence of slower segmental motion owing to the enhanced rigidity of the macromolecule possessing bulkier substituents. Under the extreme narrowing conditions, the correlation time, τ_{eff}, for each carbon atom was calculated. Results obtained are shown in Table III.

It can be said from the values for $\tau_{eff}^{CH_2}/\tau_{eff}^{CH}$ that the methylene-carbon undergoes more rapid movement than the methine-carbon in poly(oxypropylene), whereas the movement of the both carbon atoms synchronized in the molecules of poly[oxy(1-t-butyl)ethylene], suggesting more flexibility in the poly(oxypropylene) chain.

2. Regioselectivity in the Ring-Opening Polymerization of t-Butyloxirane

It was previously reported (5) that the bulk polymerization of t-butyloxirane initiated with potassium t-butoxide (t-BuOK) proceeded according to the living mechanism with the initiator efficiency being 100%.

In order to get information on the site of bond cleavage during the polymerization process, (R)-t-butyloxirane was polymerized in bulk with the t-BuOK initiator. The ^{13}C NMR spectrum of poly[(R)-t-butyloxirane] is given in Fig. 2.

Assignments of the NMR signals were made by the gated method. Each carbon signal in Fig. 2 is a sharp singlet and no extra signal due to irregular structures can be observed, indicating that the sample of poly[(R)-t-butyloxirane] obtained in 98% yield from (R)-t-butyloxirane by the bulk polymerization with t-BuOK is configurationally homogeneous (i.e., isotactic), and that the amounts of head-to-head and tail-to-tail sequences are too small, if any, to be detected by ^{13}C NMR. Therefore, the bulk polymerization of t-butyloxirane with t-BuOK was concluded to proceed to form head-to-tail sequences under the exclusive cleavage at either of the O-CH bond (α-opening) or the O-CH$_2$ bond (β-opening).

The ^{13}C NMR studies carried out on the living polymerization system of (R)-t-butyloxirane led us to conclude the β-opening to be operative in the propagation process of t-butyloxirane. This conclusion was drawn on the basis of the following observations.

In the ^{13}C NMR spectrum of the living system initiated by larger amount of t-BuOK, several new signals were observable along with the signals which were assigned previously to -CH$_2$- (74.5 ppm), -CH- (89 ppm) and -C- (35.1 ppm) of the internal units of poly[(R)-t-butyloxirane]. New signals at 64 ppm (-CH$_2$-) and 73 ppm (-C-) in the living system were assigned to structure [1],

Table I Dipole Moments of *isotactic* Poly(oxypropylene)

Solvent	Temp. in °C	μ in Debye
benzene	25	1.09 ± 0.03
	35	1.09 ± 0.03
	51	1.10 ± 0.03
cyclohexane	25	1.04 ± 0.03
	35	1.04 ± 0.03
	51	1.05 ± 0.03

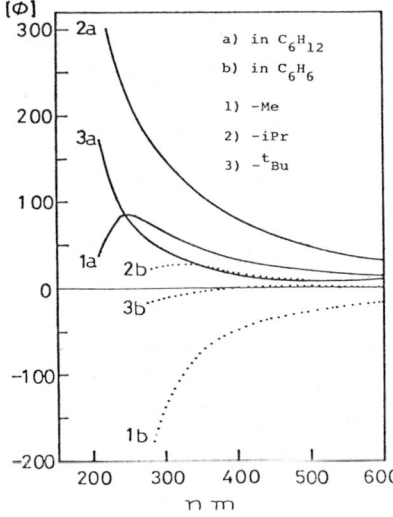

Figure 1. Optical rotatory dispersion spectra of poly(alkyloxirane)

Table II ^{13}C Spin-Lattice Relaxation Times, nT_1 (sec)

Polymer	CH	CH_2	substituent			CH	CH_2	substituent		
			C	CH	CH_3			C	CH	CH_3
Me	2.2	3.0	-	-	6.9	2.2	3.4	-	-	6.9
isoPr	0.84	0.98	-	0.84	3.9	0.70	0.84	-	0.70	3.9
t-Bu	0.52	0.54	4.0	-	2.0	0.43	0.40	3.4	-	2.2

in C_6D_6: Conc.*ca* 10 w/v%, at 60°C   in C_6D_{12}: Conc.*ca* 10 w/v%, at 60°C

because these signals were also found in the spectrum of an oligomer which was obtained after the treatment of the living system with aqueous HCl solution.

$$\begin{array}{cc} \text{CH}_3 & \text{Bu} \\ | & | \\ \text{CH}_3-\text{C}-\text{O}-\text{CH}_2- & -\text{O}-\text{CH}_2-\text{CH}-\text{OK} \\ | & \\ \text{CH}_3 & \\ \end{array}$$

    73   64   81   83
  ppm ppm   ppm ppm
     [1]   [2]

The signals at 81 ppm ($-CH_2-$) and 83 ppm ($-\overset{|}{C}H-$) observed in the living system were assigned to the structure of the growing chain end unit [2], because they disappeared when the living system was treated with aqueous HCl solution. In accord with this observation, a signal assignable to methine carbon of structure $-CH(t-Bu)-OH$ was found at 78 ppm in the spectrum of the oligomer. From these results, it was unambiguously confirmed that the bond cleavage of t-butyloxirane takes place exclusively at $O-CH_2$ bond during the polymerization process with t-BuOK as initiator.

3. Effect of t-Butyl Substituent on the Selectivity in the Polymerization of t-Butyloxirane

It was previously reported (9) that copolymerization study between R- and S-monomer is a useful tool for elucidation of the stereocontrol mechanism. When the R,S-copolymerization of t-butyloxirane was carried out starting with a monomer mixture consisting of $R/S=76/24$ using t-BuOK as initiator, R-monomer was incorporated into polymer chain preferentially over S-monomer (5). As the consequence, the optical purity in the recovered monomer became smaller than that of the starting mixture in the course of the R,S-copolymerization (Fig. 3). This is explained in terms of the growing chain control mechanism, in which the chiral structure of the growing polymer chain is responsible for the stereoselection.

$$\sim\sim\sim CH_2-CH(t-Bu)-OK \quad \begin{array}{c} \nearrow \\ (R) \\ \searrow \end{array} \quad \begin{array}{c} (R)\;\;t\text{-Bu} \\ \triangle \\ O \\ (S) \\ \triangle \\ O\;\;t\text{-Bu} \end{array} \quad \begin{array}{c} (k_{RR}) \\ \\ (k_{RS}) \end{array}$$

The curve for t-butyloxirane in Fig. 3 was analyzed in more detail by deviding the curve into j-stages. From experimental data at every stage, it was possible to calculate a parameter, α_j, which is defined as follows:

Table III Ratio of the Correlation Times $\tau_{eff}^{CH_2}/\tau_{eff}^{CH}$ for CH_2 and CH Carbons

Substituent	in C_6D_6	in C_6D_{12}
Me	0.76	0.67
isoPr	0.87	0.83
t-Bu	0.97	1.1

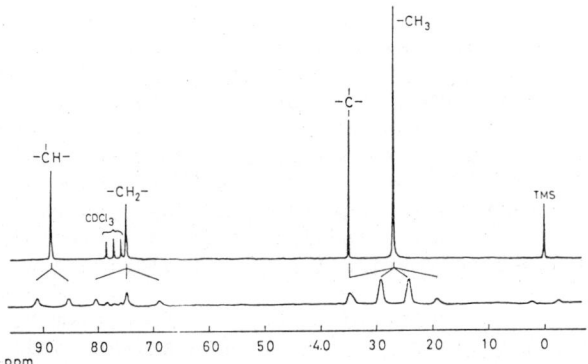

Figure 2. Pulsed FT ^{13}C-NMR spectra (25.03 MHz) of P((R)-tert-butyloxirane) in the CDCl$_3$ solution (10 w/v%) at 45°C

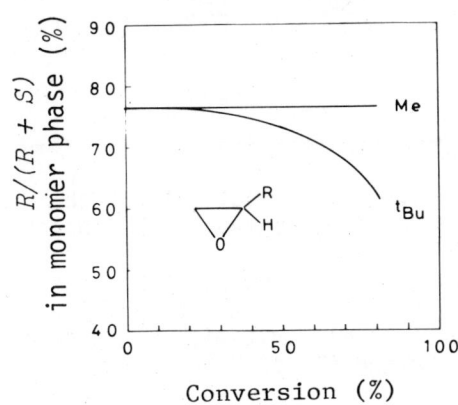

Figure 3. R-Content in unchanged monomer vs. conversion

$$[dR/dS]_j = \alpha_j [R/S]_j$$

The parameter α_j is regarded as a measure of the stereoselectivity which is associated with a growing polymer chian at j-stage. The parameter, α_j, was found to become greater as the conversion increases. Since the polymerization of t-butyloxirane initiated by t-BuOK was proved previously to form a living system (5), the above results indicate that longer chain of growing polymer to possess the greater stereoselectivity, sudden increase of α_j-value being observed when the degree of polymerization of growing chain attains about 20. Therefore, we can conclude that not only the chiral structure of growing chain end but also the chiral secondary structure of the polymer chains should be responsible for the stereoselection of incoming monomers.

Methyloxirane behaves absolutely different way from t-butyloxirane, no change at all in the optical purity of the monomer phase being observed in the course of R,S-copolymerization under similar reaction conditions (Fig. 3). It was also confirmed that the main chain of poly(methyloxirane) formed possesses randomly distributed R- and S-monomeric units with the same R/S ratio as that in the monomer phase regardless of the conversion of polymerization. These results indicate that no stereoselection takes place in the R,S-copolymerization of methyloxirane with KOR initiator (9).

C.C. Price (10) reported the IsoSyn mechanism for the stereochemistry of polymerization of RS-t-butyloxirane initiated with t-BuOK. For the formation of IsoSyn poly(t-butyloxirane), -RRSSRRSSRRSSRRSS-, the following conditions should be established

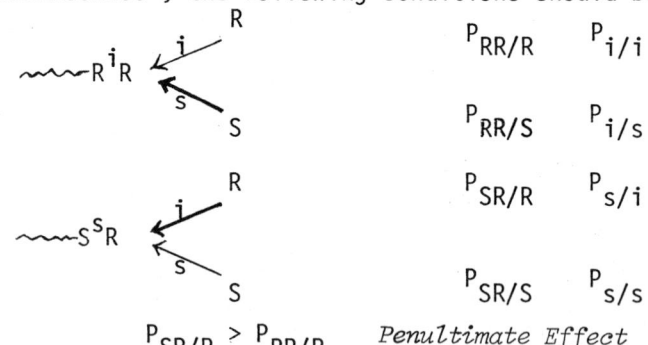

$P_{SR/R} > P_{RR/R}$ *Penultimate Effect*

where P means a conditional probability. The reactivity of the terminal unit R is expected to be definitely controlled by the penultimate unit.

In order to elucidate the nature of the growing chain control mechanism, a series of NMR studies was carried out with poly(t-butyloxirane) which was prepared by the polymerization of racemic monomer initiated by t-BuOK (11). Pulsed FT ^{13}C-NMR spectra of the polymer are shown in Fig. 4. By comparison with NMR spectra of poly[(R)-t-butyloxirane] shown in Fig. 2, it was possible to

estimate triad population in the poly[(RS)t-butyloxirane] as
follows: ii(SSS+RRR) 32%; is+si[(SSR+RRS)+(SRR+RSS)] 48%;
ss(SRS+RSR) 20%. Assuming the First Order Markov Chain mechanism,
the following values are calculated for $P_{i/s}$ and $P_{s/i}$:

$$P_{i/s}=(is+si)/[2(ii)+(is+si)]=0.43$$
$$P_{s/i}=(is+si)/[2(ss)+(is+si)]=0.55$$

These results suggest the terminal unit reacts with an incoming
monomer without any significant influence from the penultimate
unit, because $P_{s/i}$ was found almost equal to $P_{i/i}$:

At the very initial stage of the R,S-copolymerization (R/S:
76/24) of t-butyloxirane, any ordered secondary structure has not
yet been established in the growing polymer chain, so that it is
reasonable to use the value k_{RR}/k_{SR} (or k_{SS}/k_{SR})=1.3 [obtained
from 0.55/0.43 (see above)] as the reactivity ratio at the earliest stage of the R,S-copolymerization. Starting with $[R]_0/[S]_0$
=76/24, α-value can be calculated as follows:

$$\frac{d[R]}{d[S]} = \frac{[R]}{[S]} \cdot \frac{k_{RR}[R^*] + k_{SR}[S^*]}{k_{RS}[R^*] + k_{SS}[S^*]}$$

$$\frac{d[R]}{d[S]} = \frac{[R]}{[S]} \cdot \frac{(k_{RR}/k_{SS})[R^*]/(S^*) + 1}{[R^*]/[S^*] + (k_{SS}/k_{SR})} = \alpha \frac{[R]}{[S]}$$

where $k_{RR}=k_{SS}$, $k_{RS}=k_{SR}$ and $[R^*]/[S^*] = [R]_0/[S]_0$

The value, 1.14, for the α_i at the initial stage falls in the
range of value which is anticipated from the experimental curve.
Since the copolymerization was started with a mixture of R-content
being 76%, there is an enhanced chance to form isotactic enchainment, which will result in the formation of a chiral secondary
structure. Therefore, the preferential incorporation of R-monomer
will be much more amplified in the later stage as observed in the
experiment.

In order to get clearer picture as to the chiral secondary
structure of the growing chain end, the living polymerization
system [3] was studied by partially relaxed (PR) FT ^{13}C NMR in
terms of the spin-lattice relaxation behaviors.

		(3)		(2)		(1)		
(4)		Me$_3$C		Me$_3$C		Me$_3$C		
Me$_3$C-O-CH$_2$	(-CH	— O — CH$_2$ —)$_n$	CH	— O — CH$_2$	— CH	— OK	[3]
73	64	89	74.5	89.1	81	83 ppm	
(T$_1$) (sec)	(1.3)	(0.8)	(0.35)	(0.3)	(0.15)	(0.25)	

The PRFT spectra of the living system [3] showed that values of the spin-lattice relaxation time (T_1) for the carbon atoms located in the growing end units, whose signals appeared at 83 ppm, 81 ppm and 89.1 ppm, were observed to be much shorter than T_1 values for other carbon atoms of the living system. T_1 values of the methyl-carbons (1) and (2) were observed again to be shorter than those of the methyl-carbons (3) and (4).

The observed T_1 values for the living system make it possible to consider that the segmental motion of the ultimate and penultimate units of the growing chain are restricted to a significant extent.

We interpret the restricted segmental motion in terms of the formation of a rather rigid structure in which ether-oxygen atoms of the growing end units are bound together through coordination bonds with potassium cation. If all of the asymmetric centers involved in the growing end units have the same sense of stereoisomerism, the growing end moiety will exhibit a significant chiral character around the counter cation K^+. We consider the nature of the observed enhancement of the stereoselectivity, α_j, to be ascribable to the formation of the chiral secondary structure around the potassium cation. Under the reaction conditions for the asymmetric-selective R,S-polymerization stated above, probabilities for formation of such chiral structure will be rather low before the degree of polymerization of the growing chain attains about 20.

4. Specific Formation of Cyclic Tetramer by Cationic Oligomerization of (R)-t-Butyloxirane

A unique and significant effect of the bulky substituent has recently been found also in a cationic oligomerization of t-butyloxirane with $BF_3 \cdot OEt_2$ as initiator (6). When (R)-t-butyloxirane was treated with 5 mol% of $BF_3 \cdot OEt_2$, a snow-white solid was formed in 89% yield (by weight). After purification, the crystalline product (melting point 168°C) was analyzed. IR, no existence of OH group; Molecular Weight (by vapor pressure osmometry), 410; Anal, Calcd: as a cyclic tetramer, C 71.95, H 12.07; Found: C 71.91, H 12.45. $[\alpha]_D^{25}$ 50.8 (C 1.10, in C_6H_{12}); 53.8 (C 1.28, in C_6H_6).

From these results, the crystalline product was concluded to be a cyclic tetramer of (R)-t-butyloxirane. ^1H NMR and ^{13}C NMR show that the tetramer consists of four identical monomeric units. Therefore, the oxirane ring of (R)-t-butyloxirane must have opened exclusively at either of the CH-O bond (α-opening) or the CH_2-O bond (β-opening). In order to obtain information on the site of ring-opening, a model reaction of (R)-t-butyloxirane with $BF_3 \cdot OEt_2$ was carried out in the presence of t-butyl alcohol in one to one mole ratio to the oxirane. After the unchanged alcohol was removed, the reaction mixture was examined by ^{13}C NMR. Three signals were observed at 78 ppm, 73 ppm and 64 ppm besides the signals assigned to the internal oxy(1-t-butyl)-ethylene units. The first signal (78 ppm) was assigned to methine carbon of -CH-OH, while the second and third signals to the structure [1]. When a similar reaction was conducted in the presence of methanol, signals assignable to -CH_2-OCH_3 group, instead of [1], were observed at 75 ppm and 58 ppm. From these results of the model reaction, it was concluded that the cyclic tetramerization of (R)-t-butyloxirane with $BF_3 \cdot OEt_2$ catalyst proceeds under the cleavage of the CH_2-O bonds (β-opening), which resulted in the formation of (2R, 5R, 8R, 11R)(2,5,8,11-tetra-t-butyl-1,4,7,10-tetraoxacylododecane with the retention of the configuration at the asymmetric carbons.

The specific formation of the tetramer of (R)-t-butyloxirane forms a sharp contrast with the results obtained in the reaction of (S)-isopropyloxirane with $BF_3 \cdot Et_2O$, in which the reaction product was an oily compound consisting of at least seven major compounds of low molecular weight.

The bulky t-butyl substituent is expected also to cause more severe restriction in the possible preferred conformation of the cyclic tetramer in comparison with the corresponding linear polymer, poly((R)-t-butyloxirane). The chemical shift and coupling constant values in the ^{1}H NMR spectra of the cyclic tetramer in deuterated benzene and chloroform are listed in Table IV.

In Table IV, the signals around δ 2.8 ppm were assigned to the methine proton Hc, those around δ 3.6 - 3.8 ppm and δ 3.8 - 4.0 ppm to H_B and H_A methylene protons, respectively. The Newman projection for the possible three conformers around the methylene-methine carbon bond is shown in Fig. 5. The assignment of H_A and H_B are made from ^3J's and deshielding effect of the ether oxygen atoms. On the basis of the Karplus-type dependency of ^3J on dihedral angle (12), it was concluded that Hc should come approximately on the bisection plane of H_A and H_B because both of the observed $^3J_{AC}$ and $^3J_{BC}$ were about 3 Hz (Table IV). Therefore, the predominant conformation around the main chain CH-CH_2 bond should be G^+ in Fig. 5.

The most plausible spatial structure for the cyclic tetramer is that shown in Fig. 6, where the main chain of the tetramer predominantly takes a G^+G^+T conformation:

$$-O-CH(t-Bu)-CH_2-$$
$$\quad\quad G^+ \quad\quad\quad G^+ \quad T$$

According to this structure, H_A should be more deshielded than H_B because two ether oxygen atoms O(i-2) and O(i+4) come close to H_A than H_B of the methylene group (i), and so are assigned the H_A and H_B protons to the observed two methylene signals. The ORD spectra of the cyclic tetramer in benzene, in cyclohexane and in chloroform are shown in Fig. 7. In contrast to the corresponding linear poly(alkyloxirane)s, the ORD curve of the cyclic tetramer in cyclohexane is exactly the same as that in benzene and very

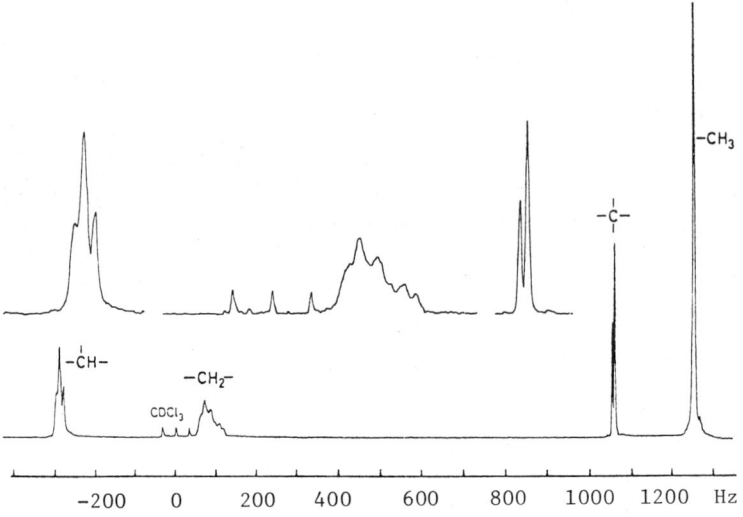

Figure 4. FT ^{13}C-NMR spectra (25.03 MHz) of P((RS)-tert-*butyloxirane*) initiated by t-BuOK in bulk. Solvent, $CDCl_3$ (conc. 30 w/v%); Temp., 50°C.

Table IV Chemical Shifts and Coupling Constants for the Cyclic Tetramer, (2R,5R,8R,11R)-2,5,8,11-tetra-*tert*-butyl-1,4,7,10-tetraoxacyclododecane, in Deuterated Benzene and in Deuterated Chloroform at Various Temperatures

Solvent	Temp,°C	Chemical shift,[a] ppm			Coupling constant, Hz		
		δ_A	δ_B	δ_C	$|^3J_{AB}|$	$|^3J_{AC}|$	$|^3J_{BC}|$
C_6D_6	30	3.70	3.64	2.66	11.7	2.6	3.0
	50	3.76	3.68	2.71	11.8	2.7	3.3
	75	3.81	3.71	2.75	11.7	2.6	3.3
$CDCl_3$	30	3.90	3.78	2.83	12.0	3.0	3.2
	50	3.90	3.79	2.83	11.9	2.9	3.3

a Chemical shift is given in ppm unit with plus value for downfield shift (δ scale) from the internal standard TMS.

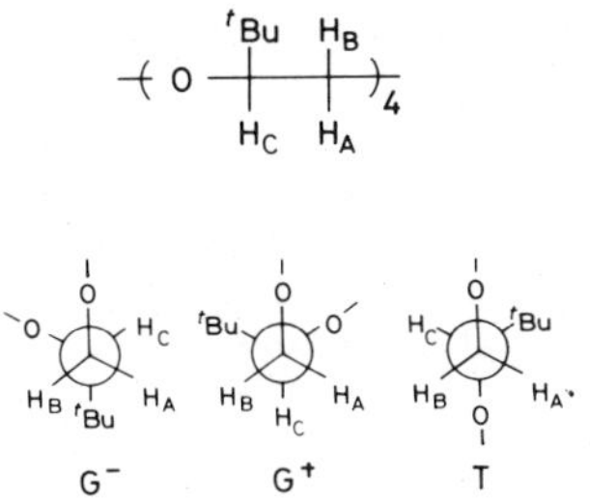

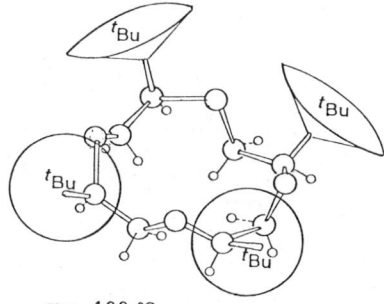

Figure 5. Three possible rotational isomers around the main chain CH–CH$_2$

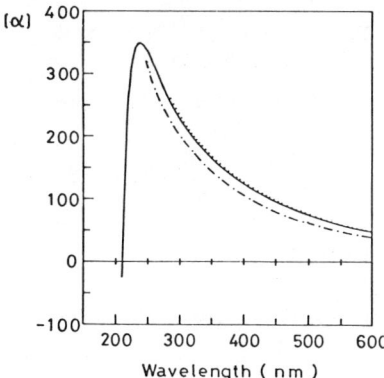

mp 168 °C
$[\alpha]_D^{25}$ 50.8 ($\underline{c}$ 1.10, C$_6$H$_{12}$)
53.8 ($\underline{c}$ 1.28, C$_6$H$_6$)

Figure 6. A proposed spatial structure of the tetramer of (R)-tert-butyloxirane

Figure 7. Optical rotatory dispersion (ORD) spectra of the cyclic tetramer of (R)-tert-butyloxirane at 25°C. (———) In cyclohexane, ($\cdot\cdot\cdot$) in benzene, ($-\cdot-$) in chloroform.

similar to that in chloroform. The similarity of the ORD spectra of the cyclic tetramer in benzene and in cyclohexane indicates the absence of an appreciable interaction of solvent molecules with the main chain, as was discussed in Section 1. The proposed structure of the cyclic tetramer may explain the similarity of the ORD curves in Fig. 7, because ether oxygen atoms are insulated from the solvent by the bulky hydrocarbon groups.

Literature Cited

(1) Price,C.C., Osgan,M., *J. Am. Chem. Soc.* (1956) 78, 690; 4787.
(2) Kumata,Y., Furukawa,J., Fueno,T., *Bull. Chem. Soc. Japan* (1970) 43, 3663; 3920.
(3) Tsuji,K., Hirano,T., Tsuruta,T., *Makromol. Chem.* (1975) Suppl. 1, 55.
(4) Sato,A., Hirano,T., Tsuruta,T., *Makromol. Chem.* (1976) 177, 3059.
(5) Sato,A., Hirano,T., Tsuruta,T., *Makromol. Chem.* (1975) 176, 1187.
(6) Sato,A., Hirano,T., Suga,M., Tsuruta,T., *Polymer Journal*, (1977) in press.
(7) Pham,K.H., Hirano,T., Tsuruta,T., *J. Macromol. Sci.-Chem.* (1971) A5, 1287.
(8) Hirano,T., Pham,K.H., Tsuruta,T., *Makromol. Chem.* (1972) 153, 331.
(9) Tsuruta,T., *J. Polymer Sci.* (1972) D6, 179.
(10) Price,C.C., Akkapedi,M.K., DeBona,B.T., Furie,B.C., *J. Am. Chem. Soc.* (1972) 94, 3964.
(11) Sato,A., Hirano,T., Tsuruta,T., *Makromol. Chem.* (1977) 178, in press.
(12) Karplus M., *J. Chem. Phsy.* (1959) 30, 11.

Acknowledgment: The author wishes to express his gratitude to Dr. T. Hirano and Dr. A. Sato for their cooperation with him to promote this study.

14

Stereoselective and Stereoelective Polymerization of Oxiranes and Thiiranes

NICOLAS SPASSKY

Laboratoire de Chimie Macromoléculaire, Associé au CNRS,
Universite Pierre et Marie Curie, 4, Place Jussieu, 75230 Paris Cedex 05, France

Ionic polymerization of oxiranes and thiiranes is usually divided in three main types : anionic, cationic and stereospecific-"coordinated". While anionic and cationic polymerization produce random amorphous polymers, at least when starting from racemic monomers, stereospecific initiators may give isotactic crystalline polymers.

Depending on the initial reagents and resulting products, several types of stereospecific processes could be considered :
- stereoselective polymerization,
- stereoelective also called asymmetric-selective polymerization,
- asymmetric-polymer synthesis.

The first two processes are dealing with monomers which are a mixture of stereoisomers, while the last one considers symmetric monomers having two asymmetric carbons of opposite configuration.

Most of the work in the field of stereoselective and stereoelective polymerization of oxiranes and thiiranes was carried out on monosubstituted monomers and was reviewed in some publications (1-6).

The stereochemical aspects of the polymerization of some disubstituted oxiranes and thiiranes were described in the work of Vandenberg (7,8).

The aim of this paper is to give a review of the more recent results concerning such reactions, to include some new unpublished data and to make proposals for mechanisms.

1 - Stereoselective polymerization

A "stereoselective" polymerization is a process in which macromolecules containing only one type of configurational unit are formed by incorporation of one stereoisomer from a mixture into a growing polymer chain. There are as many different types of macromolecules as different stereoisomers present in the initial monomer mixture.

In the case of cyclic compounds with one chiral center such

a perfect process produces a stereoregular polymer composed of two distinct isotactic poly R and poly S chains.

For example in the case of propylene oxide one must have :

$$\text{racemic} \quad \begin{array}{c} CH_3 \\ \diagdown \\ H \diagup \end{array} C \!\!-\!\! CH_2 \quad \xrightarrow[\text{initiator}]{\text{stereoselective}} \quad \begin{array}{c} CH_3 \\ | \\ \{O - C - CH_2\}_n \\ | \\ H \end{array} \text{poly R} \quad + \quad \begin{array}{c} H \\ | \\ \{O - C - CH_2\}_n \\ | \\ CH_3 \end{array} \text{poly S}$$

Generally the process is not as perfect and one obtains a strong predominance of one type of configurational unit in the chain.

Since the first example described by Pruitt & Baggett (9) in 1955, many catalytic systems have been used for stereospecific polymerization of propylene oxide and others epoxides and some of them were recently reviewed (10,11). Most of these catalytic systems contain metal-oxygen bound. Among metals iron, magnesium, zinc, aluminium, alcaline-earth and boron were the most used.

The stereoselectivity, i.e. the % of crystalline fraction, could be evaluated using the criterion of insolubility of the isotactic polymer in a solvent or determined from DTA or X-Rays measurements.

Several catalytic systems derived from the reaction of diethylzinc with compounds containing an active hydrogen were extensively studied, among them $ZnEt_2-H_2O$ and $ZnEt_2-CH_3OH$ systems, and the results discussed and largely reported (1,3,4).

It was found that the best stereospecificity is obtained for $ZnEt_2-H_2O$ system when the initiator is prepared in situ in a non-polar solvent using equimolar amounts of reagents (10). The efficiency can be increased by freeze-drying the catalyst (12).

For a given catalytic system, the stereoselectivity is depending on the enantiomeric composition and on the nature of the monomer.

For example, propylene oxide of different optical purities, was polymerized using $ZnEt_2-H_2O$ (1 : 0.7) system prepared in situ (13). As shown in table I the % of crystalline fraction as well as the tacticity are increased with an increase of the optical purity.

On the other hand, with the same initiator, the stereoselectivity is very different depending on the nature of the monomer. Almost purely isotactic products are obtained with t-butyl-thiirane while less than 30 % of crystalline fraction is isolated in the case of propylene oxide (table I).

A mechanism assuming the existence of two groups of enantiomorphic sites having more or less R and S character was proposed in order to explain the formation of polymers of different stereoregularities (1).

Table I

Stereoselectivity of $ZnEt_2-H_2O$ (1/1) initiator depending on the nature of the monomer used and its enantiomeric composition.

Monomer	o. p. initial monomer %	crystallinity %	tacticity (dyads) i %	rêf.
Methyl oxirane	0 47.5 96	27 38 82	61 72 > 90	(13)
t-butyl oxirane	0	90	-	(14)
Methyl thiirane	0 50	60-80 > 70	76 > 80	(5)
t-butyl thiirane	0-90	100	> 90	(15)

The dependance on the nature of the monomer and its enantiomeric composition seems to indicate that chiral active sites are formed after the reaction of the monomer with the initiator.

The mechanistic aspects related to the stereoselectivity of these initiators are difficult to study owing to their insolubility and overall low efficiency.

An interesting approach was tried in the case of propylene sulfide using cadmium and zinc thiolates which gave homogeneous solutions when fully consumed by the monomer (16).

The polymerization is of "living-type" process,the polymers having one living end per metal atom. Depending on the temperature and the solvent used, crystalline or amorphous polymers are obtained (table II). Zinc thiolates in contrary to zinc alcoholates were unable to give crystalline products. The increase in stereoregularity with lowering of the temperature and the negative effect of polar solvents (HMPA) may be explained by coordination equilibrium of the monomer on the active site.

Cadmium salts are the best catalysts for the polymerization of thiiranes giving polymers of the highest stereoregularity, while they are unable to polymerize oxiranes. This behaviour can be explained by the hard-soft acid-base classification in which sulfur and cadmium are closer in their character than oxygen which belongs to "hard elements".

Table II

Influence of the temperature and of the solvents on the stereoregularity of polymethyl thiiranes obtained by initiation with zinc and cadmium allyl thiolates.

Thiolate initiator	solvent	temperature °C	crystallinity	tacticity i %(dyads)
Cd	-	20	+	> 90
	benzene	20	-	50
	toluene	10	+	58
	toluene	0	+	75
	toluene	-20	+	84
	tetrahydrothiofen	0	+	78
	tetrahydrofuran	0	+	76
	HMPA	0	-	50
Zn	toluene	20	-	50
	toluene	0	-	50

2 - Stereoelective polymerization

A "stereoelective" (17) or "asymmetric-selective" (3) polymerization is a process in which a single stereoisomer of a mixture is polymerized giving macromolecules containing one type of configurational base units. For example an optically active catalyst will choose one enantiomer from a racemic mixture and form a macromolecule containing only one type of enantiomeric units. Such an ideal reaction should stop at 50 % yield after consumption of the corresponding stereoisomer.

racemic monomer (R=S) →[optically active initiator / choosing R]→ polymer poly R + unreacted monomer R/S ⟶ 0

In most of the cases the choice is not as perfect and one speaks of "stereoelective" process when a preferential polymerization of one of the enantiomers from a mixture is observed.

Thus, the enantiomorphic choice of the catalyst is a predominant element in this process and should be defined by its electivity and its selectivity.

The "stereoelectivity" could be simply defined as the relative rate of consumption of the enantiomers in the presence of chiral initiator. In the course of the reaction the unreacted monomer is continously enriched in one enantiomer and therefore the "stereoelectivity" could be simply determined from the optical purity of unreacted monomer at a given conversion.

On the contrary, in spite of the preferential constant choice of one enantiomer by the catalyst, the optical activity of the polymer is decreasing with conversion due to progressive enrichment into the opposite antipode.

The "stereoselectivity" concerns the possibility of introduction into the polymer chain of only one type of enantiomer or both of them and informations on stereoselectivity are obtained from studies of the polymer structure.

We shall discuss both of these aspects and compare results obtained with oxiranes and thiiranes on the basis of recent works reported in the litterature and based on our own experimental data.

The interaction between the monomer and the optically active initiator is determining the preferential choice of the latter for one of enantiomers. This choice is characterized by the sign, i.e. the configuration of the elected antipode and by the magnitude i.e. the optical purity of the resolved monomer.

These parameters are depending mainly on the configuration and the nature of the chiral ligand of the initiator and therefore we shall examine first the influence of the nature of the catalyst on stereoelective processes.

Then we shall demonstrate that for a given initiator, the nature of monomer and its enantiomeric composition can deeply influence the stereoelectivity.

Other parameters like temperature and solvent effects will be also discussed.

2-1) Influence of the nature of the initiator Initiators resulting from the reaction between an organometallic derivative and a chiral compound containing an acidic hydrogen were the most usually employed in the stereoelective polymerization of oxiranes and thiiranes.

The chiral ligand associated to the metallic atom is playing an important role in the configuration and the magnitude of the enantiomeric choice.

2-1-1) Effect on the configurational choice Several types of chiral compounds containing an acidic hydrogen such as alcohols, diols, aminoacids were used as coreagents with organometallic compounds ($ZnEt_2$, $CdEt_2$ and $CdMe_2$ are the most employed).

Configurational relations could be established in several cases. If one considers the absolute configuration of the chiral ligand and that of the cyclic monomer, the choice of the initiator would correspond to an "homosteric" type process if the chosen enantiomer has the same configuration as the chiral ligand used in the initiator (18). Homosteric configurational relations are illustrated in the next scheme.

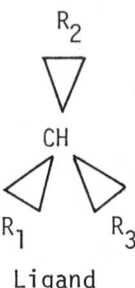

Ligand

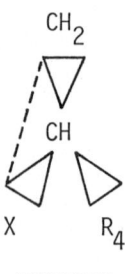

monomer

The following two series were found to give homosteric correlations

Ref.

(I) $\begin{cases} R_1 = OH \\ R_2 = CH_2OH \\ R_3 = alkyl \end{cases}$ $\begin{cases} X = O, S \\ R_4 = Me, Et, iPr, tBu \end{cases}$ our group

(II) $\begin{cases} R_1 = NH_2 \\ R_2 = CO_2H \\ R_3 = alkyl, aryl \end{cases}$ $\begin{cases} X = O, S \\ R_4 = Me \end{cases}$ (19) (20)

It was also possible to establish some correlations with other series of chiral compounds e.g. alcohols, but one must be careful in the choice of groups of comparison for a given configuration. We think that many results may be satisfactory explained by these correlations and some unknown configurations predicted from stereoelective experiments (21).

The chemical composition of the initiator could also play a decisive role in the enantiomeric choice.

We were able to establish in the case of methylthiirane that when the initiator is prepared in such conditions that alkylalcoholate species predominate over dialcoholate species, the initiator system elects the antipode the configuration of which is opposite to that of its chiral ligand. Such type of election was called "antisteric". The results were established in the case of 1,2 diols of serie (I) and for several alcohols which were reacted with three different organometallic derivatives (18).

Thus the ratio $I_s = R-M-OR^*/RO-M-OR^*$ had to be considered ($-OR^x$ being the chiral alcoholate ligand) and the following rule was found:

$I_s \leq 2$ homosteric choice

$I_s > 3$ antisteric choice

For example, when $ZnEt_2$ is reacted at room temperature with $R(-)tBu-CHOH-CH_2OH$ in (1:1) amount this system chooses preferentially the dextrorotatory methylthiirane (homosteric choice ;

$I_s = 0.44$). When the same reagents are reacted in (1:0,5) amounts the choice is opposite ($I_s = 4$; antisteric process).

The chemical composition is therefore depending on the reactivity of reagents and conditions of preparation. The reactivity of organometallic compounds decrease in the order $ZnEt_2 > CdEt_2 > CdMe_2$ and as example $CdMe_2$ gives with R(-) tBu-CHOH-CH$_2$OH in (1:1) conditions an antisteric initiator ($I_s > 3$). We have recently verified these findings on the example of two other thiiranes (C_2H_5 and CH_3OCH_2 substituted).

We were able to isolate species of both types in the case of diethylzinc-(+) 3,3 dimethyl 2 butanol initiator system. The antisteric species has a composition close to $Et_6Zn(OR)_8$ or $Zn(OR)_2$.(EtZnOR)$_6$ (-OR being the 3,3 dimethyl 2 butoxy group), while an homosteric initiator had a $Zn(OR)_2$.EtZnOR composition. Both species were soluble in benzene and were studied by ^{1}H-NMR (18). It was possible to transform one specie into the other by adding $ZnEt_2$ or by drying or heating (loss of $ZnEt_2$ and disproportionation).

Recently Ishimori and al (22) showed that $Zn(OMe)_2$.(EtZnOMe)$_6$ have a centrosymmetric structure formed of two enantiomorphic disturbed cubes. This complex had no catalytic activity at room temperature, but polymerized methyloxirane at 80°. A process of dissociation at 80° could explain such a reactivity.

We shall now consider only homosteric type initiators for simplicity.

2-1-2) <u>Effect on stereoelectivity</u> If one considers initiators prepared in homosteric conditions ($I_s < 1$) it is possible to compare the efficiency of resolution depending on the chiral hydroxy ligand associated with the organometallic compound.

The optical purity of recovered monomer at half reaction could be used as criterion of efficiency.

Thus, when racemic methylthiirane is polymerized using different initiators derived from the reaction of diethylzinc with chiral alcohols and glycols one finds the following order of efficiency :

ligand : tBu-CHOH-CH$_2$OH > tBu-CHOH-CH$_3$ > tBu-CHOH-CH$_2$OCH$_3$
(α/α_o)x/2 :   30 %   12 %   2.5 %

It was confirmed on several examples that chiral 1,2 diols gave the best resolution results. This can be due to the possible formation of rigid cyclic or perhaps polymeric species (18).

In an homologous serie of chiral hydroxy compounds the bulkiness of the substituent is favorising stereoelectivity.

For example in the 1,2 diol serie associated to diethylzinc one finds for methylthiirane following efficiencies :

1,2 diol substituent :   tBu ≃ iPr > Ph > Me
(α/α_o)x/2   :   30 %   28 %   20 %   8 %

Similar results were obtained by Furukawa et al (20) when using diethylzinc-L-aminoacid systems in polymerization of methyl oxirane.

Up to now we have found that the best stereoelective initiator system for polymerization of oxiranes and thiiranes resulted from the reaction of diethylzinc and (-)3,3 dimethyl 1,2 butane diol (DMBD) taken in (1:1) proportion.

2-2) <u>Influence of the nature of the monomer</u>   Oxiranes and thiiranes could be polymerized by the same type of initiators which makes easy a way of comparison of their behaviour. We shall now use our standard homosteric initiator $ZnEt_2$-(-)DMBD (1:1) and study the influence of the nature of the monomer on the stereoselectivity and the stereoelectivity of the process.

2-2-1) <u>Effect on stereoelectivity</u>   Let us consider optical yields obtained at half reaction with several thiiranes and oxiranes.

Thiiranes
Substituent : tBu ~ iPr > Et ~ Me > CH_3CH_2O
(α/α_o)x/2 :   46 %   30 %   16 %

Oxiranes
Substituent : CH_3CH_2O > CH_3
(α/α_o) x/2 :   25 %   20 %

It appears that in general the stereoelectivity is higher for thiiranes than for oxiranes, but also that in the case of the former the stereoelectivity is increased with the bulkiness of the substituent. The optical yield at half-reaction allows a simple comparison between all types of monomers. However, this value is not reflecting the kinetic scheme of resolution and for this purpose a study of the optical purity of recovered monomer during full course of polymerization i.e. on all conversion scale is necessary.

The experimental data taken on the whole range of conversion indicate that there are differences in kinetic behaviour between monomers. Two classes of monomers could be defined corresponding to two types of theoretical kinetic equations.

First order consumption equation

A first class of monomers obeys equation with first order in enantiomer consumption. One can write for each enantiomer :

$- d |R|/dt = K_R |R|$ $- d |S|/dt = K_S |S|$

which give
$$\frac{d|R|}{d|S|} = \frac{K_R |R|}{K_S |S|} = r_R \frac{|R|}{|S|} \qquad (1)$$

K_R and K_S are global rate constants relative to active species and r_R the stereoelectivity ratio relative to R choice. r_R was found to be constant during the full course of polymerization and therefore equation (1) could be integrated.

If one introduces experimental data which are α-optical activity of recovered monomer, α_o-optical activity of pure enantiomer, x - conversion, $|R|_o$ and $|S|_o$ initial concentrations of enantiomers :

$$\alpha = \alpha_o \frac{|R| - |S|}{|R| + |S|} \qquad x = 1 - \frac{|R| + |S|}{|R|_o + |S|_o}$$

One obtains :
$$(1-x)^{r-1} = \frac{1 + (\alpha/\alpha_o)}{|1 - (\alpha/\alpha_o)|^r} \cdot \frac{2^{r-1} |S|_o^r}{|R|_o (|R|_o + |S|_o)^{r-1}} \qquad (2)$$

which simplifies if the initial mixture is racemic ($|R|_o = |S|_o$) into :
$$(1 - x)^{r-1} = \frac{1 + (\alpha/\alpha_o)}{|1 - (\alpha/\alpha_o)|^r} \qquad (3)$$

Experimental data found for racemic CH_3(18),C_2H_5(23),CH_3CH_2O (24) thiiranes and CH_3(25),CH_3CH_2O(24) oxiranes, using our standard initiator were fitting with equation (3) with respective stereoelectivity values (r) equal to 2.4, 2.4, 1.6, 1.8 and 2.0. In Fig. 1 are plotted experimental data for methyl thiirane.

The results obtained with isopropyl and t-butyl thiirane were not fitting with equation (3) and therefore another kinetic equation of second order was proposed (6,26).

Second order consumption equation

The consumption in enantiomer is of second order and the kinetic equation becomes :
$$\frac{d|R|}{d|S|} = \rho_R \frac{|R|^2}{|S|^2} \qquad (4)$$

which could be integrated as previously after introduction of α and x and gives for the racemic monomer :

$$\frac{1}{(1-x)(1 + \alpha/\alpha_o)} = \frac{\rho_R}{(1-x)(1-\alpha/\alpha_o)} + 1 - \rho_R \qquad (5)$$

where ρ_R is the stereoelectivity constant of second order.

It appears that for a complete conversion the limit value for optical activity is no more α_o as in the case of equation (3) but $\alpha_o \cdot \frac{\rho - 1}{\rho + 1}$ and therefore optically pure monomer could not be obtained in this process.

The experimental data found for t-butyl thiirane fitted well with equation (5) as shown on Fig. 1 with ρ_R value equal to 8 at 20°C.

Isopropyl thiirane gave almost the same value. Presently the distinction between both groups seems mainly based on the bulkiness on their substituent, the second order process being observed for the bulkier compounds. Other examples are studied for better understanding.

2-2-2) <u>Effect on stereoselectivity</u> Informations on stereoselectivity are obtained by studying the stereoregularity of polymers. Generally they could be fractionated into a crystalline isotactic fraction and into an amorphous heterotactic fraction, both of them optically active.

With our standard initiator at room temperature the percentage of crystalline isotactic fraction was 20 % for methyloxirane (25), (25), 35 % for methyl thiirane (21) and practically 100 % for t-butyl thiirane (27). The isotactic fraction comes from sites of almost pure R and S character and therefore the distribution between different types of sites for one typical initiator is again depending on the nature of the monomer.

It is interesting to notice that poly(t-butyl thiirane) obtained in stereoelective experiments could be separated by selective solubility in two franctions, one of which was identified as pure poly R polymer m.p. 157°C $|\alpha|_D^{25}$ = + 164 (CHCl$_3$) and the other, as the racemate (poly R + poly S) m.p. = 204° (27). These fractions were compared with authentic samples prepared previously from pure levorotatory monomer and racemic monomer (15).

The latter results show that the stereoelective polymerization is a potential method for obtention of optically pure polymers from racemic monomers.

2-3) <u>Influence of the enantiomeric composition of the monomer. Super stereoelective processes.</u> The enantiomeric composition of the initial monomer may have a strong effect on the stereoelectivity. It was shown on the example of methyl thiirane that the value of the stereoelectivity ratio (r) could be raised up to 7 when using enriched monomers (28). The same phenomenon was recently observed with ethyl thiirane (23) and methoxymethyl thiirane (29).

The stereoelectivity is directly depending on the initial R/S composition and obeys the general first-order equation (2).

Practically, in order to obtain more and more enriched monomers the following procedure was used. The recovered unreacted monomer from one polymerization was reused as initial monomer for the next step. As shown in table III and Fig. 2 it is then possible in 3-4 steps to isolate monomers having optical purities higher than 95 %.

Such a "superstereoelective" process is therefore interesting for the preparation of small amounts of almost optically pure monomers from racemic mixtures. As an example methyl thiirane o.p. = 98 % was obtained in three steps with an overall yield of 11.5 % from the racemic compound (table III).

Table III

Effect of the enantiomeric purity of the monomer on the stereoelectivity

$ZnEt_2$-(-)DMBD (1:1) was used as initiator system

Monomer	initial monomer (α/α_o)	Conversion %	Recovered monomer (α/α_o)	r
methyl thiirane (a)	0	59	0.35	2.2
	0.35	39	0.68	4.4
	0.68	50	0.98	6.8
ethyl thiirane (b)	0	80	0.33	1.5
	0.33	20	0.44	3.2
	0.44	13	0.54	4.5
	0.54	61	0.95	6.0
methoxymethyl thiirane (c)	0	74	0.36	1.6
	0.36	24	0.55	4.2

(a) polymerization carried out at room temperature in toluene solution.
(b) polymerization carried out at -30°C in bulk.
(c) polymerization carried out at room temperature in bulk.

A monomer of the same optical purity could be obtained in a simple stereoelective experiment only at conversions higher than 98 % i.e. with less than 2 % yield.

One must add that such superstereoelective processes are also interesting as a source of polymers of high optical purity. Indeed, if one uses mixtures enriched in the enantiomer chosen in the process, one can obtain at low conversion (10 %) polymers of high optical activity. For example, when using methyl thiirane or methyloxirane of 50 % o.p. (in R enantiomer) one gets polymers 90 % enriched in this enantiomer (28).

For monomers of the second group the stereoelectivity was not affected by the initial enantiomeric composition (26).

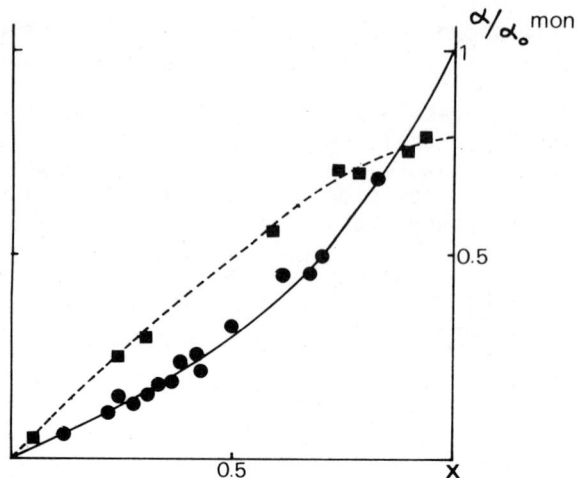

Figure 1. Stereoelective polymerization of racemic thiiranes using $ZnEt_2\text{-}(-)DMBD$ (1:1) initiator. (———) first-order curve (●, exp. data for methylthiirane); (– – –) second-order curve (■, exp. data for tert-butyl-thiirane).

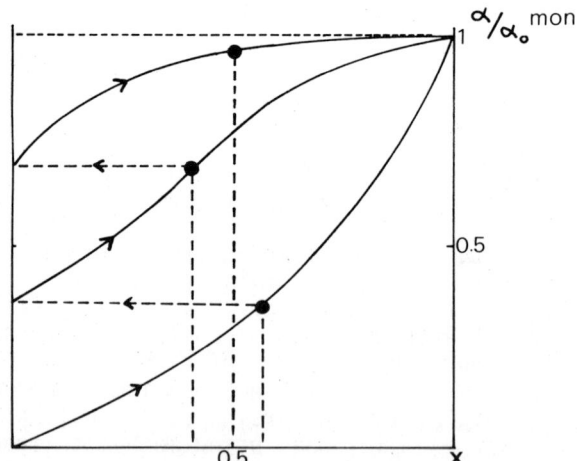

Figure 2. Superstereoelective procedure applied to the polymerization of methylthiirane using $ZnEt_2(-)\text{-}DMBD$ (1:1) initiator

2-4) Effect of the temperature on stereoelectivity Few results only were reported on the effect of the temperature on stereoelection. In the case of monomers of first class the stereoelectivity was not modified by changing the temperature, while the stereoselectivity of the process increased by lowering the temperature of polymerization as demonstrated in the case of methyl oxirane (25).

The temperature showed a strong effect in the polymerization of t-butyl thiirane (26) The stereoelectivity ρ_R doubled in value when temperature lowered from 20° to -3° and on the contrary ρ_R decreased with raising of T and at temperatures higher than 115° the choice of the enantiomer was inverted as shown in table IV. The limit value of the optical purity of monomer was also modified.

A linear correlation was found between log ρ_R and 1/T and the overall difference in energy of activation relative to the stereoelective process for both enantiomers could be calculated (5 Kcal/mol).

Table IV

Influence of the temperature on stereoelectivity in the polymerization of t-butylthiirane with $ZnEt_2$-(-)DMBD (1:1) initiator system

t°C	ρ_R	(α/α_o) recovered monomer at x % conversion	
		x = 50	limit x = 100
-3	14	58	87
20	8	48	78
63	2.5*)	25*)	43*)
135	0.9*)	2*)	4.7*)

*) S enantiomer is preferentially elected

The difference in temperature dependance between both class of monomers could be explained in terms of mechanism of polymerization as shown in chapter 4.

2-5) Effect of solvents and additives The role of solvent may be important in such "anionic-coordinated" polymerizations. It was shown for example in the case of methyl thiirane that addition of tetrahydrofuran decreased the stereoelectivity, a competition occuring between the monomer and the solvent for the coordination on the metallic atom (30).

Very recently Sepulchre (31) has shown the possibility to increase substantially the stereoelectivity by modifying the initiator with chiral agents(optically active thioethers, amines) or by using chiral solvents (limonene). Such a way seems very promising and studies are in progress in our laboratory.

3 - Asymmetric polymer synthesis

In the previous chapter we have seen that an optically active catalyst was able to make a preferential choice between two stereisomeric molecules differentiated by the opposite configuration of their asymmetric centers.

Now we wish to report our investigations on the behaviour of the same initiators in the presence of symmetric monomers having two asymmetric centers of opposite configuration in neighbouring position.

Cis-2,3 dimethyl thiirane (DMT) and cyclohexene sulfide (CS) were studied for this purpose.

A few preliminary results concerning their polymerization are given in table V. Optically active crystalline polymers were obtained which could be separated by selective solubility in fractions of different optical activity and crystallinity.

Table V
Asymmetric polymer synthesis using $ZnEt_2-(-)DMBD$ (1:1) initiator

Cis 2,3 dimethyl thiirane (DMT) and cyclohexene sulfide (CS) were polymerized in bulk at room temperature.

	Conversion %	Polymer fractions							Ref
		Sol. toluene room temp.			Sol. $CHCl_3$ room temp.				
		%	$[\alpha_p]$ ($CHCl_3$)	m.p. °C	%	$[\alpha_p]$ ($CHCl_3$)	m.p. °C		
DMT	30	29	+20	60	71	+50	125		32
	100	33	+24	40/90	67	+66	126		
CS	45	35	+ 3.8 (a)	65	10	+8.4 (b)	80		33

(a) + 20 in trichlorobenzene   (b) + 39 in trichlorobenzene.

The direction of ring-opening is oriented by the chiral choice of the catalyst which attacks preferentially one of the asymmetric carbons with inversion of configuration of the latter. The resulting polymer is optically active due to the prevalence of one type of configurational units |for example $\Sigma(-RR-) > \Sigma(-SS-)$|.

According to Vandenberg (7,8) the ring-opening of cis compounds involves a process with inversion of configuration of the attacked carbon and depending on the type of enchainment one can obtain diisotactic (I) or disyndiotactic(II) structures.

	Structure	enchainment
		head-to-tail
	diisotactic (I)	- RR - RR -
		- SS - SS -
		head-to-head
	disyndiotactic (II)	- RR - SS -

```
   C             C
    \(R)    (S)/
     C  —  C          ⟶
    / \   / \
   H    X    H
```

X = O, S

According to such a scheme, the observed optical activity must be due to the prevalence of diisotactic chains (I) of one type of chirality.

It was not possible up to now to ascertain the optical purity of prepared polymers. Using ^{13}C NMR it was found that resulting polymers presented different types of stereosequences.

In poly cis 2,3 dimethyl thiirane the peak located at 45.8 ppm ($CDCl_3$ solvent, reference to TMS) was clearly assigned to the methine chain carbon of diisotactic structure as it was directly increasing with the optical activity of the polymer. Three other peaks corresponding to chain carbons were found, showing that other structures than the simple disyndiotactic one (II) are also present.

An interesting source of informations on the structure of such polymers may be obtained from their cationic degradation as reported recently by Goethals (34). In the case of cis-dimethyl thiiranes cyclic oligomers of different structure were identified, namely trithiepanes and tetramers (35).

The degradation of optically active poly cis DMT produced optically active trithiepanes and tetramers of opposite sign. The magnitude of their optical activities is directly depending on the optical activity of the polymer used and, thus, reflects its structure. Further studies are in progress (36).

The ^{13}C NMR study of optically active poly(cyclohexene sulfide) showed also a complicated structure for the methine carbon of the main chain with the presence of five or six peaks.

Surprisingly, neither cis 2,3 dimethyl oxirane nor cyclohexene oxide have furnished products with significant optical activity when using the standard chiral initiator. The reason of this behaviour is not yet known and structure of the polymers are under the study.

One must mention that a poly cis 2,3 dimethyl oxirane of low optical activity was obtained by Vandenberg (37) when using an aluminum initiator modified by l-menthol.

Thus, new types of polymers could be obtained by means of this asymmetric synthesis where a chiral enantiomeric polymeric material is created from an achiral monomer. Such a process may be called "achiral enantiogenic" as suggested by Professor J.P. Guetté.

4 - Mechanistic aspects of the stereospecific polymerization of oxiranes and thiiranes

From the above considerations some general mechanistic features could be proposed for stereoselective and stereoelective polymerization using modified organometallic catalysts.

In a first step the monomer reacts with the initiator to form a full spectrum of sites having different R and S character. Some of these formed species have a complete selectivity and produce crystalline isotactic polymers. The proportion of such selective species for a given initiator is depending on the nature of the monomer. We have seen that for monomers with bulky substituents like t-butyl thiirane almost all the sites are purely selective, while for other monomers like methyl oxirane only 20 % of the active species are selective. If the initiator is optically active there is an unbalanced amount of R type and S type species and therefore stereoelection will occur when polymerizing a racemic monomer mixture.

Other species have a much lower selectivity and produce the amorphous part of the polymer. Again if the initiator is optically active, a predominance of one type of species occurs and therefore an heterotactic optically active polymer is obtained.

In chapter 2 we have distinguished two classes of monomers according to their kinetic behaviour in stereoelection.

This difference in behaviour can now be justified by some mechanistic considerations.

With monomers of the first class, methyl oxirane for example, the active sites are formed in an irreversible way after the reaction (or very strong complexation) of the initial monomer with the initiator. As a proof, one finds that the stereoelectivity (r), i.e. the enantiomorphic distribution of sites, is not modified by a change of the temperature of polymerization, but (r) is strongly depending on the enantiomeric composition of the initial monomer.

On the contrary, in the case of monomers of the second group (t-butyl thiirane), the active centers are formed after complexation of the monomer on the initiator species. Thus, the stereoelectivity should depend on the temperature which is indeed observed in a significant way. The second order law could be explained by a two step process : the complexation -a reversible step, then the propagation step.

Simplified representative scheme of both mechanisms could be illustrated as follows :

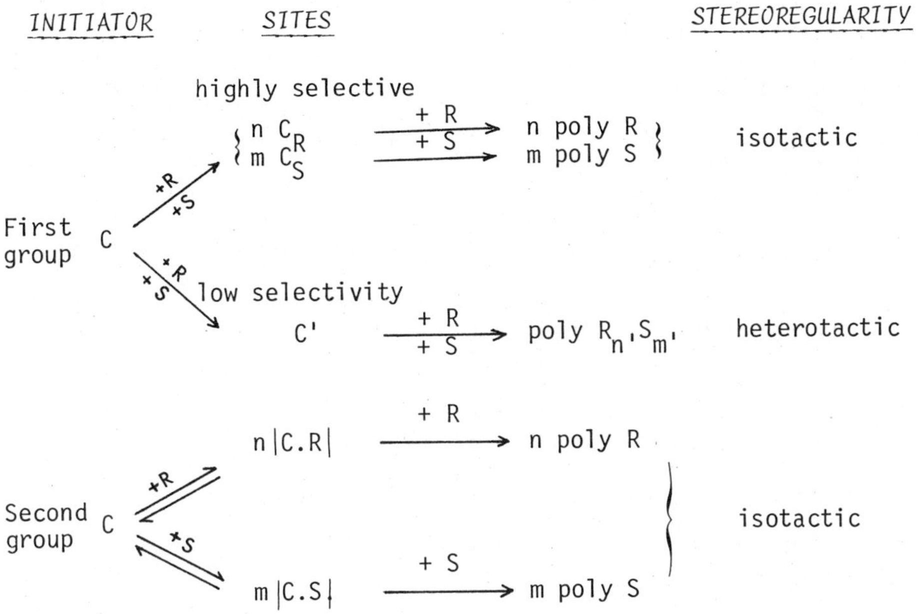

INITIATOR   SITES   STEREOREGULARITY

First group: C

- highly selective: $\{ \begin{array}{l} n\ C_R \\ m\ C_S \end{array}$ $\xrightarrow[+S]{+R}$ $\begin{array}{l} n\ \text{poly R} \\ m\ \text{poly S} \end{array} \}$ isotactic

- low selectivity: C' $\xrightarrow[+S]{+R}$ poly $R_{n'}S_{m'}$ heterotactic

Second group: C

$n|C.R|$ $\xrightarrow{+R}$ n poly R

$m|C.S|$ $\xrightarrow{+S}$ m poly S

$\}$ isotactic

When C is optically active $n \neq m$ $n' \neq m'$.

There is a mutual recognition between chiral sites and enantiomers of similar configuration. Moreover on the basis of asymmetric synthesis reactions involving symmetric monomers, we may assume that the chiral initiator can distinguish one peculiar asymmetric carbon in the molecule in the course of the ring-opening reaction. This ring-opening proceeds with inversion of configuration at the carbon as previously established by Vandenberg (7). In the case of monosubstituted monomers, the catalyst attacks the primary methylenic carbon and this does not affect the configuration of the asymmetric carbon. For symmetrically disubstituted monomers, however, when using chiral initiators, optically inactive polymers are obtained if starting from trans compounds and optically active products may be prepared from cis compounds.

More investigations are still necessary for complete understanding of stereospecific processes.

5 - Conclusion

A great variety of products could be prepared using stereospecific initiators.

Three main directions seem promising :

First, isotactic optically pure or heterotactic products of low optical activity may be obtained starting from racemic monomers.

Second, monomers of high optical purity could be isolated in limited amounts starting from racemic mixtures. In such a case the stereoelective polymerization can be considered as an original resolution method of special interest for monomers which are not easily prepared by conventional synthetic ways under their optically active form. Increase in stereoelectivity is observed when using chiral media, i.e. enantiomerically enriched monomers or external chiral additives.

Third, new optically active polymers are obtained by asymmetric transformation of symmetric monomers.

The author is grateful to Drs. Sepulchre, Dumas, MM. Coulon, Deffieux, Khalil, Momtaz, Pourdjavadi and Reix for their contribution to this work and communication of unpublished data. The author thanks Professor Sigwalt for the critical reading of the manuscript and stimultating discussions.

Literature cited

(1) Tsuruta T., Stereochemistry of Macromolecules, Ed. by Ketley A.D., Vol. 2, p. 177, M. Dekker, Inc., N.Y.,1967
(2) Sigwalt P., Int. J. Sulfur Chem. (1972) C7,83
(3) Tsuruta T., J. Polym. Sci. (1972) D, 179
(4) Tani H., Adv. Polym. Sci. (1973) 11, 57
(5) Spassky N., Dumas P., Sepulchre M. and Sigwalt P., J. Polym. Sci., Symposium n° 52 (1975) 327.
(6) Sigwalt P., Pure and Applied Chemistry (197) (in press).
(7) Vandenberg E.J., J. Polym. Sci., (1969) A 1, 7, 529
(8) Vandenberg E.J., J. Polym. Sci. (1972) A 1, 10, 329
(9) Pruitt M.E. and Baggett I.M. (to Dow Chemical Co) U.S. Pat. 2, 706, 181 (1955)
(10) Furukawa J. and Saegusa T., Polymerization of Aldehydes and Oxides, J. Wiley et Sons, N.Y. 1963
(11) Ishii Y. and Sakai S., Ring-Opening polymerization, Ed. by Frisch K.C. and Reegen S.L., Vol. 2, p. 13, M. Dekker, N.Y., 1969
(12) Oguni N., Watanabe S., Maki M. and Tani H., Macromolecules (1973) 6 (2), 195
(13) Inoue S., Tsukawa I., Kawaguchi M. and Tsuruta T., Makromol. Chem. (1967) 103, 151
(14) Price C.C., Akkapeddi M.K., Debona B.T. and Furie B.C. J. Amer. Chem. Soc. (1972) 94 (11), 3964
(15) Dumas P., Spassky N. and Sigwalt P., Makromol. Chem. (1972) 156, 55
(16) Guérin P., Boileau S. and Sigwalt P., European Pol. J. (1974) 10, 13
(17) Pino P., Adv. Polym. Sci. (1965) 4, 236
(18) Deffieux A., Sépulchre M., Spassky N. and Sigwalt P., Makromol. Chem. (1974) 175/4, 339
(19) Furukawa J., Kawabata N. and Kato A., J. Polym. Sci. (1967) B,5, 1073

(20) Furukawa J., Kumata Y., Yamada K. and Fueno T., J. Polym. Sci. (1968) C (23), 711
(21) Sépulchre M. and Spassky N., to be published
(22) Ishimori M., Hagiwara T., Tsuruta T., Kai Y., Yasuako N. and Kasai N., Bull. Chem. Soc. Jap. (1976) $\underline{49}$ (4), 1165
(23) Khalil A., Sépulchre M. and Spassky N., to be published
(24) Spassky N., Pourdjavadi A. and Sigwalt P., European Polym. J. (1977) (in press).
(25) Coulon C., Spassky N. and Sigwalt P., Polymer (1976) $\underline{17}$, 821
(26) Dumas P., Spassky N. and Sigwalt P., to be published.
(27) Dumas P., Spassky N. and Sigwalt P., J. Polymer Sci., Polymer Chem. Ed. (1974) $\underline{12}$, 1001
(28) Sépulchre M., Coulon C., Spassky N. and Sigwalt P., 1-rst International Symposium on Ring-Opening Polymerization, Jablonna (1975), Preprints p. 80
(29) Reix M., Sépulchre M. and Spassky N., to be published
(30) Spassky N. and Sigwalt P., European Polym. J. (1971) $\underline{7}$, 7
(31) Sépulchre M., Sigwalt and Spassky N., IUPAC International Symposium on Macromolecules, Dublin (1977), Preprint
(32) Momtaz A. and Spassky N., unpublished results.
(33) Reix M. and Spassky N., unpublished results.
(34) Goethals E.J., Adv. Polymer Sci. (1977) $\underline{23}$, 103
(35) Van Crayenest W. and Goethals E.J., European Polymer J. (1976) $\underline{12}$, 859
(36) Van Crayenest W., Goethals E.J., Momtaz A. and Spassky N. Unpublished results, in collaboration.
(37) Vandenberg E.J., J. Polymer Sci. (1964) B, $\underline{2}$, 1085

15

Rate and Stereochemistry of the Anionic Polymerization of α,α-Disubstituted-β-propiolactones

ROBERT W. LENZ, CHRISTIAN G. D'HONDT, and EBRAHIM BIGDELI

Polymer Science and Engineering Program, Chemical Engineering Department, University of Massachusetts, Amherst, MA 01003

Previous investigations in this laboratory (1) and elsewhere (2) have shown that polyesters prepared from chiral α,α-disubstituted-β-propiolactones by the following reactions:

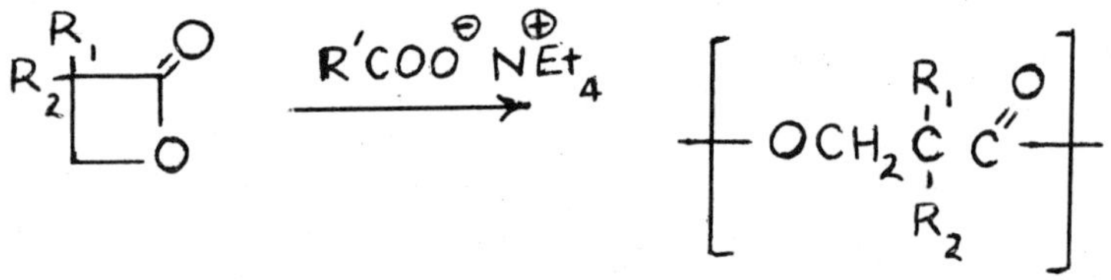

$$\underset{R_2}{\overset{R_1}{\square}}\!\!\square\!\!\overset{O}{\underset{O}{\square}} \xrightarrow{R'COO^{\ominus} \; NEt_4^{\oplus}} \left[OCH_2 \underset{R_2}{\overset{R_1}{\underset{|}{C}}} \overset{O}{\underset{}{C''}} \right]$$

are crystalline even when the two substituents, R_1 and R_2, are considerably different in size (e.g., R_1 = CH_3, R_2 = C_3H_7). This observation is surprising because the anionic polymerization reaction used is homogeneous in character, and no heterogeneous, stereoregular catalysts are required to achieve the polymer crystallinity observed. Furthermore, crystalline structure determinations by wide-angle x-ray diffraction analysis of poly-α-methyl-α-propyl-β-propiolactone showed that two crystalline forms were possible depending upon sample preparation and treatment. These two forms consisted of unit cells in which the polymer was present as either a 2_1 helix or as a fully-extended, planar zigzag conformation (1).

One purpose of the present investigation was to obtain some additional information on structure-crystallinity relationships in this family of polymers by the preparation of stereoregular isotactic polyesters from a single asymmetric isomer of the chiral monomer; that is, from an optically-active α,α-disubstituted-β-propiolactone. Because the polymerization reaction mechanism operates through scission of the alkyl-oxygen bond and does not involve bond reorganizations at the asymmetric center, it was fully expected that polymerization of the optically-active monomer would occur with complete retention to yield an isotactic polymer in which the majority of the repeating units had the same absolute configuration depending upon the optical purity of the monomer. A comparison of the crystalline properties of that polymer with the one prepared from the racemic monomer should help to explain the structural basis for the crystallinity observed for the latter (1).

Another goal of this study was to determine the important parameters which determine the rates of polymerization of these chiral β-lactones. This study is presently directed at investigating the effect of reaction variables (solvent and counterion) on polymerization rate, and in the future, attempts will be made by rate studies to ascertain if stereoelection exists in this homogeneous, anionic polymerization reaction.

Polymerization of Optically-Active Lactone

α-Phenyl-α-ethyl-β-propiolactone, PEL, was chosen as the monomer for investigation because the intermediate amino ester had previously been resolved (3). Both the racemic and optically-active monomers were converted into their polyesters in homogeneous systems using tetraethylammonium benzoate as the initiator in tetrahydrofuran solvent at room temperature. The polymerization of PEL was quite slow under these conditions and several days were required to achieve high conversions of the monomer. Because this is a "living polymer" system, high reaction conversions were required for the formation of high molecular weight polymers.

Polymers obtained from PEL monomers of different optical purity were characterized as follows: (1) for crystalline properties by differential scanning calorimetry (DSC) and wide-angle x-ray diffraction; (2) for relative molecular weights by solution viscosity; (3) for structure by IR and NMR spectroscopy; and (4) for chiroptical properties in solution by optical rotatory dispersion (ORD) and circular dichroism (CD). Molecular weight and melting point data for both the racemic and optically-active PEL polymers are collected in Table I.

Table I. Properties of Poly-α-Phenyl-α-Ethyl-β-Propiolactone

Monomer Composition		Polymer Properties	
(−)	(+)	$\bar{M}_n$	T_m, °C
50	50	9500	110
27	73	7000	116
90	10	7700	260

The large difference between the melting points in Table I of the highly optically pure, isotactic PEL polymer, on the one hand, compared to those of either the racemic or partially optically pure PEL polymer, on the other, must be a direct result of the effect of differences in stereoregularities of the two types of polymers on crystalline properties. Molecular weight was apparently not an important factor for this property because equivalent values were obtained for all three polymers in Table I. Unfortunately, the tacticity differences expected for these polymers could not be determined quantitatively by either the IR spectra of polymer films or the NMR spectra at 90 MHz of the polymers in solution. However, wide-angle x-ray diffraction measurements showed considerably different crystalline patterns and 2θ values for the optically-active PEL polymer of high optical purity on the one hand, as compared to either the racemic polymer or that with low optical purity on the other, as shown by the data in Table II, indicating that the crystalline structures of these were quite different.

In the as-prepared form, after precipitation from solution, both types of polymers appeared to be highly crystalline by this method of analysis, and both were comparable in this property to polypivalolactone, which is known to be a very highly crystalline polyester. In addition, both the optically-active and racemic polymers had considerably higher degrees of crystallinity than those previously observed for racemic poly-α-methyl-α-propyl-β-propiolactone (1). Also of importance, in addition to the different x-ray diffraction patterns of the racemic and optically-active polymers, was that the racemic polymer did not readily crystallize from the melt in the DSC characterization while the optically-active polymer of high optical purity did. Hence, the higher stereoregularity also imparts a more favorable rate of crystallization to the polymer as would be expected.

It seems likely that the observed differences in crystalline properties between the optically active and racemic PEL polymers clearly indicates that the crystalline regions in the latter are not simply formed from physically-separated blocks of R units and S units. Nevertheless, such block arrangements of chiral units could still be present in the polymer and form a different type of crystal lattice than the separate R or S polymers as was found in the crystal structure determination of isotactic, racemic poly(t-butylethylene oxide)(4).

Another possibility to account for the different crystalline properties of the racemic polymer is that, because of strong asymmetric selective effects resulting from steric interactions, the polymerization of the racemic monomer favors the enchainment of alternating R and S units; that is the formation of a highly syndiotactic polymer. Both x-ray diffraction and kinetic studies are in progress to attempt to elucidate this question, but the high degree of crystallinity and ease of recrystallization of the 80% optically-active polymer does not support this possibility because an alternating asymmetric selectivity would reduce the stereoregularity of that polymer. Indeeed, the superior crystalline properties of the optically-active polymer are surprising for a monomer of this optical purity, even if it is assumed that propagation is non-selective or Bernoullian in character. That is, this degree of monomer optical purity should lead to a polymer of only approximately 73% isotactic triad content (or 82% isotactic diads).

Rate Investigations

The rate of polymerization of racemic PEL was determined in two different solvents, tetrahydrofuran (THF) and dimethyl sulfoxide (DMSO), at two different temperatures with tetraethylammonium benzoate as initiator, and the results are collected in Table III. The reactions were followed by infrared spectroscopy based upon the carbonyl group absorption intensities for the monomer and polymer, and the data was treated according to the following pseudo-first order rate equation:

$$\ln[M] = \ln[M]_0 - k_a t$$

The absolute propagation rate constant, kp of Table III, was calculated from the apparent rate constant, k_a, by dividing by the initiator concentration.

The kinetic results for the racemic PEL monomer reveal that the value of kp is somewhat higher in THF than in DMSO at 35°C. The average value for kp in THF was 9.5 M^{-1} min^{-1} compared to 8.3 M^{-1} min^{-1} in DMSO at this temperature, indicating a significantly lower activation energy for the former reaction. Very similar results were obtained previously in this laboratory for the anionic polymerization of α-methyl-α-butyl-β-propiolactone (5), and the present authors have also confirmed the existence of this solvent effect in the equivalent polymerization reactions of the α-ethyl and α-propyl monomers of this series.

The cause of this unexpected solvent effect is not yet known, but it may be related to specific solvation characteristics of the ion pair endgroups (5). That is, it is possible that the lower rate constant for the solvent of higher polarity, DMSO, may be an indication of the formation of a structured ion pair between this solvent, the carboxylate anion and the ammonium counterion, as follows:

$$\sim OCH_2 \underset{R_2}{\overset{R_1}{C}} \overset{O}{\underset{}{C}} \underset{}{\overset{}{O^{\delta\ominus}}} \cdots \overset{CH_3}{\underset{CH_3}{S^{\delta\oplus}}} = O^{\delta\ominus} \cdots N\overset{\oplus}{Et}_4$$

Table II. Peak Angles and Intensities of
X-ray Diffraction Spectra of Poly-α-
Phenyl-α-Ethyl-β-Propiolactone

Racemic Polymer		Optically-Active Polymer	
Angle	Intensity	Angle	Intensity
7.8	m		
9.2	s		
		9.5°	s
		14.5°	m
14.5°	s		
15.0°	m		
		15.3°	s
16.0°	m		
		17.5°	w
19.5°	m	19.5°	m
21.3°	w	21.4°	w

Table III. Rate Constants for the Polymerization of
α-Phenyl-α-Ethyl-β-Propiolactone

SOLVENT	T,°C	$[M]_0$ M	$[I]_0$ M	k_a min^{-1}	k_p M^{-1} min^{-1}
THF	24	0.0123	0.0057	0.0243	4.3
"	"	0.1254	0.0059	0.0233	4.0
"	"	0.1147	0.0068	0.0279	4.1
DMSO	24	0.1410	0.0076	0.0322	4.2
"	"	0.1320	0.0098	0.0379	3.9
"	"	0.1382	0.0087	0.0341	3.9
THF	35	0.2076	0.0056	0.0551	9.8
"	"	0.2153	0.0068	0.0633	9.3
"	"	0.2412	0.0056	0.0533	9.5
DMSO	35	0.2307	0.00698	0.0551	7.9
"	"	0.2152	0.00696	0.0614	8.8
"	"	0.2417	0.0082	0.068	8.3

This ion pair could, conceivably, be of lower reactivity than that present in the solvent of lower polarity, THF, which might have greater free ion character.

These investigations are continuing to more fully verify the solvent effects observed, and in addition, similar studies are in progress on the effect of substituent size as well as reaction solvent on the anionic polymerization of a series of α-methyl-α-alkyl-β-propiolactones. An equivalent series of β-propiolactam monomers was recently investigated in this laboratory with the surprising result that the rates of propagation within this series increased with increasing size of the α-alkyl substituent (6). This result was also rationalized on the basis of changes in ion pair structure which, in that case, were believed to be induced by specific steric interactions.

Acknowledgement

The authors are grateful to the National Science Foundation, Grant No. GH-38848, for the support of this work. Use of the facilities of the NSF-sponsored Materials Research Laboratory is also gratefully acknoledged.

Literature Cited

1. Cornibert, J., Marchessault, R. H., Allegrezza, Jr., A. E., and Lenz, R. W., Macromolecules, (1973), 6, 676; Lenz, R. W., Bull. Soc. Chim. Beograd., (1974), 39, 395.
2. Thiebaut, R., Fischer, N., Etienne, Y., and Coste, J., Ind. Plast. Mod., (1962), 14, 1.
3. Fontanella, L. and Testa, E., Liebigs Ann. der Chemie,(1958), 616, 148.
4. Sakakihara, H., Takahaski, Y., Tadokoro, H., Oguni, N., and Tani, H., Macromolecules, (1973), 6, 205.
5. Eisenbach, C. D. and Lenz, R. W., Makromol. Chem., (1976), 177, 2539.
6. Eisenbach, C. D. and Lenz, R. W., Macromolecules, (1976), 9, 227.

16

Specific Interactions of Lithium Chloride in the Anionic Polymerization of Lactams

GIORGIO BONTÁ, ALBERTO CIFERRI, and SAVERIO RUSSO

Centro Studi Chimico-Fisici di Macromolecole Sintetiche e Naturali C.N.R. and Istituto di Chimica Industriale, University of Genoa, 16132 Genoa, Italy

The ranking of simple anions and cations according to their ability to affect the properties of two- and three-component systems began to be reported in the earlier days of colloid chemistry (1). In binary salt-water mixtures, for instance, a well defined order of effectiveness of anions and cations in altering the entropy of dilution, or in shifting the maximum of infrared band toward the position corresponding to the vapor phase was observed (2-4). In ternary systems such as salt, water, and a polar solute (e.g., an aminoacid), an order of ion effectiveness in altering the solubility of the polar solute was reported (5). Quite generally, the solubility was increased at low salt concentration (salting-in) and decreased at relatively high salt concentration (salting-out). With some ions however, notably SCN^-, Br^-, Li^+, the salting-in effect was always more pronounced, and occurring in a larger salt concentration range, than for the other ions, the so called salting-out agents such as F^-, SO_4^{2-}, Mg^{2+}. Corresponding effects on the activity coefficient of poorly soluble salts in electrolytic solutions were, of course, observed in earlier verifications of the Debye-Hückel theory (6,7). However, no theoretical elaboration based on purely electrostatic considerations, nor alternative approaches based on such concepts as ion hydration, water structure, compressibility of solutions or salt activity, were really successful in offering a unified description of the various salt effects described above (6,7).

When the role of salts on polymeric substances was considered, a similarity with the effects observed with simpler solutes was generally found . For instance, the shrinkage temperature of collagen tendons swollen in aqueous salt solution is depressed by increasing the concentration of salts such as KSCN or LiBr and increased by increasing KF or K_2SO_4 concentration (8-10).(Figure la). These effects are obviously equivalent to an increase, and to a decrease, of the solubility of amorphous collagen, respectively (9). Nagy and Jencks (11) investigated the role of salts in the polymerization of G-actin in aqueous solution and found that salting-in agents, such as LiBr, inhibit the polymerization which, again, is a manifestation of a decrease of the activity coeffi -

cient of soluble G-actin due to the salt.

Earlier investigators (12) dealing with salt-polymer interactions introduced an interpretation of the observed effects which-apparently- was not considered by the investigators who dealt with the behavior of simpler solutes. The interpretation is the classical one of solubilization with binding of a solvent component to the solute species. It now appears that this interpretation may have general validity (13-16). However, the alternative interpretation of an indirect salt effect- mediated by the role of salt on the water structure (2,3)- stimulated debates in the early twenties (1), and it is still not definitively abandoned (17).

It was in order to obtain a compelling evidence in favor of the binding model, and against the water structure role, that we decided to investigate the role of salts on polar polymers in the complete absence of water (18-20). The results have demonstrated that water structure - or water itself - has no prevalent role in salt-polymer interaction. In fact those salts which are able to depress the melting temperature of polymers in the presence of water are also able to cause a similar effect in the absence of water. This is illustrated, for instance, in Figure 1 by the comparison of the role of salts in depressing the shrinkage temperature of swollen collagen tendons and in depressing the melting temperature of polypyrrolidone (nylon 4).

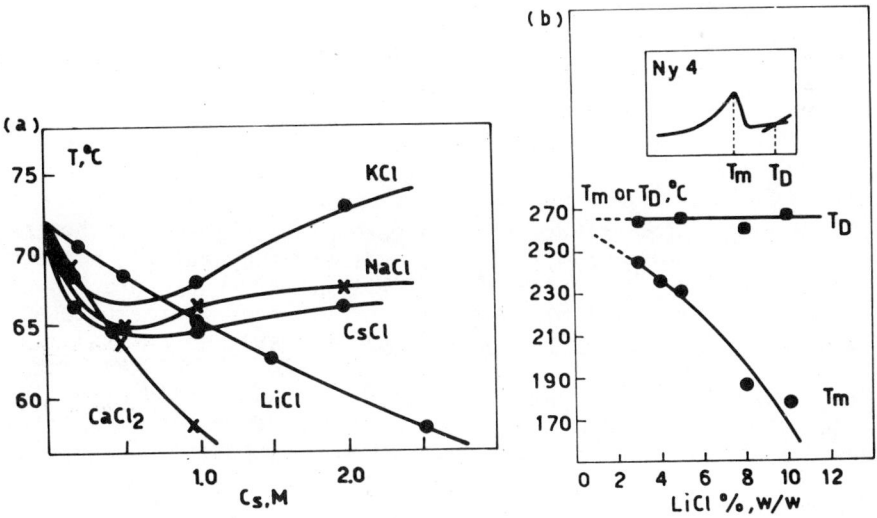

Figure 1. (a) Effect of various salts on the melting temperature of collagen swollen in water; (b) Effect of LiCl on the melting and decomposition temperatures of polypyrrolidone

While a direct binding of salting-in agents to swollen collagen could quantitatively be determined (13), the final proof of the binding mechanism should be based on the corresponding demonstration of salt - polymer adducts in the molten state, in the absence of water. The occurrence of specific interactions in binary system is demonstrated in the present investigation in the case of LiCl and lactams such as ε-caprolactam and pyrrolidone.

The study of binary polymer-salt systems - in addition to its interest as a tool for elucidating the physical chemistry of the interaction - has indicated relevant implications bearing on the processing of synthetic polar polymers. The depression of the melting temperature for polyamides of the nylon series (21) due to 0.05 LiCl mole fraction is exhibited in Figure 2. The increa-

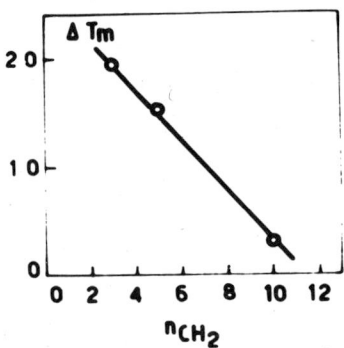

Figure 2. *Difference between the melting temperature of pure polymer and that of LiCl–polymer mixtures containing 0.05 salt mole fraction as a function of the number of methylene groups per repeating unit*

sing depression obtained with the more polar members of the series implies the possibility of processing temperatures considerably below the conventional melting temperature of pure polymers. The beneficial effect of salts is also evidenced in the alteration of other properties of the pure polymers - such as crystallization rate (19) and melt viscosity (20)-which control the processing behavior. In the particular case of polypyrrolidone, it is known that the proximity of decomposition and melting temperature has posed formidable difficulties to the processing of the pure polymer. The addition of small amounts of LiCl does not affect the decomposition temperature and allows a large depression of the melting temperature (~40°C at 5% w/w) (21), thus allowing processing uncomplicated by thermal degradation. The presence of LiCl also allows the obtainment of nylon 6 fibers with mechanical properties superior to those obtained in the absence of salt (22).

In spite of the above described beneficial effects, the necessary step of accurately mixing polymer with salt adds some difficulties, particularly if a large-scale industrial application is considered. The possibility of polymerizing the monomer in the presence of salt has been attempted in order to avoid the mixing step. Preliminary attempts to polymerize ε-caprolactam in the presence of LiCl using the conventional hydrolitic polymerization for nylon 6 have given, however, discouraging results. The salt inhibits polymerization, and very low molecular weights (<5000) and poor yields were obtained. This result is not unexpected in terms of the effects cited above, pertaining to the polymerization of G-actin, and is not in contrast with the binding hypothesis. We have nevertheless attempted the anionic polymerization of the ε-caprolactam/LiCl system in bulk, under anhydrous conditions, in the hope that the active site of polymerization might not be as strongly affected by binding of LiCl as in the case of the hydrolitic polymerization. In fact, in this case we have obtained high yields and high molecular weight polymers. The study of the latter polymerization is described in this report. It appears therefore that a rather close relationship exists between binding of LiCl to the carbonyl oxygen and the possibility of obtaining high molecular weight polycaprolactam by anionic bulk polymerization. Moreover, an interesting aspect of the investigation is that the approach used for ε-caprolactam can be applied to the polymerization of other lactams, such as pyrrolidone, which polymerize only by anionic mechanism. This was ascertained in preliminary results which will be presented in detail at a later time.

Interactions Between Lithium Chloride and Lactams

The strong interactions between lithium halide and polyamides which cause the relevant melting point depression of the polymers (Figure 2), can be better understood if we analyze the behavior of the model system lactam-halide. In this way, we can also gain additional information on mechanisms and kinetics of the anionic polymerization.

We have found that both the systems caprolactam-lithium chloride and pyrrolidone-lithium chloride, when dissolved in anhydrous methanol at room temperature, give by precipitation with a large excess of anhydrous diethyl ether a crystalline complex containing four molecules of lactam and one molecule of lithium chloride. The melting points of the complexes with caprolactam and pyrrolidone are 98.5°C and 101.7°C, respectively.

The presence of complexed LiCl in our polymerization conditions has been further supported by a detailed study of the phase diagram for the binary ε-caprolactam-LiCl system. Homogeneous solutions (melts) of LiCl-caprolactam mixtures with different salt content were prepared at 120°C and cooled down. Heating and cooling cycles were followed by DSC. The resulting equilibrium diagram is shown in Figure 3. An eutectic composition of about 5 mole

% of LiCl has been found. The 1:4 complex behaves as a compound with congruent melting point. The solid-liquid equilibria, at least in the range of 5-20 mole % LiCl, suggest complete complexation of LiCl by caprolactam.

The formation of crystalline complexes between lithium salts and lactams, such as N-methyl-γ-butirolactam, has also been observed by others. A coordination number of four is commonly encountered. In the crystal, the Li^+ ion is coordinated to the carbonyl oxygen of each of the four lactam molecules, and the NH hydrogens are hydrogen bonded to the anions (23-25).

The results described above refer to the formation of a complex between LiCl and a lactam in the crystalline state. For the purpose of the present investigation it is, however, important to establish the occurrence of a direct interaction above the melting temperature of the adduct. In fact, the binding process relevant to the melting point depression of polymers takes place prevalently in the amorphous state (13,15). No evidence of inclusion of LiCl in the crystalline lattice of nylon 6 was reported (18). We have obtained a direct evidence of such interactions by analysis of the infrared spectra at high temperature for the system caprolactam-LiCl. The infrared spectra at 120° and 155°C in the region of the carbonyl stretching bands have clearly shown the shift of the band from 1650 cm^{-1} (pure caprolactam) (26) to 1630 cm^{-1} (1:4 complex). Also 10 mole % solutions of LiCl in caprolactam show similar shifts in the direction of lower frequencies. The magnitude of the red shift is even larger, if compared to the vapor or dilute solution values, where no self-association of caprolactam is present (1672 cm^{-1}). This is indicative of bond weakening due to the electron withdrawal by the metal ion. Therefore, the binding site which occurs in the crystal is maintained also in the melt.

Ring-Opening Anionic Polymerization: The Role of Lithium Chloride

The anionic polymerization of lactams is catalyzed by strong bases capable of forming lactam anions. In the presence of a suitable initiator such as an acyl lactam, a very fast reaction occurs at temperatures in the range of 100°C to 200°C. Without the initiator, polymerization temperatures appreciably above 200°C are necessary and a noticeable induction period is present. Anhydrous conditions are an essential pre-requisite for good yields and rates and reproducible results. The most relevant reactions in the activated polymerization of lactams are the following ones:

Initiation

$$RCO-N-CO + \overset{\ominus}{N}-CO \rightleftharpoons RCO-\overset{\ominus}{N}\ CO-N-CO \qquad (1)$$

initiator   catalyst

$$\text{RCO-}\overset{\ominus}{\text{N}}\underset{\smile}{}\text{CO-N-CO} \underset{\smile}{} + \text{NH-CO}\underset{\smile}{} \rightleftharpoons \text{RCO-NH}\underset{\smile}{}\text{CO-N-CO}\underset{\smile}{} + \overset{\ominus}{\text{N}}\text{-CO}\underset{\smile}{} \qquad (2)$$

The first reaction is a ring-opening transamidation and can be depicted as a nucleophilic attack of the lactam anion on the carbonyl of the imide group. Highly electronegative substituents (such as the acyl groups) at the imide nitrogen increase the rate of the reaction. The second step is a neutralization, or proton exchange reaction, which is much faster than reaction 1).

Propagation

$$\text{RCO-NH}\underset{\smile}{}\text{CO-N-CO}\underset{\smile}{} + \overset{\ominus}{\text{N}}\text{-CO}\underset{\smile}{} \rightleftharpoons \text{RCO-NH}\underset{\smile}{}\text{CO-}\overset{\ominus}{\text{N}}\underset{\smile}{}\text{CO-N-CO}\underset{\smile}{} \qquad (3)$$

$$\text{RCO-NH}\underset{\smile}{}\text{CO-}\overset{\ominus}{\text{N}}\underset{\smile}{}\text{CO-N-CO}\underset{\smile}{} + \text{NH-CO}\underset{\smile}{} \rightleftharpoons \text{RCO-NH}\underset{\smile}{}\text{CO-NH}\underset{\smile}{}\text{CO-N-CO}\underset{\smile}{} + \overset{\ominus}{\text{N}}\text{-CO}\underset{\smile}{} \qquad (4)$$

Reaction (3) shows the prominent role of the imide linkage which is the strongest electrophilic group in the system and the actual propagation center.

The detailed mechanism of ring opening (reactions (1) and (3)) is still open to contrasting interpretations (27-31). An evaluation of their validity is out of the scope of the present paper, even if our results can throw some light on the dispute (32). A simplified kinetic scheme has been however proposed by Reimschuessel (33) and shows that the rate of polymerization is directly proportional to the concentration of imide groups (growth centers) times the concentration of the lactam anions. The latter concentration depends among others on the dissociability of the lactam salt, which in turn is a function of the nature of the cation. Šebenda (34,35) claims that the initiation and propagation rates for the anionic polymerization of caprolactam depend on the cation, in the following order:

$$\text{Li} < \text{Na} < \text{K} < \text{Cs}$$

i.e. the rates decrease by increasing the electronegativity of the cation. Similar results have been found by Sekiguchi (36), by comparing the activity of different catalysts in the polymerization of pyrrolidone.

The results of our polymerization runs are collected in Table I.

Table I

Polymerization of ε-caprolactam at 154°C (initial polymerization temperature). Initiator = N-acetyl caprolactam (1 mole %), Catalyst = alkali metal caprolactamate (1 mole %).

Sample	alkali metal component of the catalyst	LiCl weight %	LiCl mole %	Polymerization time^a, sec.	$t_{0.5}^b$, sec.	T_m^c, °C
15 G	Li		0	140	63	222.3
16 G	Na		0^d	168	114	222.5
27 G	Li		0^d	196	100	222
26 G	Na	0.35	1.00	165	117	222.5
18 G	Li	0.78	2.06	198	90	214.7
20 G	Na	1.29	3.37	235	108	215.6
21 G	Na	3.50	8.83	270	132	195.5
19 G	Li	3.86	9.68	235	135	190.4

^{a}up to equilibrium conversion.
^{b}half-conversion time.
^{c}by hot plate microscope (heating rate= 3°C/min , average values).
^{d}1 mole % of sodium chloride.

The alkali metal composing the catalyst (counterion to ε-caprolactam$^\ominus$) and the excess added salt (LiCl) are indicated in the second and third column, respectively. Polymerization kinetics data are reported in the forth and fifth column, in terms of reaction time , while the last column includes the melting temperatures of the corresponding polycaprolactam-salt system. The latter data are also illustrated in Figure 4. It appears that the polymerization catalyzed by Li$^+$ ions (sample 15 G) is faster than the one catalized by Na$^+$ ions (16 G) in the absence of added LiCl. This result is contrary to the findings of Šebenda (34,35). We also note that added LiCl reduces the polymerization rate and decreases the melting temperature of the pure polymer. However the depression of T_m due to LiCl appears to be an almost linear one only when the catalyst is the lithium salt of caprolactam (cf., in Figure 4, the difference between open and black circles).

In order to clarify the relative role of Na$^+$ and Li$^+$ counterions and to separate the effect of Li$^+$ counterion from that of added LiCl, we observe that in our polymerization experiments one could expect - considering the high dielectric constant of the reaction medium - a solubility of both LiCl and NaCl. We have found,

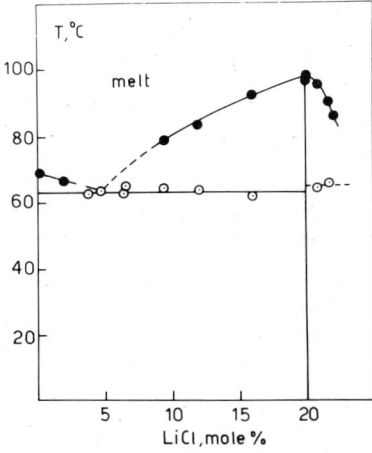

Figure 3. Phase diagram of the system, ε-caprolactam–LiCl

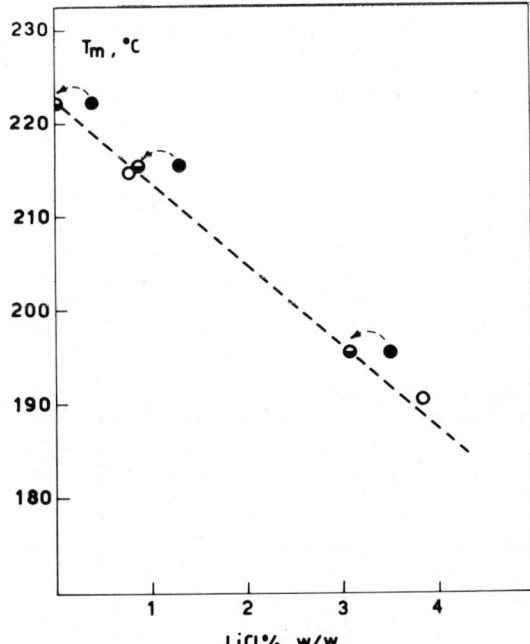

Figure 4. Effect of the alkali metal counterion on the melting behavior of polycaprolactam as a function of LiCl content. (○) = Li$^+$, (●) = Na$^+$, (◐) = assuming full exchange between Li$^+$ and Na$^+$.

however, that NaCl is completely insoluble in caprolactam at the polymerization temperature, whereas LiCl was found to be soluble at 120°C up to the concentration of 16.6% by weight (34.7 mole %). Indeed, a first evidence of an exchange between Li^+ and Na^+ ions was derived by the observation that NaCl separated from the system caprolactam-LiCl when metallic sodium was dissolved in the latter. A further support to the exchange is given by the melting behavior of salted polycaprolactam (Figure 4). Assuming a complete exchange between Li^+ and Na^+, the amount of lithium chloride is reduced up to the following values :

```
20 G        0.90% by weight
21 G        3.08% by weight
26 G        0 %
```

and 1 mole % sodium chloride is formed in the system. Replotting our data as referred to the corrected amount of lithium chloride, gives the half-full circle values which fall very close to the straight line. The melting temperature of the sample 26 G is cohincident with the melting temperature of pure polycaprolactam (samples 15 G and 16 G). This result is possible only assuming a complete exchange between Na^+ and Li^+ which are present in equimolar amounts.

The chemical nature of the cation in the catalyst has no detectable influence on the melting temperature of the pure polymer (samples 15 G and 16 G).

Corresponding results have been obtained in the anionic polymerization of pyrrolidone in bulk at 30°C. Higher initial rates have been found when lithium pyrrolidonate instead of sodium pyrrolidonate was used as catalyst. Here again, the addition of metallic sodium to the mixture of pyrrolidone and lithium chloride (1% by weight) causes the separation of sodium chloride which gives a characteristic opalescence to the solution.

It appears,therefore, that in order to study the "true" role of lithium chloride on the kinetics and mechanism of the anionic polymerization of lactams, it is advisable to use as a catalyst of the reaction a metal lactamate which does not undergo chemical exchanges with Li^+ ions.

If we compare the polymerization times of the samples 15 G, 18 G, and 19 G, we can see that lithium chloride acts as a retarder of polymerization. The presence of undissolved sodium chloride (samples 27 G, 20 G, and 21 G) provides an additional retardation, as evidenced by the data related to the polymerization of the sample 27 G. The two retardation effects are roughly additive. The retardation due to LiCl is present from the beginning of the polymerization reaction, as evidenced by the $t_{0.5}$ data (half-conversion time). It is not, therefore, caused by the modifications of the kinetic scheme due to the many side reactions which occur at the later stages of the polymerization.

A detailed study on the effect of LiCl on the polymerization kinetics and mechanism is in progress and will be published in more detail elsewhere (32).

Side Reactions and Irregular Structures

It is well known that the polyamides obtained by anionic polymerization of lactams contain some irregular structures originated from a series of side reactions. The same high reactivity of the active species (growth centers and monomer anions), which is responsible of the very fast polymerization kinetics, causes a corresponding increase in the tendency toward side reactions. The imide group, which constitutes the strongest electrophilic group in the system, undergoes a Claisen-type condensation in the strongly basic medium provided that hydrogen on the α-carbon is present. The resulting products are keto imides (N-acylated alkylamides of β-keto acids). The keto imides are unstable at high temperatures and, through a sequence of condensation and acylation reactions, are converted into keto amides. Both keto imides and keto amides are comparatively strong acids, and they contribute to decrease the concentration of lactam anions. Therefore, the global effect is not restricted to the lowering of the concentration of the growth centers (imides), but it involves also the decrease in the basicity of the medium (37).

Further condensation reactions of keto imides and keto amides may yield to many possible products: oxypyridone, isocianate, uracil, malonamide, ketones, and so on. A scheme of the most probable side reactions in the anionic polymerization of lactams with a methylene group next to the carbonyl is given in Figure 5, taken from a comprehensive review of Šebenda (38). As already mentioned, a complete and detailed kinetic scheme for the anionic polymerization of lactams cannot disregard the complex role of side reactions, which contribute also to the formation of additional growth centers. For these reasons, only simplified kinetic equations with very limited validity have been proposed so far (33).

Some of the above mentioned irregular structures show absorption peaks in the ultraviolet region between 250-300 nm. As pointed out by Šebenda (37), the absorption maxima at 277 nm (solvent H_2SO_4 50% w/w) are directly proportional to the total amount of the products derived from the imide groups. In particular, by using model compounds he found that the degradation of keto amide groups give rise to an absorption peak at 280 nm, thus suggesting that most of the irregular structures absorbing in the uv arise from the keto amide.

We have studied in detail nature and amount of the uv absorbing groups in the polycaprolactam samples synthesized in presence of lithium chloride. The solvent used for the spectrophotometric measurements was 99.5% formic acid, containing 4.5% (w/w) lithium chloride. A band maximum at 269 nm was found for all samples. Our preliminary data are collected in Table II. Apart from the difference in the concentrations of initiator and catalyst, the polymerization conditions for the two sets of data were strictly the same, and they have already been reported in Table I.

It is evident from the above data that the chemical nature of the catalyst counterion (Li^+ or Na^+) as well as its concentration has no effect on the uv absorption intensity, whereas the presence of lithium chloride in the polymerization system strongly reduces the amount of the irregular structures absorbing in the uv region. The effect seems to be lineally dependent on the halide content. These structures are presumably formed through a sequence of condensation reactions at the α-carbon atom of the N-acylated and N-carbamoylated amides. Lithium chloride should, therefore, interfere very strongly with the Claisen-type condensa-

Table II

O.D. values of 1% (w/w) polycaprolactam solutions in HCOOH containing 4.5% (w/w) of LiCl. Band maximum at 269 nm.

Sample	Alkali metal counter ion	$LiCl^c$ weight %	$O.D._{1\ cm}$
14 G[a]	Na	0	0.44
28 G[a]	Li	0	0.44
8 G[a]	Na	2.43	0.37
10 G[a]	Na	3.52	0.32
11 G[a]	Na	5.19	0.26
13 G[a]	Na	7.21	0.21
15 G[b]	Li	0	0.43
27 G[b]	Li	0[d]	0.45
26 G[b]	Na	0	0.41
20 G[b]	Na	0.90	0.38
19 G[b]	Li	3.86	0.35

[a]initiator and catalyst concentrations: 0.5 mole%; [b]initiator and catalyst concentrations: 1 mole %; [c]corrected values, on the basis of full exchange between Li^+ and Na^+; [d]containing 1 mole % NaCl.

tion in a way which is, at present, under investigation (32). Even at this preliminary stage, however, it is noteworthy to point out the relevant role of lithium chloride toward the synthesis of more regular polymers of caprolactam.

We have also carried out molecular weight fractionations of our polymer samples, in order to study the effect of lithium chloride on the MWD. The data on MW and MWD will be reported elsewhere (39), but we can mention here the results on the uv absorption of the fractionated samples, i.e. the distribution of the uv absorbing groups as a function of polymer molecular weight. As an example, O.D. values for two unfractionated (open circles) and fractionated (full circles) polymers are reported in Figure 6. Curve a) refers to polycaprolactam synthesized in the absence of LiCl (sample 14 G), while curve b) gives the O.D. values of the sample 11 G (5.19 % by weight of LiCl). It is evident that all the fractions irrespective of the chain length, show the same amount of irregular structures. Identical behavior was found for the other samples.

This relevant result is not surprising, being the side reactions which cause the structural irregularities able in part to generate new growth centers, as evidenced in Figure 5. In fact, most of the uv absorbing groups are originated from the imide end groups and are the loci for further growth. After our results, the side reactions which do not contribute to the formation of new growth centers are not very probable in our experimental conditions.

Our data for the distribution of irregular structures as a function of MW do not support recent interpretations (40) based on a close correlation between bimodal MWD and amount of fast side reactions. If irregularities were prevailing in the initial stage of polymerization, when higher molecular weights are produced, the amount of uv absorbing species would be a function of the chain size. A more detailed interpretation of the correlations between MW, MWD and LiCl will be presented at a later time (39). It is important, however, to emphasize here that lithium chloride causes only an unrelevant decrease of the polymer molecular weight, which remains in the range of 15,000 to 25,000, i.e. within the usual values of anionic polycaprolactam MW's.

Some preliminary data on polypyrrolidone synthesized in presence of lithium chloride show absorption patterns similar to those found for polycaprolactam, with the intensity of the band maximum at 274 nm strongly depressed by 1% by weight of LiCl. This similarity, which is present despite the fact that the polymerization temperatures for the two monomers are very different (>120°C for caprolactam, and 30°C for pyrrolidone), is in striking contrast with the interpretations commonly encountered in the literature that the side reactions are determined by the simultaneo-

228   RING-OPENING POLYMERIZATION

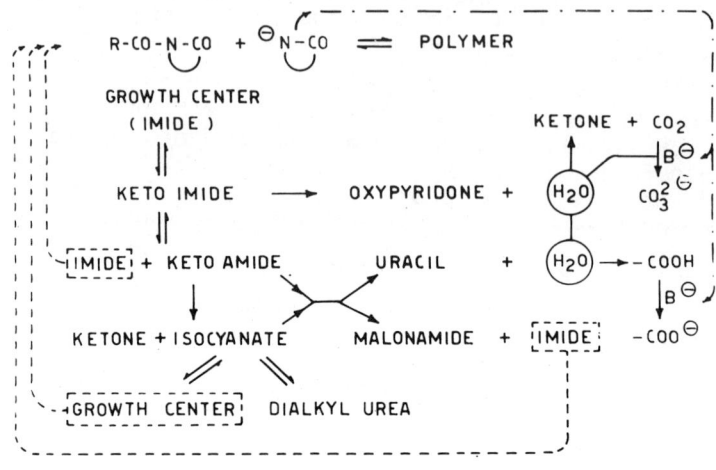

Figure 5. Side reactions occurring during the anionic polymerization of lactams (38)

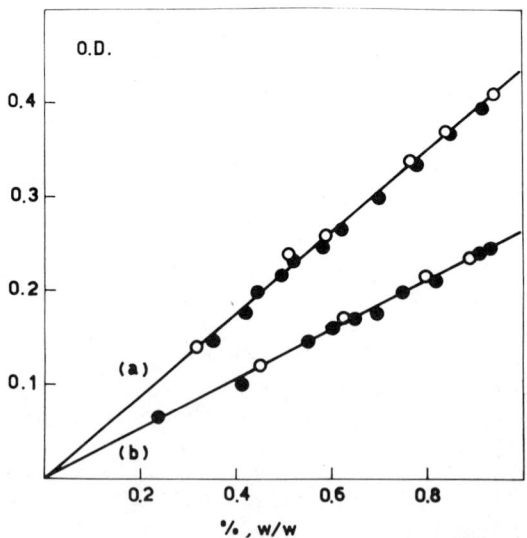

Figure 6. O.D. values vs. polymer concentration for sample 14 G (a) and 11 G (b) (path length = 1 cm)

us effect of high temperature and strong basicity of the medium (37,41,42). An attempt to clarify the true nature of the effect which cause the side reactions is under way (32). It is mainly based on the unique opportunity to use lithium chloride as an agent able to control and reduce the irregular structures.

Melting Behavior of Polycaproamide Synthesized from Caprolactam-LiCl Mixtures.

The full set of data on the melting temperatures forthe polycaproamide samples as functions of LiCl content in the polymerizing mixture are plotted in Figure 7 (curve a). The data refer to two different sets of initiator and catalyst concentrations (0.5 and 1 mole %) and are already corrected and related to the "true" concentration of lithium chloride. Our curve runs parallel to the curve (b) based on some previous data (19), obtained by dissolving lithium chloride in melted polycaproamide. The scale difference is attributed to the different experimental techniques: our data are T_m values, obtained by polarizing microscope equipped with hot stage, whereas in ref.(19) the authors obtained T_m^o values using DSC analysis and Hoffman plots. In fact, T_m values of some polycaproamide samples prepared as described in ref.(19), were determined following the present method and their melting values as functions of LiCl content fall close to the curve (a). We can therefore conclude that it is possible to "directly" synthesize salted polycaproamide, characterized by the same melting behavior found with the other methods of salt addition. This result is technologically very relevant and permits to by-pass all the practical limitations of the other mixing techniques as pointed out in the introduction section. Moreover, analogously to previous findings (18), the salt can be quantitatively removed from the as-formed salted polyamide by washing with hot water (~90°C), all the physical properties of polycaprolactam being easily restored.

Conclusions

Formation of adducts between LiCl and ε-caprolactam or pyrrolidone having 1:4 composition was observed by solubility and phase diagram studies. Specific interactions between LiCl and the lactams were also evidenced in the amorphous state, and in the absence of water, by infrared spectroscopy.

The anionic bulk polymerization of caprolactam and pyrrolidone in the presence of LiCl under anhydrous conditions is characterized by good yields and fast reactions. The MW and MWD of the polymers are comparable to those obtained by conventional hydrolitic polymerization. The extent of side reactions is reduced during the anionic polymerization in presence of LiCl, thus leading to a more regular polymer, actually superior to that obtained by the usual anionic technique. The melting temperature of the resulting polymer-LiCl mixtures is lower than that of the pure polymer, and

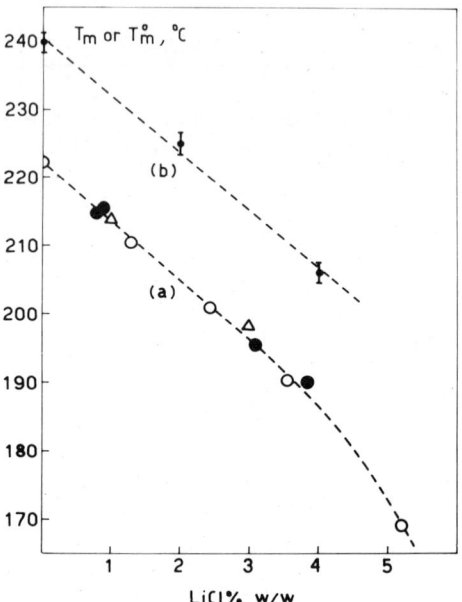

Figure 7. Effect of LiCl on the melting temperature of polycaprolactam. (○, ●) = Direct synthesis, (○) = 0.5 mol % initiator and catalyst, (●) = 1 mol % initiator and catalyst, (△) = salt added to melted polyamide.

is continuously depressed by increasing LiCl content in a way entirely similar to that observed by mixing preformed polymer with LiCl.

Binding of LiCl to the carbonyl oxygen of the lactam is postulated not to adversely interfere with the polymerization site during the anionic polymerization. Binding of LiCl to the carbonyl oxygen of the amide bond is postulated to persist in the amorphous state of the polymer, possibly involving multiple interactions leading to labile crosslinking effects (20).

Acknowledgements

The Authors are thankful to Dr.R.Braggio, Ms. E.Aglietto, Mr.M.Nencioni and Mr.E.Savà for their helpful collaboration in the experimental part of the present work.

Literature Cited

1) McBain J.W., Colloid Science, ed.D.C.Health, Boston,Mass, 1950.
2) Frank H.S., and Robinson A.L., J.Chem.Phys.,(1940), 8, 933.
3) Frank H.S., and Wen Yang Wen, Discuss.Faraday Soc., (1957), 24, 133.
4) Boswell A.M., Gore R.C., and Rodebush W.H., J.Phys.Chem., (1941) 45, 543.
5) Cohn E.J., and Edsall J.T., Proteins, Amino Acids, and Peptides, Reinhold Publishing Corp., New York, N.Y., 1943.
6) Long F.A., and McDevitt W.F., Chem.Rev., (1952), 51, 119.
7) Edsall J.T.,and Wyman J., Biophysical Chemistry, Academic Press, New York, 1958.
8) Gustavson K.H., The Chemistry and Reactivity of Collagen, Academic Press, New York, 1956.
9) Ciferri A., Rajagh L.V., and Puett D., Biopolymers, (1965), 3, 461.
10) Puett D., and Rajagh L.V., J.Macromol.Chem.,(1968), A2, 111.
11) Nagy B., and Jancks W.P., J.Am.Chem.Soc.,(1965), 87, 2480.
12) Katz J.R., and Weidinger A., Biochem.Z., (1933),259, 385.
13) Ciferri A., Garmon R., and Puett D., Biopolymers (1967),5,439.
14) Bianchi E., Conio G., Ciferri A., Puett D., and Rajagh L.V., J.Biol.Chem.,(1967), 242, 1361.
15) Orofino T.A., Ciferri A., and Hermans J.J., Biopolymers,(1967), 5, 773.
16) Conio G., Patrone E., Rialdi G., and Ciferri A., Macromolecules, (1974), 7, 654.
17) Von Hippel P.H., and Wong K.Y., J.Biol.Chem.,(1965), 240,3909.
18) Valenti B., Bianchi E., Greppi G., Tealdi A., and Ciferri A., J.Phys.Chem., (1973), 77, 389.
19) Bianchi E., Ciferri A., Tealdi A., Torre R., and Valenti B., Macromolecules, (1974), 7, 495.
20) Acierno D., Bianchi E., Ciferri A., De Cindio B., Migliaresi C., and Nicolais L., J.Polym. Sci.C,(1976), 54, 259.

21) Valenti B., Bianchi E., Tealdi A., Russo S., and Ciferri A., Macromolecules, (1976), $\underline{9}$, 117.
22) Acierno D., La Mantia F., Polizzotti G., Alfonso G.C., and Ciferri A., J.Polym. Sci., Polym.Letters Ed., (1977),
23) Madan S.K., Inorg.Chem.(1967),$\underline{6}$, 421.
24) Wuepper J.L., and Popov A.I., J.Amer.Chem.Soc.,(1969), $\underline{91}$,4352.
25) Balasubramanian D., and Shaikh R., Biopolymers, (1973), $\underline{12}$, 1639.
26) Millich F., and Seshadri K.V., in "Lactams", Ch.3 of "Cyclic Monomers", edr.Frisch, K.C., Wiley-Interscience, New York 1972.
27) Wichterle O., Makromol.Chem., (1960), $\underline{35}$, 174.
28) Wichterle O., Šebenda J., and Králíček J., Fortschr.Hochpolym. Forsch., (1961), $\underline{2}$, 578.
29) Champetier G., and Sekiguchi H., J.Polym.Sci., (1960), $\underline{48}$,309.
30) Sekiguchi H., J.Polym.Sci. A ,(1963), $\underline{1}$, 1627.
31) Sekiguchi H., and Coutin B., J.Polym.Sci.Polym.Chem.Ed.,(1973) $\underline{11}$, 1601.
32) Aglietto E., Bontà G., Ciferri A., Nencioni M., and Russo S., to be published.
33) Reimschuessel H.K., in "Lactams", ch.7 of "Ring-Opening Polymerization", edrs.Frisch K.C., and Reegen S.L., M.Dekker, New York 1969.
34) Čefelín P., and Šebenda J., Coll.Czech.Chem.Comm., (1961),$\underline{26}$, 3028.
35) Šittler E., and Šebenda J., J.Polym.Sci. C ,(1967), $\underline{16}$, 67.
36) Sekiguchi H., Rapacoulia Tsourkas, P., and Coutin B., J.Polym.Sci. C, (1973), $\underline{42}$, 51.
37) Šebenda J., Masař B., and Bukač Z., J.Polym. Sci. C (1967), $\underline{16}$, 339.
38) Šebenda J., J.Macromol.Sci.-Chem. A , (1972), $\underline{6}$, 1145.
39) Ciferri A., Russo S., and Savà E., to be published.
40) Roda J., Králíček J., and Šanda K., Eur.Polym.J.,(1976), $\underline{12}$, 729.
41) Bukač Z., and Sebenda J., Coll.Czech.Chem.Comm.,(1967), $\underline{32}$, 3537.
42) Bukač Z., Tomka J. and Šebenda J., Coll.Czech.Chem.Comm., (1968), $\underline{33}$, 3182.

17

Isomerization Polymerization of Lactams

H. K. REIMSCHUESSEL

Chemical Research Center, Allied Chemical Corp., Morristown, NJ 07960

The isomerization polymerization of lactams is a rather recently discovered phenomenon. (1) It pertains to substituted lactams in which a particular substituent is or contains a carboxylic group capable of interacting with the amide function of the lactam. Whereas the ordinary ring opening polymerization of lactams yields polyamides, the isomerization polymerization results in the formation of polyimides. Either process is characterized by competition between an intramolecular reaction of cyclization and the intermolecular polymerization reaction. In the ordinary ring opening polymerization of lactams the former reaction is part of a polymer-monomer equilibrium, and the product of cyclization is the particular lactam itself. The chemical structure of the repeating unit of the corresponding polymer molecule is in this case identical to that of the opened lactam ring. This applies, of course, also to any cyclic oligomers formed during the polymerization process. A rather different situation, however, characterizes the isomerization polymerization for which no polymer-monomer equilibrium is indicated. The structures of both the propagating species entailed in this polymerization reaction and the product of any cyclization reaction differ from that of the particular lactam. Furthermore, no structural identity exists in this case between the lactam and the repeating unit of the polymer molecule.

Whether polymerization or cyclization is the dominating reaction depends for either process on thermodynamic and kinetic factors, and on the total molecular strain energy of the particular ring structure. In case of lactams, the six-membered δ-valerolactam, for instance, is the most stable ring structure and exhibits the least tendency to polymerize. Regardless of the ring size, introduction of substituents generally increases both the rate of ring closure and the stability of the ring, it results consequently in a decrease of the polymerizability of the particular lactam. This is reflected in a lower heat of polymerization and a higher monomer equilibrium concentration for lactams such as methyl caprolactam and the respective equilibrium polymers. (2) Relative to the unsubstituted lactam, ring closure of the corresponding substituted one is characterized by a lower enthalpy and higher entropy. Both effects shift the equilibrium toward the cyclic monomer at the expense of the extent of polymerization. This is true, however, only for lactams containing substituents that are incapable of reacting with other functions present in the system. If the particular substituent is capable of inter- or intramolecular interaction with the amide function of the lactam to the extent that either a cyclic transition state or a cyclic intermediate results, because the formation of such a particular cyclic structure is highly favored both geometrically and thermodynamically, then a drastically altered reactivity of the parent ring system could be the consequence. It is well known that carboxyl groups are capable of reacting with amide functions. Transamidation and acylation are rather prominent examples of processes entailing this interaction. Lactams containing carboxyl groups as substituents or as principal moiety of substituents were therefore synthesized and investigated.

Substituted Lactams

The lactams investigated thus far may be divided into three groups: 1) carboxymethyl lactams; 2) carboxy lactams; 3) carboxy lactams containing non-reactive substituents. The first group consists of α-carboxymethyl caprolactam, (αCM7), β-carboxymethyl caprolactam, (CM7), 4-carboxymethyl-2-piperidone, (CM6), and 4-carboxymethyl-2-pyrrolidone, (CM5). The second group is represented by 4-carboxy-2-piperidone (C6) and 4-carboxy-2-pyrolidone, (C5), and the third group comprises 6,6-dimethyl-4-carboxy-2-piperidone, (DMC6), 5,5-dimethyl-4-carboxyl-2-pyrrolidone, (DMC5), 4-carboxy-6-methyl-2-piperidone, (MC6), 4-carboxy-6-ethyl-2-piperidone, (ME6), 4-carboxy-5-methyl-2-pyrrolidone, (MC5), and 4-carboxy-5-ethyl-2-pyrrolidone, (EC5).

The α-carboxymethyl caprolactam was obtained from α-bromocaprolactam via nucleophilic substitution employing sodium diethyl malonate (3), whereas both β-carboxymethyl caprolactam and 4-carboxymethyl-2-piperidone were synthesized via nucleophilic addition of the malonate anione to the corresponding α, β unsaturated lactam (4,5). The 4-carboxy-methyl-2-pyrrolidone was obtained by hydrolysis of the corresponding ethyl ester which was synthesized according to the procedures given by Henecka et al. (6,5).

The synthesis of 4-carboxy-2-piperidone entailed addition of hydrogen cyanide to dialkyl itaconate, and reductive cyclization of the resulting dialkyl cyanomethyl succinate to 4-alkoxycarbonyl-2-piperidone followed by saponification (7). The other member of this group, 4-carboxy-2-pyrrolidone, was obtained from its methylester which was synthesized via esterification of amino methyl succinic acid (8). The 6,6 dimethyl-4-carboxy-2-piperidone was synthesized by reductive cyclization of methyl 3-methoxycarbonyl-5-methyl-5-nitrohexanoate, followed by saponification of the methylester group (8). The 5,5-dimethyl-4-carboxy-2-pyrrolidone was prepared by the same sequence of reactions starting from ethyl 3-ethoxy carbonyl-2-pyrrolidone (8). The synthesis of 4-carboxy-6-alkyl (methyl and ethyl)-2-piperidone consisted of nucleophilic addition of nitroethane (or nitro propane) to dimethyl itaconate, reductive cyclization of the resulting alkyl 3-methoxycarbonyl-5-nitrohexanoate (or-heptanoate), and saponification of the ester groups (9). The synthesis of 4-carboxy-5-alkyl (methyl and ethyl)-2-pyrolidone entailed Michael reaction with nitromethane (or 1-nitropropane) on either diethyl maleate or fumarate, reductive cyclization of the ethyl 3-ethoxycarbonyl-4-nitropentanoate (or hexanoate), and saponification of the ester function (9). The addition of nitroalkanes to dimethyl itaconate and diethyl maleate (fumarate) yielded in both cases mixtures of disastereoisomers. Thus, reductive cyclization of the corresponding mixtures afforded in case of alkyl 3-methoxy carbonyl-5-nitroalkanoate the formation of both 4-methoxycarbonyl-6-alkyl-2-piperidone and 3-carboxymethyl-5-alkyl-2-pyrrolidone, whereas the reduction of alkyl 3-ethoxycarbonyl-4-nitroalkanoate produced 4-ethoxycarbonyl-5-alkyl-2-pyrrolidone and an appreciable amount of noncrystallizable material (9).

General reaction schemes for the syntheses of the particular substituted lactams have been summarized in Table I.

Principal Reactions, Structure and Properties of Reaction Products

Numerous reactions can be envisaged that are peculiar to the functions that characterize the considered lactams. For the present review, however, only those reactions are of interest that are thermally induced by heating the particular lactams in an inert atmosphere to temperatures above their respective melting points. Under this condition either or both polymerization and rearrangement may occur. Whereas the latter may or may not entail the formation of water the former always does. The structures of both the lactam derivatives and the corresponding reaction product (polymers or/and rearrangement products) are shown in Table 2. The 4-carboxymethyl-2-pyrrolidone (CM5) was found to be the only member of the present series of lactams that neither polymerized or rearranged when heated above its melting point. The three other carboxymethyl lactams polymerized. The polymer derived from α-carboxymethyl caprolactam (αCM7) was a colorless transparent material that did not melt or decompose below 300°C (10). It was insoluble in all solvents and did not contain soluble compounds or unreacted monomer. This polymer was obviously highly crosslinked. The polymerization of both 4-carboxymethyl-2-piperidone (CM6) and β-carboxymethyl caprolactam (CM7) resulted in high molecular weight, linear, crystallizable polymers that were soluble in solvents such as formic acid, m-cresol, trifluoroethanol, and sulfuric acid but insoluble in all common organic solvents. The polymer derived from CM6 did not melt below 400°C but showed signs

Table 1 General Reaction Schemes

STRUCTURE

MONOMER	PRODUCTS
CM5	NO REACTION
αCM7	
CM6	
CM7	
C6	
C5	
DMC6	
DMC5	
MC6 EC6	
MC5	

Table 2. Structure of Lactams and Reaction Products

of decomposition at temperatures starting from 300°C. The conversion of monomer to polymer was about 80 to 85%. The polymerization of CM7 resulted in essentially complete conversion to a polymer that melted at 281°C and had a glass transition temperature of about 90°C. Both 4-carboxy-2-piperidone (C6) and 4-carboxy-2-pyrrolidone (C5) polymerized upon heating to temperatures in the range of 200°C to form linear amorphous polymers (7,8) that were soluble in formic acid, m-cresol, sulfuric acid, and trifluoroethanol but insoluble in the common organic solvents. Essentially complete conversion of monomer to polymer was also a characteristic of either of these two cases. The gem-dimethyl lactam derivatives did not polymerize but rearranged. Upon heating to temperatures above their respective melting point (232°C, 206°C), 6,6-dimethyl-4-carboxy-2-piperidone (DMC6) isomerized quantitatively to 5,5-dimethyl-3-carboxymethyl-2-pyrrolidone whereas 5,5-dimethyl-4-carboxy-2-pyrrolidone (DMC5) rearranged with elimination of water to isopropylidene succinimide. The occurrence of both polymerization and isomerization upon thermal treatment distinguished the behavior of the 4-carboxy-6-alkyl-2-piperidones. Heating either the methyl (MC6) or the ethyl (EC6) derivative to 230°C resulted in both polymerization and the formation of the corresponding 3-carboxymethyl-5-alkyl-2-pyrrolidone. There was no indication of polymerization when 4-carboxy-5-methyl-2-pyrrolidone (MC5) was heated to temperatures in the range of 200°C. The sole reaction product was in this case ethylidene succinimide, which was formed with elimination of water.

The reaction products were identified by conventional analysis. The chemical structure of the polymers was deduced mainly from the information obtained from infra red and NMR analyses, and solubility characteristics. The infra red spectra of all of the polymers showed strong absorptions related to the imide moiety. Absorptions in the 1675 to 1705 cm^{-1} and 1725 to

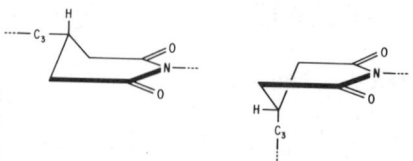

Figure 1. Conformations of PCM7

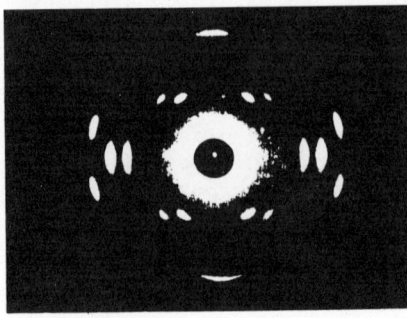

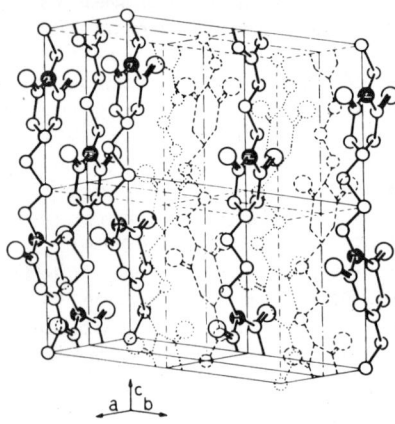

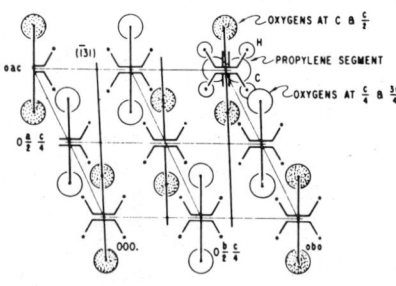

Figure 2. X-ray pattern of oriented monofilament; configuration of polymer chains in the unit cell; schematic projection on the AB plane

1790 cm^{-1} ranges were attributed respectively to asymmetrical and symmetrical carbonyl vibrations, whereas C-N-C anti-symmetrical stretching and imide group vibrations accounted for absorptions in the 1350 to 1390 cm^{-1} and 1140 to 1150 cm^{-1} ranges.

A rather detailed structural analysis was performed on the poly [(2,6-dioxo-1,4-piperidinediyl)trimethylene] which, as has been shown, is the polymer obtained by thermal polymerization of β-carboxymethyl caprolactam.([11](#)) Depending upon the conditions of polymerization, this polymer could be obtained either as a predominantly crystalline or an essentially amorphous material. Differences observed in the infra red spectra of the crystalline and amorphous polymers have been considered indicative of the existence of different conformations of the dioxopiperdine ring, which in turn can give rise to two different chain conformations. As indicated in Figure 1, one is characterized by an equatorial position of the trimethylene moiety with respect to the plane of the imide group, whereas the other pertains to a structural unit in which the trimethylene group is positioned axially to the plane of the ring.

NMR analysis indicated that the latter conformation is the predominant structure of the polymer in formic acid solution. The former has been ascribed to the crystalline modification. This was supported by x-ray analysis which indicated a triclinic unit cell containing eight structural units as shown in Figure 2. The unit cell had the following parameters: a=9.64Å, b=11.32Å, c=15.80Å, α=98°, β=96°, γ=114°. The polymer was found to crystallize in positively birefringent spherulites. In Table 3 are summarized the characteristic frequencies of the infra red spectra of both the poly [(2,6-dioxo-1,4-piperidinediyl)trimethylene] (PCM7), and poly [(2,5-dioxo-1,3-pyrrolidinediyl)dimethylene] (PC6), whereas Table 4 lists the nmr data obtained on these two polymers.

The response of the amorphous polymer derived from CM7 to annealing was rather different from that of polymers such as the nylons and poly(ethylene terephthalate) who readily crystallize when heated above their respective glass transition temperatures. Even prolonged annealing of the unoriented amorphous polyimide resulted in only insignificant crystallization. Annealing of amorphous polymer which was oriented by cold drawing resulted in rapid and extensive crystallization. However, no crystallization was induced during the orientation process; this is quite different from the behavior of most linear polymers such as polyamides and polyesters. This phenomenon can be rationalized by the assumption that the glassy state of this polyimide is characterized by the presence of chain segments of either of the two considered conformations. Crystallization depends therefore not only upon the mobility of the macromolecules but also on rate and extent of conformational changes of the dioxopiperidine moiety during the orientation process.

Except for the poly[2,6-dioxo-1,4-piperidinediyl)dimethylene] which was highly crystalline, all other linear polymers were amorphous. This is of course readily explained by the presence of an asymmetric center entailing the carbon atom in the 3-position in the ring moiety of the repeating unit. Additional support for the chemical structure of the polymer PCM7 was obtained from mass spectroscopical studies.(12) Subjecting a polymer sample (η_{red} = 2.7) to 70 eV at 270°C yielded as a major fregment [-CH$_2$-CH$_2$-CH$_2$-CH(CH$_2$CO)$_2$N-]$_2$ H, m/e 307, which corresponds to a polymer molecule segment containing two repeating units. When the monomer was introduced at 270°C into the preheated source, signals appeared at m/e 154, 307, 460, and 613, after unreacted monomer had flushed off. These values correspond to fregments of the general structure [-CH$_2$-CH$_2$-CH$_2$-CH(CH$_2$CO)$_2$N-]$_n$ H where n is 1, 2, 3, and 4, respectively. A possible mechanism for the formation of these fregments may entail the six-center pyrrolysis reaction followed by

Table 3 Characteristic Frequencies in The Infrared Spectra of PCM7 and PC6 (11,12)

Frequency, cm^{-1}		Assignment
PCM7	PC6	
2930, 2867	2940	stretching vibrations of $-CH_2-$ in the tri(di)methylene moiety
1728	1790	symmetrical) carbonyl vibrations
1677	1705	asymmetrical)
1460	1495	bending in $-CH_2-$ of the tri(di)methylene moiety
1437	1445	bending in ring $-CH_2-$
1358	1390	$-C-N-C-$ anti-symmetrical stretching
1139	1150	imide group vibrations

Table 4 NMR Data for PCM7 and PC6 (11,12)

Proton Peak Position (ppm, TMS)

PCM7	PC6	Assignment	Measured Relative Number of Protons
1.48 – 1.52 (m)	1.17–2.13 (m)	$-N-CH_2-CH_2-CH_2-$	4
		$-N-CH_2-CH_2$	2
3.79 – 3.83 (t)	3.26 (m)	$-N-CH_2$	2
		$-N-CH_2-$	2
2.32; 2.58; 2.74; 2.85; 3.0		ring protons	5
	2.49 (m)	ring protons	3

addition of a proton.

Viscosity - Molecular Weight Relations. The relationship between the viscosities of polymer solutions and polymer molecular weights was studied in some detail for both the poly [2,6-dioxo-1,4-piperidinediyl)trimethylene](PCM7) and poly[2,5-dioxo-1,3-pyrrolidinediyl)dimethylene](PC6). Intrinsic viscosities were determined on m-cresol solutions and evaluated by using the Huggins equation $\eta_{sp}/c = [\eta] + k'[\eta]^2 c$. The well known Kuhn - Mark - Houwink - Sakurada - equation $[\eta] = K_{w(n)} M_{w(n)}^a$ was employed for correlating the viscosities with the molecular weights. The latter were obtained either from light-scattering (M_w) or osmotic pressure measurements (M_n). Values for the Huggins constant k' and the interaction parameters K and a are listed in Table 5.

Table 5: Values for k', $K_{w(n)}$ and a in the Equations $\eta_{sp}/c = [\eta] + k' [\eta]^2 c_1$ and $[\eta] = K_{w(n)} M_{w(n)}^a$

Polymer	k'	a	K_w	K_n
PCM7	0.35	0.73	4.5×10^{-4}	7.5×10^{-4}
PC6	0.41	0.65	4.3×10^{-4}	

A Schulz - Flory distribution of molecular weights was indicated by a value of 2 for the ratio M_w/M_n.

The viscosity-molecular weight relations for the considered two systems were derived from polymers characterized by weight number average molecular weights up to 75000 (PCM7) and 150000 (PC6) respectively. Considerably higher molecular weights are obtainable for these two systems. The polymerization of the β-carboxy-methyl caprolactam for instance has resulted in polymers of molecular weights in excess of 30000.

Mechanical Properties. Tensile properties were determined according to ASTM D1708 for PCM7 and PC6 on films obtained by compression molding, and for PCM7 on monofilaments obtained by melt extrusion and subsequent drawing, employing draw ratios in the range of 4:1 to 6:1. The values obtained depended upon the molecular weights employed and the particular conditions of sample preparation. Since thus far no attempts were made to develop optimum processing conditions, the data listed in Table 6 may not be representative of ultimately attainable values. They show, however, that materials of considerable strength are readily obtained.

The corresponding polymer samples were tested at relative humidities of 0% and 50%. It was found that the moisture content did not significantly affect the tensile properties. This is explained by the rather low equilibrium moisture regain that characterizes these polymers. At relative humidities of 50% and 100%, the moisture contents of PCM7 were respectively 0.56 and 1.40% whereas in case of PC6 they were 3.6 and 11.3%.

Table 6: Tensile Properties

	PCM7	PC6
Tensile strength psi	12000 - 18000	9000 - 16000
Elongation %	70 - 90	3 - 10
Modulus (2% secant) psi	450000 - 580000	320000 - 520000
Tensile strength g/d	6 - 8.5	
Elongation %	12 - 16	
Tensile Modulus g/d	75 - 105	

Thermal Characteristics. Differential thermal analysis, loss of stress birefringence, and torsion tests were used to determine the transition temperatures for the linear polyimides. The corresponding values are listed in Table 7 together with the main decomposition temperatures as determined by both differential thermal analysis and thermogravimetric analysis at a programmed heating rate of 10°C/min.

Table 7: Thermal Analysis Data

Polymer	Melting Point °C	Glass Transition °C	Main Decomposition °C
PCM6	400	168 - 173	420
PCM7	281	85 - 91	460
PC6	-	127 - 135	400
PC5	-	205 - 210	410

Kinetics and Mechanisms of Reaction

Information on possible reaction mechanisms were obtained from kinetic studies. The conversions of β-carboxymethyl caprolactam to the polyimide and of 5,5-dimethyl-4-carboxy-2-pyrrolidone to isopropylidene succinimide have been considered representative examples for respectively the polymerization and rearrangement reactions. The evaluation of experimental data for the polymerization was based upon the concept that the extent of reaction is represented by the momentary concentration of imide linkages(I) which is related to the respective concentrations of both unreacted monomer(M) and polymer molecules(c) by the stoichiometric relation:

$$(I) \quad I = M_o - M - c = U - c \quad (1)$$

Where M_o is the initial monomer concentration, and U the monomer conversion, (14). From equation 1 follows:

$$dI/dt = -dM/dt - dc/dt = dU/dt - dc/dt \quad (2)$$

Experimental data were used to construct the plot shown in Figures 3 and 4. These plots show that both monomer conversion

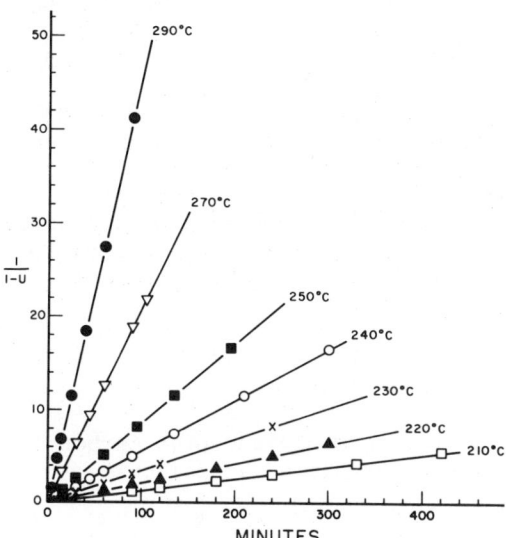

Figure 3. Second-order rate plot of conversion data for CM7

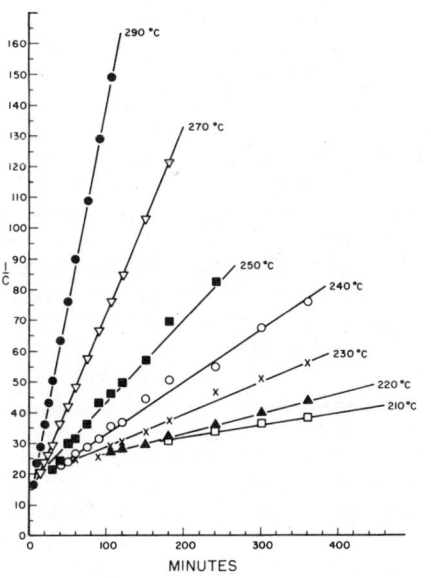

Addition and Condensation Polymerization Process

Figure 4. Second-order rate plot for chain growth (PCM7) (14)

and chain growth are second order reactions. For the main phase of the considered polymerization process, the rate of polymerization was found to be adequately represented for the temperature range of 210°C to 290°C by equation 3:

$$dI/dt = k_p(M^2 + 3.167c^2) \quad (3)$$

where $k_p = 74.9 \times 10^7 \exp(-23800/RT)$

The conversion of 5,5-dimethyl-4-carboxy-2-pyrrolidone to isopropylidene succinimide was studied at the temperatures of 225, 232, and 240°C (8). Attempts to determine the overall order of this reaction showed that also in this case linear relationships were indicated only in second-order rate plots as shown in Figure 5.

The mole fraction of x of the pyrrolidone derivative was calculated according to the relation $x = a_1/(a_1+a_2)$ where a_1 and a_2 are the areas of the peaks at 1.10, 1.35, and 1.78, 2.20 (δ), respectively, in the corresponding nmr spectrum. The slopes of the straight lines in Figure 5 represent values for the overall rate constant k_r for the particular temperatures. The temperature dependence of k_r was found to obey the Arrhenius equation, and the activation energy E and the pre-experimental

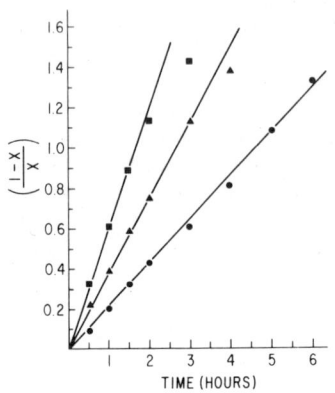

factor A were estimated according to the relationship $\log k_r = \log A - E/(4.574T)$ to be $A = 25.85 \times 10^{12}$ and $E = 34346$ cal/mole.

Comparing the structures of the lactams listed in Table 2 with those of their corresponding reaction products indicates that the conversion of the former entails isomerization processes. In case of the β-carboxymethyl caprolactam, the isomerization could be

Figure 5. Second-order rate plot for the conversion of C5

explained according to a type of the commonly accepted carbonyl addition-elimination mechanism, and formulated as follows: (15)

$$\text{(structure 1)} \rightleftharpoons [\text{(structure 2)}] \longrightarrow \text{(structure 3)} \quad (4)$$

Polymer formulation would then of course be the result of polycondensation of the isomerization product via an intermolecular reaction between the amino- and anhydride functions of the 3(3-amino propyl) glutaric anhydride. Whereas second order kinetics can be readily accommodated for this polycondensation which affects the concentration of the polymer molecules c, an intramolecular reaction such as the considered simple isomerization according to equation 4 should obey first order kinetics. The second order kinetic representation over rather extended ranges of monomer conversion suggests therefore that the isomerization process may be governed by a more complex mechanism. With respect to carboxylic acids, lactams are nucleophiles of moderate basic power and as such convert acids into their conjugate bases. In case of the considered carboxymethyl- and carboxy lactams, this means that both, an electrophilic group (the lactam carbonyl function), and a nucleophile (the carboxylate ion) are present in the same molecule. Thus, if the configuration of the particular lactam is conducive to ring formation, a bicyclic intermediate of the type shown in equation 4 will form. The competitive intermolecular reaction is in this case thermodynamically less favored since the formation of the corresponding intermediate necessitates two molecules coming together to form one species. This process would result in a loss of translational freedom and correspondingly in a large loss of entropy. Cyclization, on the other hand, affects only internal, or vibrational, freedom which is not very large for the considered lactams in any case. The entropy loss

for the formation of the bicyclic intermediate is therefore much
less than for that of the intermolecular analogue. If in addition
the particular ring systems, such as most five-, and six-membered
rings, are characterized by little or no molecular strain, then
the considered intra- and intermolecular reactions will not be
distinguished by large differences in the respective enthalpy
changes. These changes will in this instance therefore not offset
the cyclization favoring entropy effects and will thus not
contribute significantly to the free energy change.

It is therefore reasonable to postulate that the mechanisms
of the considered processes entail as the principal initial reaction the formation of a bicyclic intermediate. Once formed it
may undergo any of three possible fast reactions: 1) elimination
of the internal nucleophile from the carbon atom of the lactam
carbonyl group; 2) elimination of the amino moiety from the lactam
amide function; 3) addition of a proton. The first reaction would
be a reversal of the cyclization and thus regenerate the original
lactam derivative. The second reaction is the one depicted in
equation 4 for the β-carboxymethyl caprolactam. It is however
incompatible with the second order kinetics that was actually
observed for the conversion of this lactam. Furthermore, this
reaction could not be part of the rearrangement of the 5,5-
dimethyl-4-carboxy-2-pyrolidone to isopropylidene succinimide
since no opening of the lactam ring was indicated. The third
reaction, considered less likely in most of the ordinary addition-
elimination reactions involving carboxylic acid derivatives, entails the addition of a proton to the electron pair that stems
from the carbon-oxygen double band and has become localized on
the oxygen atom:

$$\left[\begin{array}{c} \text{bicyclic structure with } O, NH, C-O, C-O: \end{array} \right] \longrightarrow \text{bicyclic structure with } O, NH, C-O, C-OH \qquad (5)$$

This addition, as illustrated in equation 5 for the β-carboxymethyl caprolactam, results in a stabilization of the bicyclic
structure and precludes the attainment to coplanarity required for
the p-orbital overlap that characterizes the known equilibrium of
the amide moiety: $-C(O)-N< \leftrightarrow -C(O)^{(-)} = \overset{+}{N}<$
For the considered lactam derivatives the addition of a proton
seems therefore to be the most likely reaction, it apparently
results in the formation of stable bicyclic structures that have
a finite existence, and are presumed to be the principal reaction
intermediates. The overall kinetics of the conversion of the
particular lactam derivatives may thus be determined by reactions
entailing these species. The second order kinetics observed for
the conversion of both the polymerizing lactam derivatives and
those that rearrange to a different monomeric structure may be

explained by mechanisms that involve either electrophilic catalysis or interaction between two of the considered protonated species.

Conceivable mechanisms entailing catalysis by ionizable carboxy groups are shown in Figure 6 for the conversion of β-carboxy methyl caprolactam, the rearrangement of 5,5-dimethyl-4-carboxy-2-pyrrolidone, and the isomerization of 6,6-dimethyl-4-carboxy-2-piperidone.

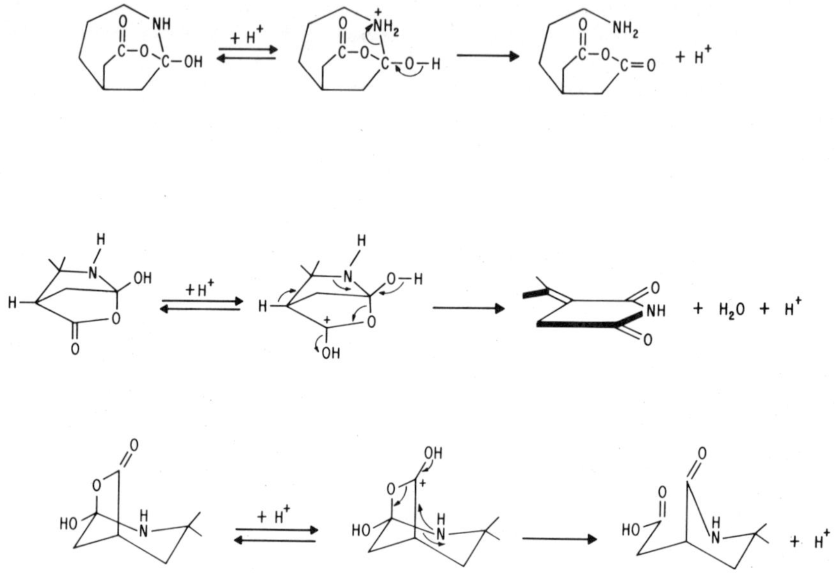

Figure 6. Reaction mechanisms entailing electrophilic catalysis

Since each of the lactam derivatives contains a carboxylic acid group, the rate of conversion may be expressed by equation 6

$$-dM/dt = k_c [COOH][M] = k_c [M]^2 \qquad (6)$$

According to which the second order kinetic representation can be readily rationalized.

Alternative mechanisms, characterized by interactions between intramolecularly protonated species, are illustrated in Figure 7 for the conversion of β-carboxymethyl caprolactam and the rearrangement of 5,5-dimethyl-4-carboxy-2-pyrrolidone. Both involve bimolecular reactions. Whereas the former is an association reaction resulting in the linear dimer of the corresponding polymer, the formularization of the latter corresponds to that of an exchange reaction resulting in the formation of the succinimide derivative and the regeneration of the original pyrrolidone derivative. Both are multi-center reactions entailing the

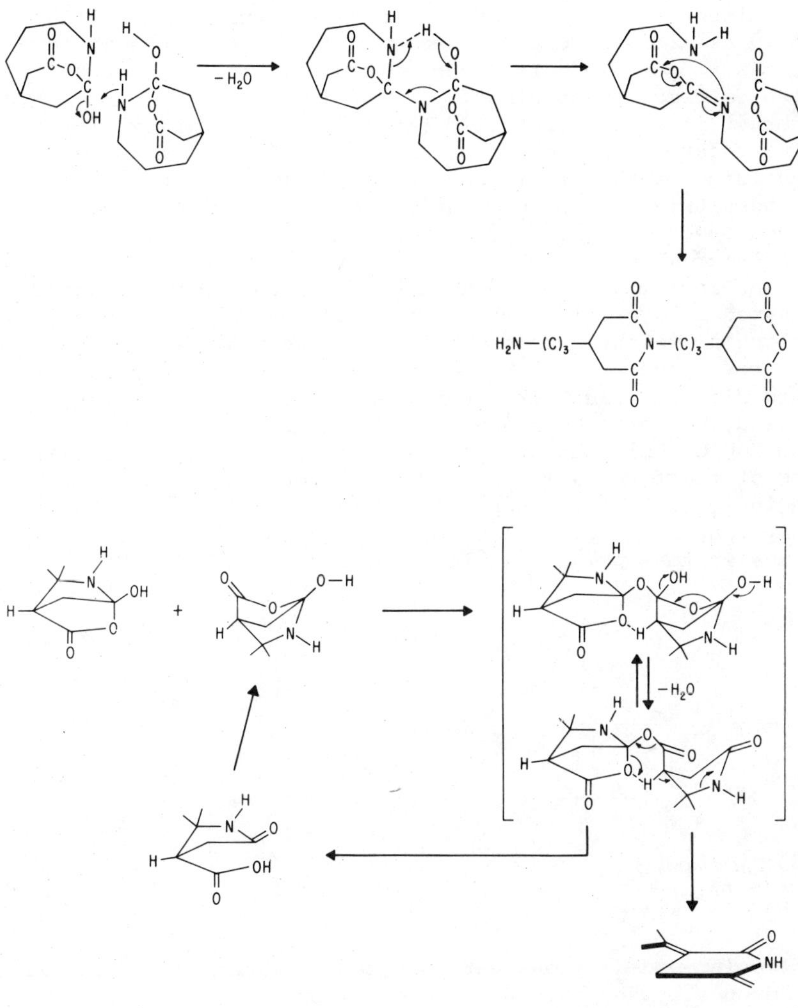

Figure 7. Reaction mechanisms for bimolecular reactions

elimination of water and rearrangement via a six-membered cyclic transition state. In both cases the arrangement yields a coplanar species that has a lower free energy than the strained and highly organized corresponding multi-cyclic structure. In case of the pyrrolidone derivative the coplanar structure is the final product in case of the caprolactam derivative the coplanar iminolactone rearranges via a four center reaction to the energetically even more favored linear dimer.

Whether it is electrophilic catalysis or a complex bimolecular reaction that constitutes the principal mechanism for the conversion of the particular lactam derivatives cannot be deduced conclusively from presently available information. However, mass spectroscopical examination (13) of low molecular weight polymer which had been obtained by polymerizing β-carboxymethyl caprolactam at temperatures between 200 and 220°C showed the presence of a cyclic dimer. This was not observed in case of high molecular weight samples obtained by polymerization at temperatures above 280°C. It could also be shown that this cyclic structure did not result from pyrolysis reactions; it actually disappeared upon prolonged heating at temperatures in the range of 275°C to 300°C while the previously mentioned linear pyrolysis products of the formula $[-(CH_2)_3CH(CH_2CO)_2N-]H$ started to appear. This is indicative of the capability of the cyclic dimer to polymerize by ring opening. A possible mechanism for its formation during the initial polymerization reaction is shown in Figure 8.

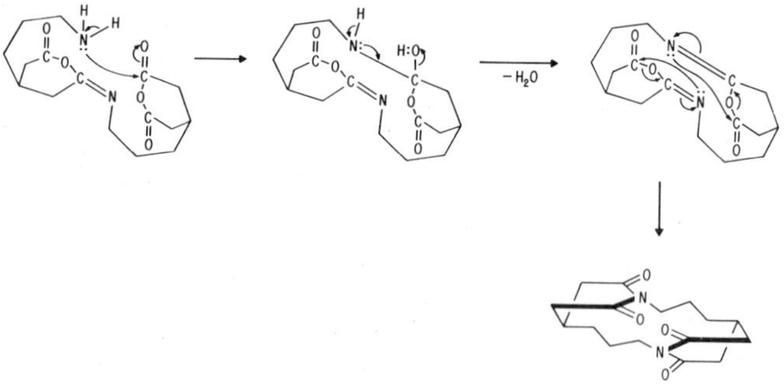

Figure 8. Mechanism of formation of cyclic dimer

Molecular models show that this cyclic dimer is characterized by the absence of any molecular deformation.

The presence of the cyclic dimer in the initial reaction products lends some support for the bimolecular reaction mechanism shown in Figure 7. The postulated formation of a bicyclic structure, involving intramolecular nucleophilic addition of the carboxylate ion to the lactam carbonyl group and addition of a proton, appears to be the initial reaction in either mechanism. This postulate is supported by the observation that neither polymerization nor rearrangement occurred when the corresponding ester lactams rather than the carboxy lactams were employed.

Furthermore, with respect to the ring size, it is well known that among the lactams, the six-membered 2-piperidone exhibits the least tendency toward ring-opening polymerization, whereas the five-membered 2-pyrrolidone polymerizes rather readily. If one would assume that the phenomena related to the carboxy group containing lactams were merely the result of some type of activation of the lactam amide bond, then the reactivities of the particular derivatives should in any instance parallel those of the corresponding parent lactams. We have seen that this is not the case: the 4-carboxymethyl-2-piperidone polymerized quite easily upon heating whereas the 4-carboxymethyl-2-pyrrolidone, under corresponding conditions, did not react at all. It has therefore been concluded that a bicyclic intermediate of the type shown in Figure 5 for the β-carboxymethyl caprolactam is the reactive species in both polymerization and rearrangement, and that the ability to form such a structure is a necessary condition for the conversion of the considered lactam derivatives. The nonreactivity of the 4-carboxymethyl-2-pyrrolidone appears thus to be a consequence of this compound's inability to form a corresponding bicyclic structure as it can be readily demonstrated with molecular models.

There appears to be a reciprocal relation between the extent of reaction and the extent of bond angle distortion in the corresponding bicyclic intermediate. Bicyclic structures characterized by essential absence of bond angle bending according to molecular models were the β-carboxymethyl caprolactam and the 4-carboxy-2-piperidone. For both compounds the extent of reaction was particularly high with respect to both the monomer conversion and the degree of polymerization. In case of the 6,6-dimethyl piperidone derivative, isomerization to the 5,5-dimethyl-3-carboxymethyl-2-pyrrolidone was essentially quantitative. On the other hand, both conversion in rearrangement reactions, and molecular weights in case of polymerization, were low when the attainment of the postulated bicyclic intermediate was accompanied by appreciable bond angle distortion, as indicated for the pyrrolidone derivatives. Whereas the β-carboxymethyl caprolactam can form the bicyclic intermediate with essentially no distortion of bond angles, stereo models indicate that a corresponding structure derived from α-carboxymethyl caprolactam is not favored, though not impossible, its formation results in a considerable molecular strain. It is conceivable that this contributes to the observed intermolecular crosslinking occurring upon polymerization which should in this case result in a polymer structure characterized by the presence of succinic imide units.

Conclusion

Isomerization polymerization is feasible with lactams containing a carboxylic group capable of interacting with the lactam amide function; it affords macromolecules whose repeating units derive from isomers of the original monomers. Thus, the polymers obtained are polyimides rather than polyamides.

Introduction of additional but non-reactive substituents into the corresponding lactams results in isomerization to other stable ring systems rather than in polymerization.

Both polymerization and isomerization presuppose the formation of a bicyclic intermediate which then reacts by a mechanism entailing either electrophilic catalysis or a bimolecular reaction.

LITERATURE CITED

1. Reimschuessel, H.K., J. Polym. Sci. Polymer Letters (1966) $\underline{4}$, 953
2. Schaffler, A., and W. Ziegenbein, Chem. Ber. (1955) $\underline{88}$, 1374, 1906
3. Reimschuessel, H.K., J. Heterocyclic Ch. (1964) $\underline{1}$, 193
4. Reimschuessel, H.K., J.P. Sibilia, and J.V. Pascale, J. Org. Chem. (1969) $\underline{34}$, 959
5. Reimschuessel, H.K., Transactions New York Academy of Sciences, Ser. II, (1971) $\underline{33}$, 219
6. Henecka, H., U. Horlein, and K.H. Risse, Ang. Chem. (1960) $\underline{72}$, 960
7. Reimschuessel, H.K., K.P. Klein and G.J. Schmitt, Macromolecules (1969) $\underline{2}$, 567
8. Klein, K.P., and H.K. Reimschuessel, J. Polym. Sci., A-1 (1971) $\underline{9}$, 2717
9. Klein, K.P., and H.K. Reimschuessel, J. Polym. Sci., A-1 (1972) $\underline{10}$, 1987
10. Reimschuessel, H.K., U.S.P. 3384625, Brit. P. 1042640 (Allied Chemical)
11. Reimschuessel, H.K., L.G. Roldan, and J.P. Sibilia, J. Polym. Sci., A-2 (1968) $\underline{6}$, 559
12. Reimschuessel, H.K., K.P. Klein, J. Polym. Sci., A-1 (1971) $\underline{9}$, 3071
13. McCarthy, E.R., J.S. Smith and H.K. Reimschuessel unpublished work.
14. Reimschuessel, H.K., Advances in Chemistry Series (1969) $\underline{91}$, 717
15. Bender, M.L., Chem. Rev. (1960) $\underline{60}$, 53

Copolymerization of ε-Caprolactam with β-(3,4-Diaminophenyl) Propionic Acid

S. W. SHALABY* and E. A. TURI
Chemical Research Center, Allied Chemical Corp., Morristown, NJ 07960

Polymeric materials having wide ranges of mechanical and thermal properties have been made by the copolymerization of ε-caprolactam with suitable aromatic monomers[1-3]. Copolyamides of ε-caprolactam and m-xylylenediammonium isophthalate or β-(4-aminophenyl) propionic acid were studied earlier in this laboratory[2,3]. While the latter group of copolyamides, based on β-(4-aminophenyl) propionic acid, were shown to be crystalline over a wide composition range, most members of the former group of copolymers were essentially amorphous. It was also shown that the incorporation of a small fraction of these aromatic moieties into nylon 6 led to some noticeable changes in its mechanical and thermal properties. Copolymers based on 90/10 and 85/15 of caprolactam and β-(4-aminophenyl) propionic acid were shown to be more rigid and somewhat more thermally stable as compared to nylon 6. This was attributed to the rigid and intrinsically thermostable aromatic moieties in these copolyamides. On the other hand, the homolymer of β-(4-aminophenyl) propionic acid was less thermally stable than poly(2,5-ethylene benzimidazole) (PEBI), which was studied earlier by the authors[4]. This, the well-documented high thermal stability and excellent tensile properties of imidazole-type polymers and our interest in improving these properties in nylon 6, led to the initiation of the present studies. In this communication, the possible formation of a copolymeric chain of ε-caproamide and ethylene benzimidazole sequences and the properties of the resulting copolymers are reported. Since it was shown earlier that poly(2,5-ethylene benzimidazole) can be formed easily by homopolymerization of β-(3,4-diaminophenyl) propionic acid[4] (DPPA) or its methyl ester (MDPP, these were chosen as comonomers for the synthesis of the copolymers subject of the present studies. The structure of nylon 6, PEBI and the copolymers of caprolactam

---

* Present address: Ethicon, Inc., Somerville, N.J. 08876

and DPPA or MDPP (P-CL-co-EBI) can be illustrated as follows:

$$m \; \underset{CL}{(CH_2)_5 \underset{NH}{\overset{CO}{\diagup}}} \longrightarrow \{(CH_2)_5-CO-NH\}_m$$

Nylon 6

$$n \; \underset{\underset{NH_2}{NH_2}}{\bigodot}^{CH_2CH_2COOR} \longrightarrow \{CH_2CH_2-\underset{\underset{H}{N}}{\bigodot}^{N}\}_n$$

CPPA = R H
MDPP = R CH$_3$

PEBI

$$CL + DPPA \; (or \; MDPP) \longrightarrow \{\cdots-CH_2CH_2-\underset{NH-CO-\cdots-NH(CH_2)_5CO}{\bigodot}^{NH_2}\}_p$$

$$\{\cdots-CH_2CH_2-\underset{\underset{H}{N}}{\bigodot}^{N}\cdots\cdots-NH(CH_2)_5CO-\cdots\}_p \longleftarrow$$

P-CL-co-EBI

EXPERIMENTAL - a) Analytical Methods

Reduced viscosities (η_{red}) were obtained for polymer solutions in sulfuric acid (0.5 g/100 ml). The infrared spectra of compression-molded films (molded at about 260°C) were obtained on a Beckman IR-9 spectrophotometer. The differential scanning calorimetry (DSC) data were recorded, in most cases, on a DuPont 990-DSC apparatus in nitrogen at a heating rate of 20°C/min. and using about 10 mg. sample. The polymers used in these experiments were annealed at 100°C in vacuo for 20 hrs. before obtaining the initial thermal analysis heating data. Alternatively, in order to achieve similar thermal history of the examined copolymers, the samples were heated to and held for a few minutes at temperatures above their melting temperatures, quenched in liquid nitrogen and then reheated as shown in Tables V, VI, and VIII. Two samples (Table VIII) were analyzed on a Perkin-Elmer DSC-1B in nitrogen using a heating rate of 20°C/min. and about 10 mg. samples. Most of the thermogravimetric analysis (TGA) data were obtained on a DuPont 951 Thermal Analyzer in nitrogen, using a heating rate of 10°C/min. and about 10 mg. samples. The TGA data of two samples (Table VIII) were obtained on a Cahn RG Electrobalance in nitrogen and air, using a heating rate of 10°C/min. and about 5 mg. samples. A Norelco diffractometer with crystal monochromatized copper x-radiation was used to obtain the X-ray powder diffraction patterns of polymer granules which were annealed at 100°C in vacuo for 20 hrs. The per cent crystallinities were determined by the method of Hermans and Weidinger[5]. The tensile properties of certain copolymers were measured on microtensile samples which were prepared by compression-molding. The data were obtained according to ASTM specifications, using an Instron head speed of 0.5 in./min.

b) General Polymerization Method

The required amounts of ε-caprolactam and DPPA or MDPP were placed in a large polymerization tube (about 20 times the volume of the polymerization mixture). The tube was purged several times with argon. The reaction was conducted under one atmosphere of argon for 5 hrs. and the tube was then sealed under reduced pressure. The polymerization was then continued under differenct conditions as shown in Table I.

The resulting polymer was ground, extracted with water in a Soxhlet extractor for 2 days and dried in vacuo at 70°C to a constant weight. After determining the % extractables (100-% conversion), the polymer granules were annealed at 100°C for 20 hrs. in vacuo.

RESULTS & DISCUSSION

Synthesis of Polymers & Determination of Their Composition

β-(3,4-Diaminophenyl) propionic acid (DPPA) and methyl β-(3,4-diaminophenyl) propionate (MDPP)[5] were prepared and purified as described in a previous report. ε-caprolactam was purified by distillation in vacuo before use. Copolymers of caprolactam and DPPA or MDPP were prepared according to the schemes outlined in Table I. Viscosity data of all polymers are also summarized in Table I. All polymers revealed reasonably high reduced viscosities. Polymers made at moderate temperatures (a maximum polymerization temperature of 255°C) exhibited higher solution viscosities than those obtained using high polymerization temperatures (a maximum polymerization temperature of 270°C). Chain degradation at the high polymerization temperature can be responsible for the observed depreciation in the reduced viscosity of the latter copolymers. For copolymers made under similar reaction conditions (V to VII and VIII to XI) the reduced viscosity seems to decrease with the increase of their aromatic content. This may be associated with the inability of the more aromatic copolymers to undergo appreciable chain extension at reaction temperatures which are slightly or moderately higher than their T_g.

Three approaches for determining the final composition of the copolymers were used in the present studies. The first

TABLE I COPOLYMERIZATION DATA

Polymer Number	I	II	III	IV	V	VI	VII	VIII	IX	X	XI	XII	XIII	XIV
Caprolactam, mole %	95	90	85	80	65	50	20	95	90	85	80	95	90	93
3,4-diaminophenyl-β-propionic acid, mole %	5	10	15	20	35	50	80	5	10	15	20	–	–	7
Methyl 3,4-diamino-β-phenylpropionate, mole %	–	–	–	–	–	–	–	–	–	–	–	5	10	–
Reaction time, hr, in the following order at														
220°C/1 Atms.	4	4	4	4	4	4	4	4	4	4	4	4	4	(2)[b]
240°C in sealed tube	–	–	–	–	20	20	20	–	–	–	–	–	–	–
255°C in sealed tube	20	20	20	20	–	–	–	–	–	–	–	20	20	20
270°C in sealed tube	–	–	–	–	–	–	–	20	20	20	20	–	–	–
240°C/1 Atms	–	–	–	–	–	–	–	–	–	–	–	–	–	–
255°C/1 Atms	5	5	5	5	5	5	5	5	5	5	5	5	5	–
255°C/2 mm	0.5	0.5	0.5	0.5	0.5	0.5	0.5	0.5	0.5	0.5	0.5	0.5	0.5	–
230°C/2 mm	3	3	3	3	3	3	3	3	3	3	3	3	3	–
Water extractables, %	7.1	6.3	2.8	4.0	2.0	2.2	0.4	6.2	5.0	4.6	8.8	9.0	9.8	11.0
Final caproamide content, mole %	94.7	89.3	84.7	79.4	64.5	49.4	20.0	94.7	89.5	84.3	78.5	89.5	89.0	92.3
$\eta_{red.}$ of copolymer	2.00	1.93	1.88	2.29	2.94	2.04	1.28	1.43	1.18	1.75	0.96	1.17	0.82	0.87

(a) Using 90/1 weight ratio of caprolactam and aminocaproic acid.
(b) Heated at 200°/1 Atm.

approach entailed the use of the per cent conversion obtained
by extraction and the assumption that the extractable fraction
consists of caprolactam or its water-soluble oligomers. In
essence, it is assumed that the aromatic comonomers in the
initial mixture were incorporated in the copolymer chain. This
assumption can be justified if one realizes that the per cent
conversion increases with the increase of the aromatic content
of the polymerization mixture and an almost quantitative conversion can be achieved in the formation of the aromatic homopolymer. Using the extraction data in Table I, the final
composition of the copolymers were calculated and shown to be
comparable to those of the initial comonomer mixtures (Table I).
In a second attempt to determine the composition of the
copolymers, typical samples were analyzed for their carbon,
hydrogen and nitrogen contents, as shown in Table II. The
elemental analysis data indicate, qualitatively, that the final
copolymer compositions are comparable to the initial compositions of the comonomer mixtures. Due to the minor differences
in the elemental contents of the copolymers and the level of
experimental error associated with elemental analysis, no
attempts were made to use the elemental analysis data, quantitatively, for determining the composition of the copolymers.
The infrared spectra of thin polymer films were used in the
third approach to determine the composition of the copolymers,
as discussed in the next paragraph.

Three characteristic absorption frequencies were used in
the IR studies. These are associated with the aromatic
out-of-plane C-H bending (ν_1), the aromatic in-plane C-H
bending (ν_2) and the $(CH_2)_5$ skeletal vibrations (ν_3) at 810,
1010 and 1170 cm^{-1}, respectively[3,4,6]. The relative absorbances
(A_1/A_3 and A_2/A_3) of the aromatic and aliphatic moieties were
calculated and used as a measure of their concentration along
the copolymer chain as shown in Table III. Thus, the A_1/A_3 data
were first plotted against the molar composition of the
copolymers, as calculated using the extraction data, to obtain
a straight line relationship as shown in Figure 1. The slope
of the straight line was then used for dividing the A_1/A_3
values of the individual copolymers to obtain a new molar composition, TCA_1 (mole % of total chain aromatic sequences
using A_1/A_3) based on IR measurements. The agreement between
the IR and extraction-based compositions was fair for most copolymers; higher IR values were recorded for copolymers I, III
and VI. Upon using the A_2/A_3 values to obtain another set of
IR-based molar composition values for the copolymers, TCA_2
(Figure 2 and Table III) it was found that the agreement between
extraction and IR-based molar compositions was poor. Copolymer
VII displayed a noticeable deviation from the straight line
relationship shown in Figure 2. All IR-based molar compositions were much higher than those based on the extraction
data.

TABLE II   ELEMENTAL ANALYSIS DATA OF COPOLYMERS

Polymer No.	Initial mole % of Caprolactam	Found %C	Found %H	Found %N	Calculated[a] %C	Calculated[a] %H	Calculated[a] %N
I	95	64.21	9.56	12.52	64.26	9.59	12.73
II	90	65.23	9.33	12.98	64.82	9.38	13.09
III	85	65.83	8.81	13.84	65.38	9.17	13.44
IV	80	66.02	8.74	13.76	65.95	8.96	13.79
V	65	68.14	7.85	15.03	67.64	8.33	14.85
VI	50	69.70	7.46	16.09	69.34	7.70	15.91
VII	20	72.22	6.44	17.64	72.72	6.43	18.02
VIII	95	64.17	9.43	12.65	64.26	9.59	12.73
IX	90	65.30	9.29	13.09	64.82	9.38	13.09
X	85	66.09	8.76	13.84	65.38	9.17	13.44
XI	80	65.87	8.73	13.65	65.95	8.96	13.79
XII	95	64.30	9.53	12.58	64.26	9.59	12.73
XIII	90	65.53	9.39	12.95	64.82	9.38	13.09

(a) Based on initial mole % of caproamide($C_6H_{11}NO$, 113.2) and ethylene benzimidazole units ($C_9H_9N_2$, 145.18).

TABLE III   INFRARED DATA AND POLYMER COMPOSITION

	Polymer Number							(b)
	I	II	III	IV	V	VI	VII	C-1
Initial Mole % of Aromatic Comonomers	5	10	15	20	35	50	80	100
IR Relative Absorbance Data[a]								
A_1/A_3	0.30	0.38	0.71	0.69	1.14	1.84	2.66	–
A_2/A_3	0.20	0.24	0.33	0.36	0.54	0.83	1.62	–
A_1/A_2	1.50	1.58	2.15	1.92	2.11	2.21	1.64	1.34
Copolymer Composition								
• Mole % of total chain aromatics using extraction data (c)	5.3	10.7	15.3	20.6	35.5	50.6	80.0	–
• Mole % of total aromatics using A_1/A_3, (TCA_1) (d)	9.2	11.7	21.8	21.2	35.1	56.6	81.8	–
• Mole % of total chain aromatics using A_2/A_3 (TCA_2)	14.6	17.5	24.1	26.3	39.4	60.6	(118.2)[f]	–
• Relative concentration of un-cyclized chain aromatics (RUCA) using A_1/A_2 (e)	1.12	1.18	1.60	1.43	1.57	1.64	1.22	(1.00)[e]
• $(RUCA/TCA_1) \, 100 = \phi$	12.2	10.1	7.3	6.7	4.5	2.9	1.5	1.0

(a) A_1 = Absorbance at 810 cm^{-1} due to CH in benzene ring.

   A_2 = Absorbance at 1010 cm^{-1} due to CH in benzene ring.

   A_3 = Absorbance at 1170 cm^{-1} due to caproamide sequence.

(b) See Reference #4, polymer no. C-1.

(c) A slope of 0.0325 (from Fig. 1) was used for this calculation.

(d) A slope of 0.0137 (from Fig. 2) was used for this calculation.

(e) By dividing A_1/A_2 of the copolymer by that of the homopolymer (C-1) in which all sequences were assumed to be benzimidazole.

(f) Polymer displayed marked deviation in relative absorbance (Fig. 2).

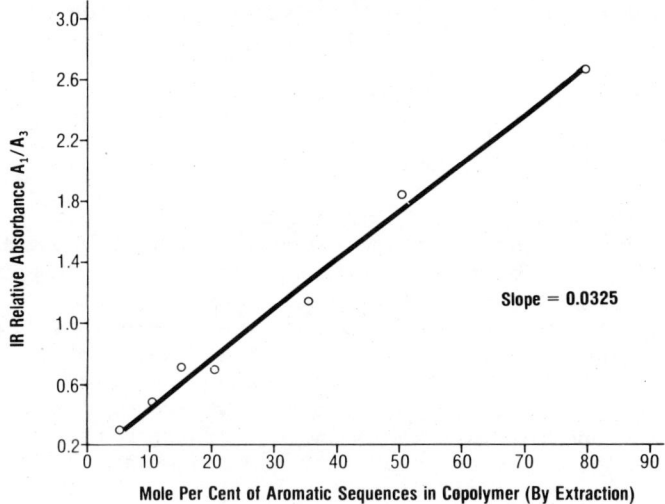

Figure 1. Effect of composition on absorbance at 810 (A_1) and 1170 cm^{-1} (A_3)

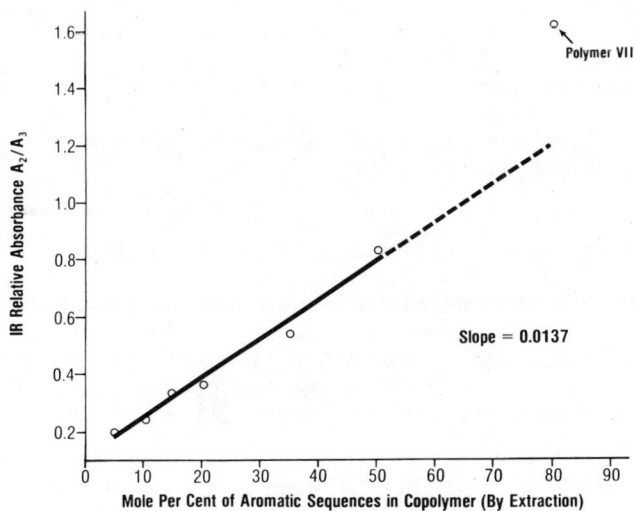

Figure 2. Effect of composition on absorbance at 1010 (A_2) and 1170 cm^{-1} (A_3)

This indicates that the effect of molar composition on the absorbance values associated with the out-of-plane (ν_1) and in-plane deformations (ν_2) is not identical for both frequencies.

$$TCA_1 = (A_1/A_3)\ (1/0.0325)$$
$$TCA_2 = (A_2/A_3)\ (1/0.0137)$$

This unexpected behavior [7,8] cannot be ascribed to the change in the aromatic content of the gross environment about the benzene ring in the chain. On the other hand, the unparalleled dependence of absorbance at ν_1 and ν_2 may be attributed to subtle differences in the micro-environment about the benzene ring which can be caused by incomplete cyclization of amino-amide groups to benzimidazole structures. Although one may expect that the relative absorbance of ν_1 and ν_2 for a benzene ring in the uncyclized amino-amide differs from that of benzimidazole, it is proposed that the change in the ν_2 absorbance, due to the difference in chemical structure, is more than that of ν_1. Hence, one expects that the relative absorbance of ν_1 and ν_2 will vary with changes in the concentration of uncyclized amino-amide groups in the aromatic fraction in copolymers having the same total molar composition of aromatic sequences. Taking this into consideration and using the relative absorbance A_1/A_2 for an aromatic homopolymer, C-1 (in which all aromatic sequences are assumed to be made practically of benzimidazole groups[4]), the relative concentration of the uncyclized chain aromatic (RUCA) was calculated by dividing the relative absorbance values of the copolymers by that of C-1 (Table III). In other words, the deviation of the A_1/A_2 values of the copolymers from the 1.34 value of the homopolymer can be used as a measure of the structural imperfections in their chains due to the uncyclized amino-amide moieties. Qualitatively, copolymers I and VII reflect a minimum and maximum level of chain imperfections, respectively.

$$RUCA = (A_1/A_2)\ copolymer/(A_1/A_2)\ homopolymer$$

For quantitative or semi-quantitative use of the RUCA values, they were normalized with respect to the aromatic content of the copolymers through dividing them by the corresponding values for the mole fraction (mole % divided by 100) of chain aromatics, TCA_1. These $(RUCA/TCA_1)100$ values are used later in the Discussion for correlating the composition of the copolymers with their thermal properties and will be referred to simply as ϕ.

Crystallinity and Thermal Properties

The effect of composition on the degree of crystallinity and the 2Θ values for the major reflections of the copolymers is illustrated by the data in Table IV. These data indicate that (a) the per cent crystallinity decreases with the increase in the copolymer aromatic content; (b) low levels of crystallinities, about 20%, were recorded for copolymers having 15 and 20 mole % of the aromatic sequences; (c) copolymers with ≥35% aromatics were amorphous to X-ray; (d) with the exception of copolymer II (which displayed 30 and 6% crystallinity due to the α- and γ-crystalline form of nylon 6, respectively), all crystalline copolymers revealed the two characteristic α-form reflection of nylon 6; and, (e) the intensity of the α-form reflection at 2 θ of about $23°$ decreased with the increase in the aromatic content of the copolymer, which can be associated with the effect of the structural imperfections on the normal lateral packing of the nylon 6 chains.

The above X-ray crystallinity data are consistent with DSC data in Table V, which indicate that copolymers having ≥35% aromatics in their chains do not undergo a first order thermal transition. The DSC data in Table V also show that (a) both the initial and reheating T_g increase with the increase in concentration of the rigid aromatic moities in the chains; (b) both the initial and reheating T_m decrease with the increase in the aromatic content of the copolymers; (c) the T_c increases as the concentration of aromatics in the polymer chains increases; (d) copolymers with 15 and 20% aromatics do not crystallize readily upon reheating their quenched melts; and, (e) an unusual difference is present between the initial and reheating T_g values. In order to compare the initial and reheating T_g and T_m values of the copolymers and their dependence on composition, the graphical illustration shown in Figure 3 was constructed. It is apparent from this Figure that (a) both the initial (T_m^o) and (T_m) melting temperatures decrease linearly with the increase in the aromatic content of the copolymers; (b) the dependence of T_m^o on composition is almost identical to that of T_m; (c) both the initial (T_g^o) and reheating glass transition temperatures (T_g) increase linearly with the increase in the aromatic content of the copolymers; (d) the T_g^o and T_g of the aromatic homopolymer (C-1) do not fall on the T_g-composition straight line of the copolymers; and, (e) the change in T_g with composition is more dramatic than that of T_g^o. The unusual increase in the T_g of the copolymers as a result of the thermal treatments associated with the reheating measurements is likely to be related to the fraction of uncyclized amino-amide aromatic sequences. To examine this possibility, the increase in glass transition temperature ($T_g - T_g^o$) was plotted against the previously calculated composition function ϕ (see Tables III and V), which is proposed to be an inverse concentration function of the amino-amide moieties along the copolymer backbone. This is achieved by constructing plot C of Figure 4, which shows a gradual decrease in $T_g - T_g^o$ with the increase of

TABLE IV X-RAY DATA*

Polymer Number	Mole % of Caprolactam[a]	Reduced Viscosity	Major Reflections $2\theta°(I/I_0)$	% Crystallinity[c]
XV	100	2.45	19.9 (92) 23.6 (100)	44
XVI	100	0.91	20.0 (94) 23.6 (100)	44
I	95	2.00	20.0 (100) 23.4 (93)	40
XIV	93	0.89	20.0	[d]
II	90	1.93	20.0 (95) 21.3 (100) 23.4 (76)	36
XIII	90	0.82	19.9 (100) 23.5 (87)	35
III	85	1.88	20.0 (100) 23.4 (92)	21
IV	80	2.29	19.8 (100) 23.4 (72)	25
XI	80	0.96	19.8 (100) 23.5 (70)	22
V	65	2.94	20.0 [b]	[d]
VI	50	2.04	19.6 [b]	[d]
VII	20	1.28	19.4 [b]	[d]

* The powdered samples were annealed for 20 hrs. at 100°C before testing.
(a) Initial composition.
(b) Center of broad reflection.
(c) With the exception of sample II (which displayed 30 and 6% crystallinity due to the α- and γ- crystalline form, respectively) all crystalline copolymers revealed the characteristic α-form reflections of nylon 6.
(d) Essentially amorphous.

TABLE V DSC DATA[a], EFFECT OF COMPOSITION ON THE THERMAL TRANSITIONS OF COPOLYMERS

	Polymer Number								
	N-6[b]	I	II	III	IV	V	VI	VII	C-1[c]
Composition Mole % of CL[d]	100	95	90	85	80	65	50	20	0
$\eta_{red.}$	2.45	2.00	1.93	1.88	2.29	2.94	2.04	1.28	0.34
Initial heating, $T_g°$, °C	–	–	45	54	53	61	61	67	240
Initial heating, $T_m°$, °C	219	207	194	(164)[e] 174	(154)[e] 174	–	–	–	–
Reheating[f], T_g, °C	40	56	67	85	89	124	156	205	242
T_c, °C	69	100	154	–	–	–	–	–	–
T_m, °C	220	210	190	–	–	–	–	–	–
$T_g - T_g°$, (°C)	–	–	22	31	27	63	95	138	2
ϕ	–	12.2	10.1	7.3	6.7	4.5	2.9	1.5	1.0

(a) DuPont 990-DSC, in nitrogen, 20°C/min. heating rate.

(b) See Reference #3.

(c) See reference #4, polymer C-1.

(d) Based on initial composition of comonomers.

(e) Minor endotherm.

(f) Samples were heated to 260°C (except C-1, which was heated to 300°C), held for 2 min., then quenched in liquid nitrogen.

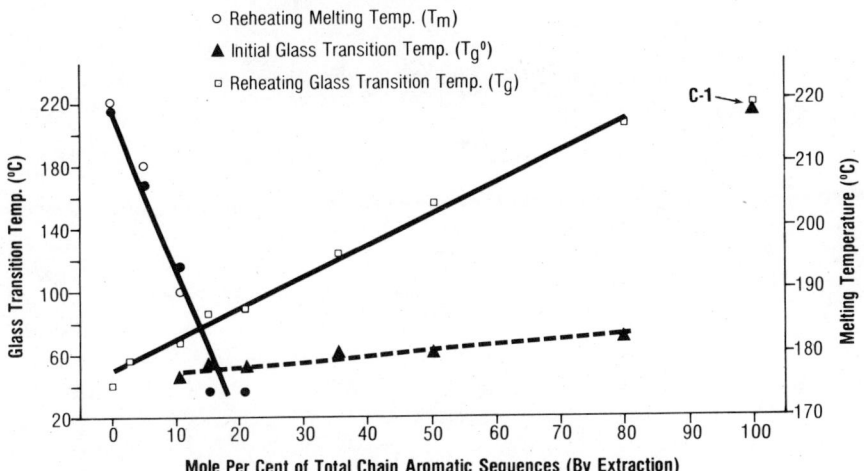

Figure 3. Effect of composition on polymer initial and reheating T_g and T_m

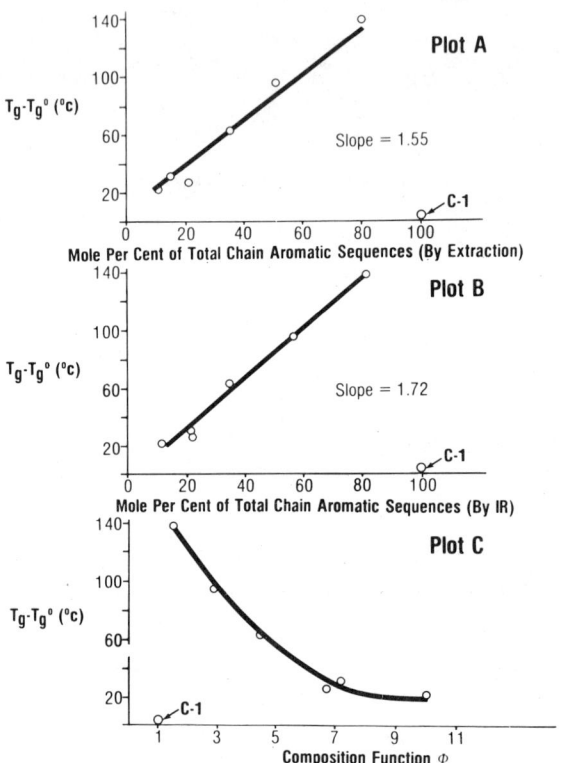

Figure 4. Effect of polymer composition on the change of T_g from thermal treatment

the ϕ values. Before attempting to explain the actual source of this relation in plot C, plots A and B of Figure 4 will be discussed. Both plots A and B indicate that the $(T_g - T_g^o)$ values increase with the increase in the aromatic content of the copolymers, whether the determined compositions are based on the extraction (plot A) or IR data (plot B). The data scattering and the slope of the straight line in both plots suggest more dependence and a better correlation of the $(T_g - T_g^o)$ data with IR determined compositions, as compared with those based on extraction data. The abnormal deviation of the $(T_g - T_g^o)$ values of the homopolymer (C-1) from the straight line relationships of plots A and B and the curve of plot C can be easily recognized upon examining Figure 4. This is proposed to indicate that a major contribution to the high reheating T_g values is associated with thermally-induced reactions of the uncyclized amino-amide groups which are present in the copolymers and not the homopolymer. If the aromatic moieties are the only participant in these reactions, one would expect an essentially linear dependence of $(T_g - T_g^o)$ on composition function ϕ. Since this is not the case, as plot C indicates, it is suggested that the caproamide sequences do take part in these reactions. Re-examination of plot C and taking into account this suggestion, allows one to conclude that the efficiency of the amino-amide moieties in increasing the T_g upon reheating does decrease with the decrease in the caproamide content. If one assumes that the amino-amide moieties undergo dehydration intramolecularly and/or intermolecularly (through reaction with carboxylic end groups), formation of new ethylene-benzimidazole units and/or branching can occur during thermal treatments similar to those used in the DSC measurements. The cyclization and/or branching would also be expected to increase with the decrease in the rigidity of the chains and/or the melt viscosity of the polymer matrix. Both of these properties depreciate gradually with the increase of the aliphatic content of the copolymer chains which, in turn, is translated to a lowering of T_g^o. Accordingly, the $(T_g - T_g^o)$ values would be expected to decrease with the decrease in the efficiency of the amino-amide group (i.e., the decrease in the composition function ϕ) in inducing branching and/or cyclization. This is indeed the case as shown in plot C of Figure 4. It also is interesting to note that the ϕ values (in Tables III and V) decrease steadily with increase in the aromatic (or increase in aliphatic) content of the copolymers, which does substantiate the above physical interpretation of the role of ϕ.

The effect of polymerization temperature and type of aromatic comonomer used on the degree of polymerization, degree of crystallinity and thermal properties of the copolymer was studied and the data are summarized in Table VI. Comparison of copolymers II, IX, and XIII which have essentially the same aliphatic content, indicate that the solution viscosity of

XIII which was made from an aromatic diamino-ester is lower than that of II derived from the corresponding acid. In addition, the use of a high polymerization temperature ($270^\circ C$) leads to (a) no significant difference in crystallinity; (b) a decrease in T_c; (c) a decrease in $(T_g-T_g^\circ)$; and, (d) an increase in thermal stability. The change in $(T_g-T_g^\circ)$ and thermal stability suggests that the polymer (IX) made at $270^\circ C$ contains less uncyclized aromatics, initially, than that made at $255^\circ C$ (II). This is logical to expect since cyclization to benzimidazole is expected to be high at the high temperature. Also, the effect of the initial content of amino-amide groups on $(T_g-T_g^\circ)$ is consistent with the discussion in the previous paragraph regarding the role of ϕ. A similar effect of the polymerization temperature on the polymer properties can be recorded as one examines the data for copolymer IV and XI. Furthermore, the data of copolymer VII reveal a higher thermal stability and $(T_g-T_g^\circ)$ value than the rest of the copolymers due to its higher aromatic content. It is important to note that the thermal treatments of VII cause increases in its T_g but $(T_g-T_g^\circ)$ changes per degree rise in the maximum temperature of the thermal treatment decrease above $230^\circ C$. This may suggest that the branching and/or crosslinking due to the amino-amide groups do not require excessively higher temperatures than $200^\circ C$.

The effect of composition (in terms of mole per cent caproamide) on the thermal stability of the copolymers is illustrated by the TGA data in Table VII. These data indicate that copolymers with high aromatic contents are generally more stable, thermally, than those with less aromatics in their chains. The effect of composition on the thermal stability is most noticeable between 450 and $500^\circ C$. At this temperature range nylon 6 is known to undergo excessive degradation[9] and, hence, it is likely that the aromatic moieties interfere with the unzipping mechanism by which the polycaproamide depolymerizes thermally.

Thermal and Tensile Properties of A Typical Copolymer and Nylon 6

The fact that the crystallizability of the partially aromatic copolymers decreases drastically with the increase of their aromatic content makes their comparison with nylon 6, as semicrystalline materials, an uneasy task. However, it was felt instructive to compare nylon 6 with a highly crystalline copolymer having a very low aromatic content (7 mole %) such as copolymer XIV. The thermal and tensile data of XIV and a nylon 6 sample (XVI) having comparable solution viscosities are summarized in Table VIII. These data show that (a) the annealed sample of XVI seems to have a higher degree of crystallinity (high ΔH_f°) than XIV, which is also associated with $19^\circ C$ difference in T_m°; (b) a reheated sample of XVI is

TABLE VI THERMAL ANALYSIS[a] & X-RAY DATA: EFFECT OF POLYMERIZATION CONDITIONS & THERMAL TREATMENTS ON THE THERMAL BEHAVIOR & CRYSTALLINITY OF COPOLYMERS

	II	IX	XIII	IV	XI	VII
Copolymer Composition,						
Mole % of CL[b]	90	90	90	80	80	20
Mole % of DPPA	10	10	-	20	20	80
Mole % of MDPP	-	-	10	-	-	-
Maximum Polymerization Temp., °C	255	270	255	255	270	255
Red. Viscosity (in H_2SO_4)	1.93	1.18	0.82	2.29	0.96	1.28
% Crystallinity of annealed (100°C/20 hr) Polymer:	36	35	-	25	22	-
DSC Data:						
Initial heating $T_g^°$, °C	45	47	45	53	58	67
Initial heating $T_m^°$, °C	194	195	193	174	175	-
Reheating data of samples quenched from 260°C, T_g, °C	67	64	59	89	85	205
T_c, °C	154	148	135	-	-	-
T_m, °C	190	190	194	-	-	-
$(T_m - T_g^°)$, °C	22	17	14	36	27	138
T_g(°C) due to different heating cycles:						
1st reheating after quenching from 150°C	-	-	-	-	-	130
2nd reheating after quenching from 200°C	-	-	-	-	-	180
3rd reheating after quenching from 230°C	-	-	-	-	-	203
4th reheating after quenching from 260°C	-	-	-	-	-	208
TGA data, % Wt. loss at 425°C	10	10	12	9	9	7
450°C	29	35	33	20	18	8
475°C	64	57	73	54	41	12
500°C	86	90	94	85	83	34

(a) DuPont 990 DSC and 951 TGA, in nitrogen, 20°C/min. and 10°C/min. heating rate, respectively.
(b) Based on initial comonomer composition.

TABLE VII TGA DATA[a]: EFFECT OF COMPOSITION ON THERMAL STABILITY OF COPOLYMERS

	Polymer Number						
	I	II	III	IV	V	VI	VII
Composition, Mole % of CL	95	90	85	80	65	50	20
$\eta_{red.}$ (H_2SO_4)	2.00	1.93	1.88	2.29	2.94	2.04	1.28
% Wt. loss in nitrogen at 200°C	1	1	1	2	2	3	3
300°C	2	2	1	3	2	3	4
400°C	4	4	3	5	4	5	6
425°C	9	10	8	9	6	7	7
450°C	24	29	20	20	13	12	8
475°C	65	64	45	54	31	33	12
500°C	95	86	86	85	70	65	34
600°C	97	90	88	88	81	74	65
700°C	-	93	91	89	82	76	68
800°C	-	98	95	93	85	80	72
900°C	-	-	99	97	90	86	77
1000°C	-	-	100	100	96	93	84

(a) DuPont-951 TGA, 10°C/min. heating rate.

TABLE VIII COMPARISON OF THERMAL & TENSILE PROPERTIES[a]

OF NYLON 6 & A TYPICAL COPOLYMER

	Polymer Number	
	XVI	XIV
Molar Composition (CL/DPPA) Initial	100/0	93/7
Final	100/0	92/8
Reduced Viscosity	0.91	0.89
DSC Data[b]: Initial heating, T_m^o (°C), ΔH_f^o (cal/g)	225, 24.6	204, 16.2
Reheating T_g(°C), T_c(°C)	43, 71	56, 101
$-\Delta H_c$ (cal/g)	6.7	7.2
T_m(°C), ΔH_f (cal/g)	220, 16.7	205, 13.2
TGA Data[c]: % Wt. loss in nitrogen		
@ 200, 300, 350	2, 3.5, 6.5	1, 2, 4.5
400, 425 and 450°C	20, 44, 81	16, 37, 77
% Wt. loss in air		
@ 200, 300, 350	2, 3, 6.5	2, 2.5, 4.5
400, 425 and 450°C	31, 61, 86	21, 42, 74
Tensile Properties: UE (%), UTS (psi)	182, 5268	40, 5407
WB (lb-in), SM (psi)	3.71, 105682	0.11, 140802
YE, YS	8, 4452	8, 5407

(a) Samples were conditioned at 50% R.H. and 25°C; UE = ultimate elongation, UTS = ultimate tensile strength, WB = work to break, SM = 2% secant modulus, YE = yield elongation, YS = yield strength.

(b) Perkin Elmer DSC-1B, in nitrogen, 20°C/min. heating rate. For reheating data, samples were heated up to 260°C, held for 2 min. then quenched in liquid nitrogen.

(c) Cahn RG electrobalance, 10°C/min. heating rate.

more crystallizable (low T_c and high ΔH_f) than XIV; (c) the reheated copolymer has a higher T_g and lower T_m than nylon 6 (this was associated with a difference of 13 and 15°C, respectively); (d) the copolymer is more stable thermally than nylon 6; (e) nylon 6 is more pliable and tougher than XIV as indicated by its higher extensibility and work to break values as compared with the copolymer; and, (f) the copolymer has higher tensile strength and modulus than nylon 6. The noticeably higher modulus of XIV as compared with XVI is attributed to the high rigidity of the copolymer aromatic sequences.

ACKNOWLEDGEMENT

The authors wish to thank Dr. P.J. Harget, Mr. R.A. Kirk, Mrs. L.S. Komarowski, Mr. A.B. Szollosi and Mr. D.W. Richardson for their valuable contributions during the different stages of these studies.

ABSTRACT

Several copolymers of ε-caprolactam and β-(3,4- diaminophenyl) propionic acid were prepared using comonomer mixtures containing 95, 93, 90, 85, 80, 65, 50 and 20 mole per cent of caprolactam. In most cases, the composition of the resulting high molecular weight copolymers were comparable to the corresponding initial comonomer compositions. Infrared spectroscopy and elemental analysis data of the copolymers suggest the presence of ethylene-benzimidazole moieties in their chains. Copolymers containing 35 mole per cent or more of the aromatic sequences were shown to be amorphous by differential scanning calorimetry and X-ray diffraction techniques. On the other hand, copolymers containing between 80 and 95 mole per cent of caprolactam moieties were semicrystalline and their degree of crystallinity ranged between about 20 and 40%. Similarly, the melting temperatures of these copolymers varied between 165 and 207°C. The glass transition temperature (T_g) of both the amorphous and crystalline copolymers were shown to increase with the increase in their aromatic content. Upon subjecting the copolymers to certain thermal treatments, noticeable changes in T_g were recorded. This was ascribed to structural imperfections in the chain and the proposed thesis was substantiated by IR and addititional thermal analysis data. A comparison of the thermal and tensile properties of a typical semicrystalline copolymer and nylon 6 is also reported.

LITERATURE CITED

1. Ajrodli, G. Stea, G., Mattiussi, A., & Fumagalli, M., J. Appl. Polym. Sci., (1973) 17, (3187).
2. Shalaby, S.W., Turi, E.A. & Pearce, E.M., J. Appl. Polym. Sci., (1976) 20, (3185), and references therein.
3. Shalaby, S.W., Turi, E.A. & Harget, P. J., J. Polym. Sci., Polym. Chem. Ed., (1976) 14, (2407), and references therein.

4. Shalaby, S.W., Lapinski, R.L. & Turi, E.A., J. Polym. Sci., Polym. Chem. Ed., (1974) 12 (2891).
5. Hermans, P.H. & Weidinger, A., J. Appl. Phys., (1968) 19 (491) J. Polym. Sci., (1949) 4 (135), J. Polym. Sci., (1950). 5, (565).
6. Conley, R.T., "Infrared Spectroscopy", Chap. 5, Allyn & Bacon, Boston, Mass., 1966.
7. Eglinton, G. in "Physical Methods in Organic Chemistry", Chap. 3, Schwarz, J.C.P., Ed., Holden-Day, Inc. San Francisco, Calif., 1964.
8. Korshak, V.V., Teplyakov, M.M. & Fedorova, R.D., J. Polym. Sci., Part A-1, (1971), 9, (1027), and references therein.
9. Pearce, E.M., Shalaby, S.W., & Barker, R.H., in Flame Retardant Polymeric Materials", Chap. 6, Lewin, M., Atlas, S.M. & Pearce, E.M., Eds., Plenum Press, New York, N.Y., 1975.

19

Anionic Polymerization of Fluorocarbon Epoxides

JAMES T. HILL and JOHN P. ERDMAN

E. I. du Pont de Nemours and Co., Inc., Elastomer Chemicals Department, Experimental Station, Wilmington, DE 19898

The objective of these investigations was to prepare perfluorinated polyether elastomers having both good low temperature flexibility and high thermal stability. The ring opening polymerizations of hexafluoropropylene epoxide (HFPO) and octafluoroisobutylene epoxide (OFIBO) were examined as a potential route to such materials.

$$CF_3CF\overset{O}{\overset{/\backslash}{-}}CF_2 \qquad (CF_3)_2C\overset{O}{\overset{/\backslash}{-}}CF_2$$

$$\text{HFPO} \qquad\qquad \text{OFIBO}$$

A number of nucleophiles are capable of opening the epoxide rings in these monomers (1,2,3). Fluoride ion opens the rings rapidly while preserving the perfluorinated nature of the products. Nucleophilic attack occurs exclusively at the more substituted carbon forming isolable perfluoroalkoxides.

$$CF_3CF_2CF_2O^-M^+ \qquad (CF_3)_2CFCF_2O^-M^+$$

Under some conditions the salts lose the elements of M^+F^- and form the corresponding acyl fluorides. The source of the fluoride ion and nature of its gegenion are important both in the ring opening reaction and to the stability of the alkoxide product. Of more than 40 fluoride salts investigated cesium fluoride was found to be most effective for initiating ring opening and forming stable alkoxides.

The alkoxides can react with additional epoxide to form straight chain oligomers (4).

$$CF_3\overset{O}{\overset{/\backslash}{C}FCF_2} + F^- \rightarrow CF_3CF_2CF_2O^- \xrightarrow{HFPO} CF_3CF_2CF_2O(\overset{CF_3}{\underset{|}{C}FCF_2}O)_n\overset{CF_3}{\underset{|}{C}FCF_2}O^-$$

A medium is required for cesium fluoride to react with either HFPO or OFIBO at room temperature. Tetraglyme (TG) was found to be the best of a number of polar and non-polar media tested. Initiation or ring opening by cesium fluoride in tetraglyme is difficult because of the low solubility of the salt in either the solvent or the liquid monomers. Polymerization is slow to initiate, difficult to control and is accompanied by a chain transfer reaction which yields only low molecular weight oligomers. From the reaction of HFPO we were able to isolate and characterize oligomers up to the tetradecamer. Higher oligomers were not detected. As shown below the chain transfer reaction formally can be represented by the elimination of solvated cesium fluoride from the growing alkoxide followed by fluoride attack on epoxide to generate new polymer chains.

$$R_fCF_2O^-Cs^+ \rightarrow R_f\overset{O}{\overset{\|}{C}}F + [CsF]$$

$$[CsF] + CF_3\overset{O}{\overset{/\backslash}{C}F-CF_2} \rightarrow CF_3CF_2CF_2O^-Cs \rightarrow \text{new polymer chains}$$

To prepare high molecular weight polymers requires the use of soluble cesium perfluoroalkoxide initiators. The purified acyl fluoride oligomers can be reacted with cesium fluoride in tetraglyme to form well defined, stable, homogeneous initiator solutions. In the case of HFPO these initiators are more useful than fluoride salts since they are more compatible with the monomer, smoothly initiate the polymerization, permit good temperature control and allow the synthesis of fluid polymers containing up to 100 monomer units. We have been able to demonstrate polymerizations from the monofunctional and difunctional initiators (5), shown below.

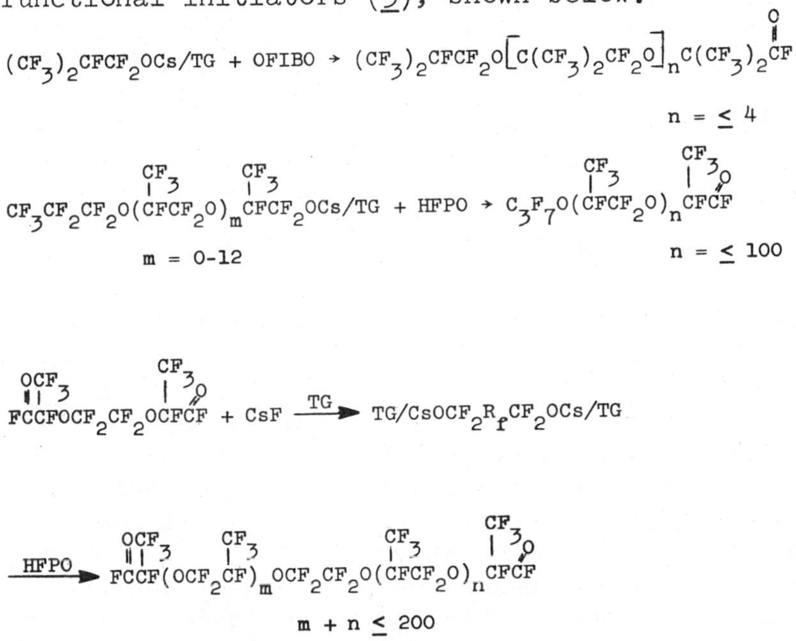

$$(CF_3)_2CFCF_2OCs/TG + OFIBO \rightarrow (CF_3)_2CFCF_2O[C(CF_3)_2CF_2O]_nC(CF_3)_2\overset{O}{\overset{\|}{C}}F$$

$$n = \leq 4$$

$$CF_3CF_2CF_2O(\underset{|}{\overset{CF_3}{C}}FCF_2O)_m\underset{|}{\overset{CF_3}{C}}FCF_2OCs/TG + HFPO \rightarrow C_3F_7O(\underset{|}{\overset{CF_3}{C}}FCF_2O)_n\underset{|}{\overset{CF_3}{\overset{|}{C}}}F\overset{O}{C}F$$

$$m = 0-12 \qquad\qquad n = \leq 100$$

$$\underset{FCCFOCF_2CF_2OCFCF}{\overset{OCF_3\quad\; CF_3}{\overset{\|\;\;\;\;\;\;\;\;\;\;\;\;\;\;\;\;|\;\;\;\;}{\;}}} + CsF \xrightarrow{TG} TG/CsOCF_2R_fCF_2OCs/TG$$

$$\xrightarrow{HFPO} \underset{FCCF(OCF_2CF)_mOCF_2CF_2O(CFCF_2O)_nCFCF}{\overset{OCF_3\;\;\;\;\;CF_3\;\;\;\;\;\;\;\;\;\;\;\;\;\;\;\;\;\;CF_3\;\;\;\;\;CF_3}{\overset{\|\;\;\;\;\;\;\;\;\;\;\;|\;|\;\;\;\;\;\;\;\;\;|}{\;}}}$$

$$m + n \leq 200$$

A wide variety of diluent solvents, coordinating solvents, counter ions and perfluorinated acyl fluorides were examined in an effort to suppress the chain transfer reaction that limits the degree of polymerization (DP). Because of this limitation the highest number average molecular weights (M_n) we have observed for poly HFPO are 15,500 from monofunctional initiators and ~25,000 from difunctional initiators. In contrast, we have not been able to polymerize OFIBO beyond DP 4. This suggests that HFPO is about 25 times as reactive as OFIBO with regard to ring opening polymerization initiated by solvated perfluoroalkoxide salts (6). Since the attack is at the tertiary carbon atom of both epoxide rings we conclude that steric hindrance in OFIBO is mainly responsible for its lower propagation rate, making the polymerization less competitive with the chain transfer reaction. Increased inductive stabilization of the negatively charged alkoxide by the additional trifluoromethyl group in OFIBO may also be a minor factor contributing to its lower reactivity.

If the homogeneous initiators are heated to greater than 100°C the alkoxides revert to acyl fluoride, cesium fluoride precipitates and the tetraglyme separates from solution. On cooling and remixing the homogeneous initiator is reformed quantitatively. This suggests that the mobile equilibrium,

$$R_fCF_2OCs/TG \rightleftharpoons R_fC(=O)F + CsF/TG$$

is perhaps responsible for the chain transfer reaction.

Two observations discount this hypothesis. In Figure 1 the ^{19}F NMR spectra of the difunctional acyl fluoride and its corresponding cesium alkoxide initiator are compared. Resonances for the fluorines on the end groups in the two species are well separated and distinct. In temperature studies of the alkoxide solutions over the range -40 to +80°C we observe neither acyl fluoride resonances nor changes in chemical shifts that could be ascribed to rapidly equilibrating species. That the chain transfer rate is very sensitive to the polymerization temperature over the range -35 to +25° suggests that the equilibrium, if it plays a significant role, should be readily observable.

The second point is that in the presence of the alkoxide initiators additional unreacted cesium fluoride does not increase the rate of chain transfer during HFPO polymerizations. As we know that the metal salt is by itself capable of initiating oligomerization when it is mixed with tetraglyme, it must have some finite solubility in the medium. We must conclude then that fluoride ion initiation is not competitive with alkoxide initiation and further that the alkoxide-acyl fluoride equilibrium is not important to the chain transfer reaction.

Our experiments show that the size of the gegenion, the solvent's coordinating ability, and the solublity of the initiator system in the reaction medium are major variables determining the rate of chain transfer. With small counterions, lithium, sodium, or potassium, or with insoluble initiator salts high yields of only low molecular weight polymer are obtained. Cesium salts routinely give the least amount of chain transfer. Similarly the transfer rate diminishes as the affinity of the coordinating solvent toward the large alkali metal ions is increased.

These data suggest that the equilibrium between contact, solvent separated, and free ion pairs is related to the rate of chain transfer. We believe that chain propagation occurs via reaction of the monomer either directly with free ions or by insertion of the epoxide oxygen into the coordination sphere of the metal ion followed by alkoxide attack and ring opening. The chain transfer probably arises from a bimolecular reaction of the monomer with contact ions via a six-membered cyclic process.

Polymerization Technique

For successful polymerizations scrupulus maintenance of anhydrous materials and conditions are required as even a few ppm of water are deleterious to the degree of polymerization.

The fluid initiator is prepared in a dry box under nitrogen by mixing acyl fluoride, cesium fluoride, and tetraglyme in approximately 1:1:>2 molar ratio. A slight excess of salt insures complete conversion of the acyl fluoride to alkoxide. Cesium fluoride dissolves rapidly and when the excess is removed by centrifugation the single phase supernatant is crystal clear and is found to contain the theoretical amount of cesium.

The initiator is weighed into the reaction vessel and is cooled to -35° at which point a solvent for the polymer is added. In most cases this is liquid hexafluoropropylene (HFP). We have found that solvent in addition to tetraglyme is necessary for successful polymerization. HFP is a solvent for the perfluorinated ends of the growing chains. We believe that its main functions are to keep the viscosity of the polymerizing mass low and to provide for efficient transfer of the heat of polymerization away from the growing polymer ends. There is some evidence that HFP is a trap for stray fluoride ions which helps to suppress chain transfer. However, alternate solvents incapable of accepting fluoride ions have been used successfully as diluents and have been shown to have no measurable effect on the rate of chain transfer.

As soon as all the diluent is added the mixture is stirred rapidly to disperse the semisolid initiator as well as possible; it is not completely soluble at this stage. HFPO is then added through calibrated meters at a constant rate and condensed. Reaction temperatures are maintained between -30 to -35°C. Lower temperatures freeze the initiator to a completely inactive solid while higher temperatures result in the loss of solvent and monomers and sharply increasing chain transfer rates. As the monomer is consumed the initiator thins out and becomes evenly dispersed throughout the milky mass. At ~ 5000-6000 M_n the reactor's contents suddenly becomes clear. As additional monomer is added the mixture again becomes cloudy as high MW fluorocarbon polymer begins to phase out of solution. At the same time the solution viscosity increases markedly and the mass becomes difficult to stir.

When the desired quantity of monomer has been added the reaction is stirred for another hour, after which no further increases in MW are observed. HFP is removed under vacuum as the reactor is warmed to room temperature. Material balances are > 99%. No unreacted monomer is found in the recovered solvent indicating that conversions of HFPO are close to 100%. Though quantitative recovery of the acyl fluoride terminated polymer is possible in most cases it is more convenient to quantitatively convert the end groups to the ethyl ester by quenching with absolute ethanol and subsequently removing the salt and tetraglyme which separates from the polymer.

Polymer Characterization

Number average molecular weights (M_n) for these polymers can be determined by a variety of methods; vapor phase osmometry, IR, UV and saponification with alcoholic potassium hydroxide. Purified oligomers up to the tetradecamer were used to establish IR extinction coefficients for the acid fluoride (5.35 μm) and the ethyl ester (5.6 μm). Number average MW's up to ~ 25,000 are accessible with these techniques; all gave equivalent results. For polymers initiated by difunctional catalysts the IR method affords equivalent weights which are corrected to number average molecular weights with an NMR measurement of the average functionality.

When chain transfer occurs during a difunctional polymerization monofunctional polymers result which have at one end a perfluoropropyl ether group.

$$CF_3\underline{CF_2}CF_2O \sim\sim\sim \underline{CF}(CF_3)C(O)OEt \quad \text{monofunctional polymer}$$

s, 133.33 ppm   t, 131.56 ppm

$$EtOC(OCF_3)\underline{CF} \sim\sim\sim \underline{CF}(CF_3)C(O)OEt \quad \text{difunctional polymer}$$

t   t

The amount of chain transfer can be measured by ^{19}F NMR. The secondary fluorines (s) absorb at 133.33 ppm ($CFCl_3$) and the tertiary (t) at 131.56 ppm. We find that these well defined, separated, and easy to integrate peaks occur in the ratio 2:1 for monofunctional polymer and that no "s" absorbances exist for purely difunctional polymers. The relative quantities in a mixture of mono- and difunctional materials can be determined from the integral ratio, t/s, by applying the equation:

$$\text{mole fraction difunctional polymer, } D = \frac{t/s - 1/2}{t/s + 1/2}$$

Examples of spectra for a 33% difunctional polymer (bottom) and a 95% difunctional polymer are illustrated in Fig. 2. For any degree of difunctionality the M_n is equal to $1 + D$ times the IR equivalent weight.

M_n, Functionality, and Time

One polymerization was conducted using a tetraglyme solution of HFPO trimer alkoxide for initiation and HFP as the diluent; monomer was added slowly and continuously over 50 hrs. Samples were removed during the polymerization to follow the M_n as a function of time and the amount of monomer added. Figure 3 shows this relationship and illustrates the severe limitation on M_n by chain transfer. At the end of 50 hrs the M_n was only 5000 compared to 59,400 M_n expected if no chain transfer had occurred. At 7 hrs into this run the chain transfer parameter, MW theory/M_n, was 8.0, indicating that each molecule of initiator had undergone an average of 7 transfer reactions. This parameter increased to only 11.9 at the end of 50 hrs suggesting that most of the chain transfer takes place early in the polymerization - perhaps even at the onset of polymerization. We find also that the rate of chain transfer increases with increasing monomer addition rates. Under reaction conditions identical to those used in the above experiment tripling the monomer addition rate results in a 5-fold increase in the chain transfer rate while a 6-fold monomer rate increase results in a 16-fold

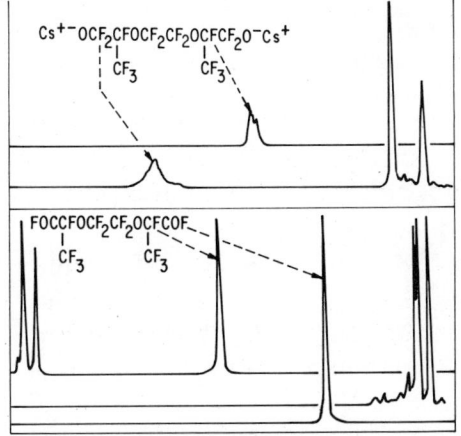

Figure 1. ^{19}F NMR of the difunctional acyl fluoride and its corresponding cesium alkoxide

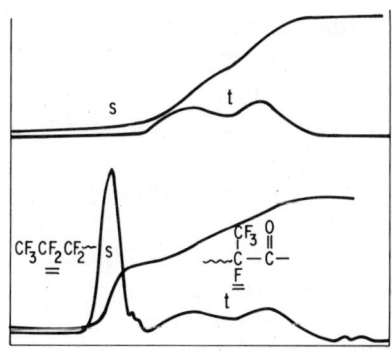

Figure 2. Partial expanded scale ^{19}F NMR of 33% difunctional poly HFPO (bottom) and 95% difunctional poly HFPO (top)

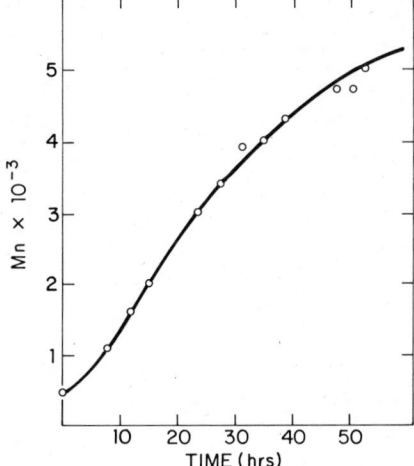

Figure 3. HFPO polymerization using cesium HFPO trimer alkoxide initiator

increase in the transfer rate.

We obtained more information from a similar experiment using the difunctional initiator in which both functionality and M_n were followed as a function of time. Figure 4 shows the molecular weight response. It is obvious that there is an induction period for the polymerization. The M_n lags considerably behind the theoretical M_n up to ~ 200 min and then it increases, for a brief period faster than the rate of monomer addition, indicating that there had been a large accumulation of unreacted monomer during the induction period. This slow initiation is probably caused by HFPO having to diffuse through the mass of very viscous tetraglyme to find alkoxide groups. Subsequent thinning and dispersion of the catalyst increases the rate of monomer reaction.

The other point to observe in this figure is the apparent discontinuity at 720 min. Here was one of the places that the living anionic nature of these polymerizations was demonstrated. After 12 hours of reaction monomer addition was stopped and the polymerization mixture cooled to -78°C and kept there for 8 hr. After rewarming the mixture to -32°C, monomer addition was restarted and the polymerization continued. The discontinuity is not unusual as a slight M_n increase was observed whenever a polymerization was shutdown and later restarted. The observation suggests that relative to the alkoxide concentration HFPO is always present in excess.

Figure 5 illustrates the difunctionality response for this experiment. The induction period and discontinuity are also present. From the difunctionality measurements it is possible to calculate a pseudo-first-order rate constant for chain transfer, K_t. The average K_t for this run was $7.9 \times 10^{-6} \sec^{-1}$. In the rate expression, the assumption that the number of alkoxide ends is constant at all times leads to low values for K_t as calculated from the first four points. This might be expected for during the induction period not all the alkoxide ends are

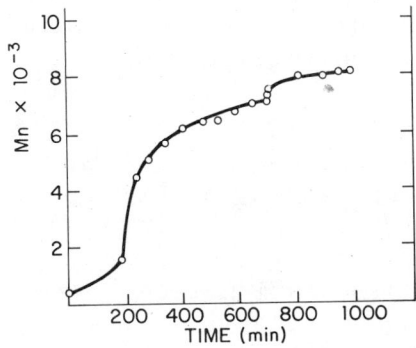

Figure 4. HFPO polymerization using difunctional initiator

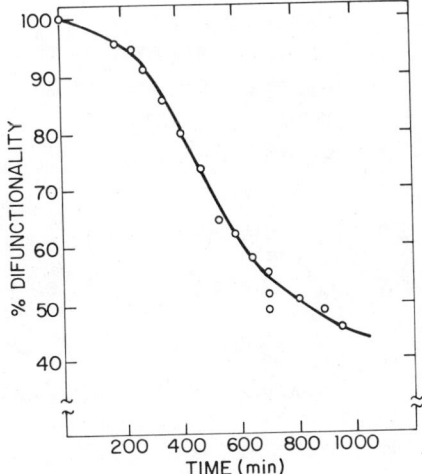

Figure 5. HFPO polymerization using difunctional initiator

available for propagation. Later on as the initiator is more evenly dispersed the assumption becomes valid. Calculation of K_t from 20 other polymerizations run under similar conditions gave an average value of $8.4 \times 10^{-6} \sec^{-1}$.

Armed with the chain transfer constant and the time dependence of the molecular weight and difunctionality we developed the appropriate kinetic expressions to estimate the pseudo-first order rate constant for the ring opening propagation step. We find that a value of $8 \times 10^{-4} \sec^{-1}$ gives the best fit to all the data. It is interesting to note that the values we obtained for K_p and K_t indicate that the maximum DP possible for a polymer initiated with a monofunctional alkoxide is about 100 (16,500 M_n). The difunctional molecules can be expected to grow to no more than DP 200 (33,000 M_n) but this material would be contaminated with a substantial fraction of lower DP monofunctional polymer.

Using the kinetic parameters derived from our experiments we were able to calculate reaction times, monomer addition rates and the correct amount of difunctional initiator to use to prepare moderate MW difunctional polymers containing no detectable monofunctional contaminants. With careful control we can routinely prepare 100% difunctional polymers of the structure

$$\underset{\|}{\overset{OCF_3}{F\overset{\|}{C}CF}}(OCF_2\overset{CF_3}{\underset{|}{C}F})_m-OCF_2CF_2O(\overset{CF_3}{\underset{|}{C}FCF_2}O)_n\overset{CF_3}{\underset{|}{C}F}\overset{O}{\overset{\|}{C}F}$$

$$m + n \cong 35$$

Chain Extended Polymers

Because of the chain transfer we were unable to prepare elastomeric HFPO homopolymer. The highest MW materials prepared were very viscous liquids at room temperature. The chemical inertness of the backbone and the high reactivity of the acyl fluoride and ethyl ester end groups enables the chain extension with a wide variety of reagents. A few examples are shown below.

$$R_f(\overset{O}{\underset{\|}{C}}F)_2 \xrightarrow{\text{diols}} \text{polyesters}$$

$$\xrightarrow{\text{diamines}} \text{polyamides}$$

Polyesters were prepared by condensation of the diacid fluorides with a variety of fluorinated and non-fluorinated diols. Alternatively they could be prepared by metal catalyzed ester interchange using the poly HFPO diesters. Polyamides can be prepared either by solution or interfacial techniques. These materials were deficient in both thermal and hydrolytic stability.

Aromatic chain extension links were prepared in an effort to increase the high temperature stability of the polymer. Benzimidazoles were synthesized by condensing either the acid fluorides or esters with diaminobenzidine and subsequently dehydrating with heat. For 1,3,4-oxadiazoles the precursor polyhydrazides could not be prepared by the direct reaction of the acyl fluorides with hydrazine as intractable unstable corsslinked materials were formed. They were best prepared by converting the prepolymer end groups to phenyl esters and then reacting with hydrazine to form the dihydrazides. Subsequent reaction with a variety of diacid halides afforded attractive elastomers. Dehydration with phosphorous pentoxide gave the co-polyoxadiazoles but these were no longer rubbery.

Benzimidazole

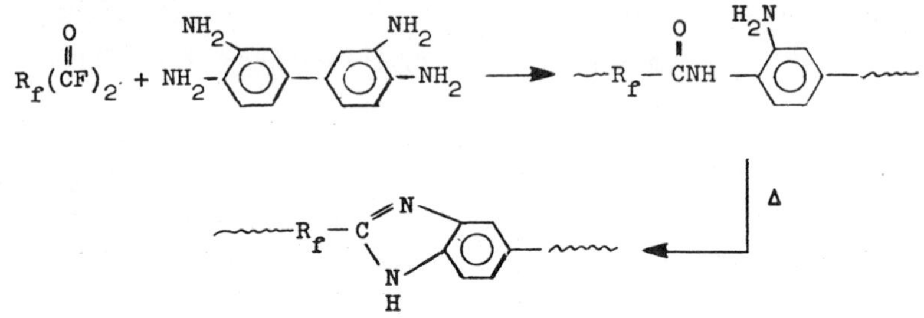

$$R_f(\overset{O}{\overset{\|}{C}}F)_2 + NH_2-\underset{NH_2}{\bigcirc}-\underset{NH_2}{\bigcirc}-NH_2 \longrightarrow \sim\sim R_f-\overset{O}{\overset{\|}{C}}NH-\underset{H_2N}{\bigcirc}-\sim\sim$$

$$\sim\sim R_f-C\underset{\underset{H}{N}}{\overset{N}{\diagdown}}\bigcirc\sim\sim \quad \overset{\Delta}{\longleftarrow}$$

1,3,4 Oxadiazoles

$$R(COF)_2 + C_6H_5O^-K^+ \longrightarrow R_f(\overset{O}{\overset{\|}{C}}OC_6H_5)_2 \xrightarrow{N_2H_4} R_f(\overset{O}{\overset{\|}{C}}NHNH_2)_2$$

$$\xrightarrow{\underset{CCl}{\bigcirc}\overset{O}{\underset{\|}{CCl}}} \sim\sim R_f\overset{O}{\overset{\|}{C}}NHNH\overset{O}{\overset{\|}{C}}-\bigcirc-\sim\sim \xrightarrow{P_2O_5} R_f C\underset{N-N}{\overset{O}{\diagup\diagdown}}C-\bigcirc-\sim\sim$$

↓ diamine

1,3,4 triazole crosslink

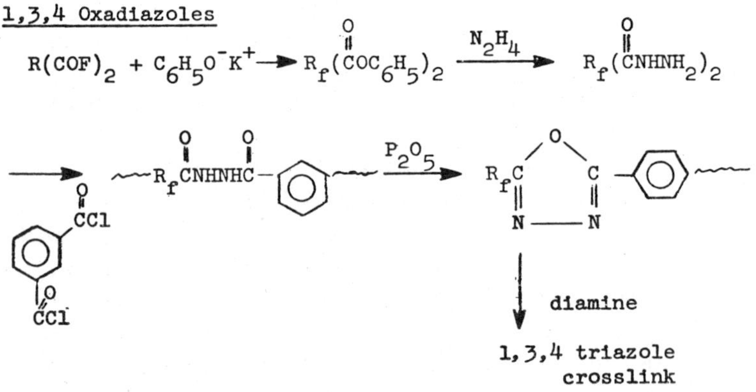

We were <u>unable</u> to chain extend the difunctional HFPO beyond $\overline{DP}$ = 10 (60,000 M_n). As predicted the polymers exhibited Tg's close to -50°C. The aromatic links exhibited only fair stability at 350°C. Though the fluorinated chains impart some resistance to aqueous bases we find the materials chain extended with 5 membered heterocycles do not survive such exposure. The strong electron withdrawing effect of the fluorinated substituents render the ring carbons very susceptible to nucleophilic attack. In aqueous base the polymers are degraded to the original molecular weights.

The most interesting chain extended polymers were the s-triazines. The diesters can be converted to diamides with ammonia and thence to dinitriles

$$R_f(COEt)_2 \xrightarrow{NH_3} R_f(CNH_2)_2 \xrightarrow{P_2O_5} R_f(C\equiv N)_2$$

$$R_f(C\equiv N)_2 \xrightarrow{NH_3} R_f(C-NH_2)_2 \xrightarrow[-NH_3]{\Delta}$$

$$\xrightarrow{AgO, \Delta}$$

with phosphorous pentoxide. The nitriles can be converted directly to s-triazines at high temperature in the presence of silver oxide catalyst or can be converted to bisamidines with ammonia. When these fluids are heated ammonia is evolved and the triazine networks form. Like the linear polymers these crosslinked materials exhibit glass transitions in the range -50 to -60°C. They have good hydrolytic stability and are virtually unaffected by prolonged heating at 350°C. The modulus, tensile strength, and elongation at break of the best triazine polymers, however, are too low for general use in molded goods. We believe, however, that the poor physical properties are not due to inherent weaknesses of the HFPO backbone or the triazine links but rather to loose chain ends and short distances between the trifunctional crosslinks.

Literature Cited

1. Hill, J. T., J. Macromol. Sci.-Chem., (1974), $\underline{A8}$(3), 499.
2. Eleuterio, H. S., U. S. Patent 3,358,003 (1967).
3. Sianesi, D., Pasetti, A. and Tarli, F., J. Org. Chem. (1966), $\underline{31}$, 2312
4. Moore, E. P., U. S. Patent 3,322,826 (1967)
5. Fritz, C. G. and Moore, E. P., U. S. Patent 3,250,807 (1966)
6. Hill, J. T., Eighth Int. Symp. Fluor. Chem., Kyoto, Japan, Aug. 24, 1976.

Ring-Opening Polymerization via C–C Bond Opening

H. K. HALL, JR., H. TSUCHIYA, P. YKMAN, J. OTTON,
S. C. SNIDER, and A. DEUTSCHMAN, JR. (1)

Department of Chemistry, University of Arizona, Tucson, AZ 85721

Ring-opening polymerization usually involves compounds containing strained C-O, C-N, or C-S single bonds. Polymerizations involving strained C-C single bonds are less familiar. Cyclopropane and cyclobutane do not give clean results, because reagents sufficiently vigorous to open the ring also attack the resulting chain. Two types of strained bicyclic compound undergo ring-opening polymerization via C-C bond opening. The first group consists of compounds with a strained polycyclic structure. Examples include a variety of bicyclobutanes 1, (2) bicyclopentane-[2.1.0]carbonitrile 2 (3), benzocyclopropenecarbonitrile 3 (4), benzocyclobutene 4 (5,6), two tetracyclooctanes 5 and 6, (7,8), and 1,3-dehydroadamantane 7 (9,10).

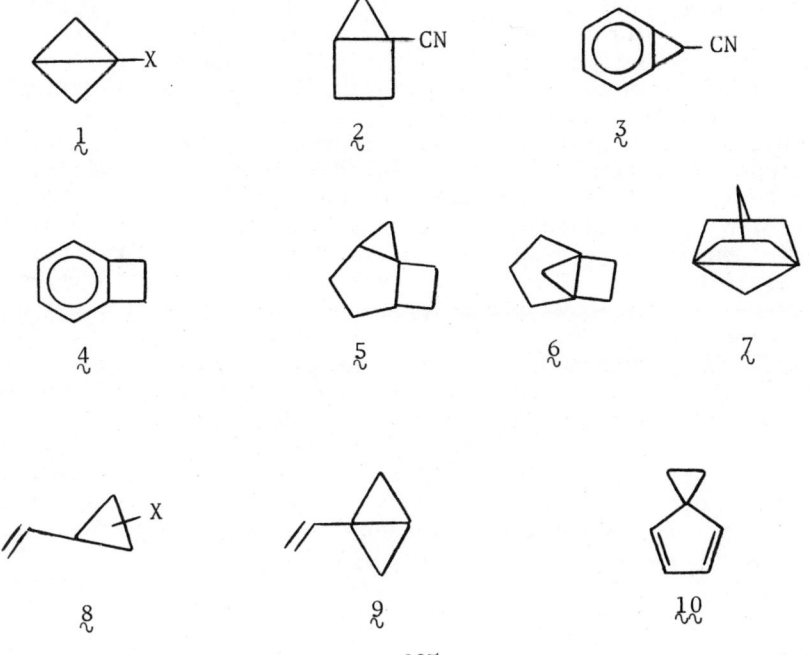

Two factors account for the ability of such polycyclic compounds to polymerize. First, the strain energy is very high, and may be relieved on polymerization. Secondly, the compact cage molecule shows low steric hindrance to attack. Breaking a C-C single bond always encounters hindrance from substituents on the adjacent carbons, but this repulsion is minimized in compact ring structures.

The second group of monomers contains those in which a double bond is "conjugated" with a strained single bond. A number of vinylcyclopropanes 8 (11,12) belong in this category, as do 1-vinylbicyclobutane 9 (2) and [2,4]-spiroheptadiene 10 (13). For these compounds, the double bond offers a point of reaction for the growing polymer chain. Again, the projecting $CH_2=$ group offers minimal steric hindrance.

Unlike other ring-opening polymerizations, most C-C single bond polymerizations have been carried out by free radical initiation, even though examples of cationic, anionic, and coordination polymerizations have been presented. The bicyclobutane monomers, such as bicyclobutane-1-carbonitrile (1a X = CN), are as reactive in free radical polymerization as vinyl monomers (2).

Anionic Polymerization of Bicyclobutane-1- arbonitrile - We inquired whether anionic polymerizations of 1a could also be carried out. The most successful anionic polymerizations of methacrylonitrile (the vinyl analog of 1a), have been those of Joh and his colleagues (14,15), who used dialkylmagnesium and magnesium dialkylamide initiators. Therefore we utilized them with bicyclobutane-1-carbonitrile.

Bicyclobutane-1-carbonitrile polymerized very readily with these organomagnesium initiators (Table I). The magnesium amides and "ate" compounds gave highest yields, with the mercaptides close behind. Dialkylmagnesiums gave lower yields. Dioxane and toluene as solvents gave the highest yields, and the polymer precipitated from these media. Unstirred, magnetically stirred, and mechanically stirred polymerizations gave comparable results. Homogeneous polymerizations were performed in dimethylformamide, sulfolane and tetramethylene sulfoxide solution. The yields under these conditions were very low or zero.

Polybicyclobutanecarbonitrile obtained in this way was a white powder, unlike the fibrous material obtained by free radical initiation. The inherent viscosities in dimethylformamide were usually about 0.1 dl.g.$^{-1}$ and rarely exceeded 0.5 dl. g.$^{-1}$. The nmr spectra resembled those of the radical-initiated polymer. When magnesium di(isopropylmercaptide) was used as the initiator, isopropylmercapto end groups were visible in the nmr spectra. The infrared spectra also resembled those of the radical-initiated polymer, and supported the 1-cyano-1,3-cyclobutanediyl structure. One absorption which did not conform to this structure was visible in every ir spectrum. This absorption at 1700 cm^{-1} is ascribed to a ketone carbonyl group. This assignment was confirmed by stirring the polymer with sodium borohydride overnight in sulfolane-water (4:1), whereupon this band disapeared. The absorption intensity of this carbonyl band correlated inversely with inherent viscosity, indicating that it was involved in the termination reaction.

We propose the following mechanism:

Initiation - Addition of organometallic RM to the strained 1,3 bond is known to occur even for the more sterically crowded

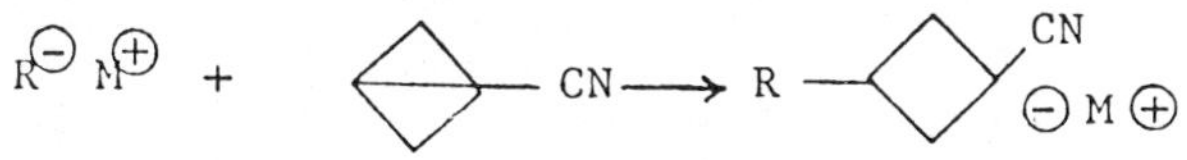

$$R^{\ominus} M^{\oplus} + \text{[bicyclobutane]}-CN \longrightarrow R-\text{[cyclobutane]}(CN)(M^{\oplus})$$

3-methyl-1-bicyclobutanecarbonitrile (16), and is supported by detection of the corresponding end groups in the NMR spectra of polymers initiated by magnesium di(isopropylmercaptide) (and other initiators).

Propagation -

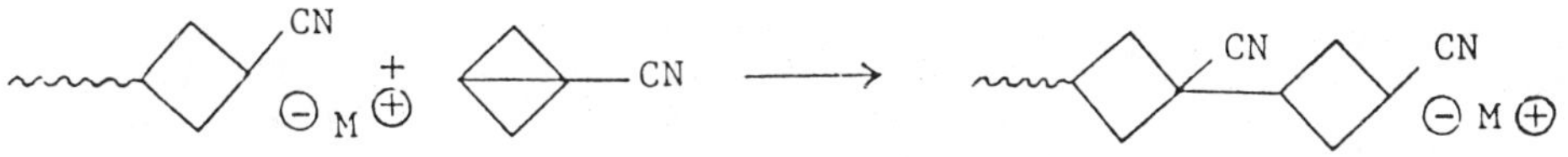

$$\sim\sim\text{[cyclobutane]}(CN)(M^{\ominus\oplus}) + \text{[bicyclobutane]}-CN \longrightarrow \sim\sim\text{[cyclobutane]}(CN)-\text{[cyclobutane]}(CN)(M^{\ominus\oplus})$$

Termination -

$$\sim\sim\text{[cyclobutane]}(CN)(M^{\ominus\oplus}) + \sim\sim\text{[cyclobutane]}(C\equiv N) \longrightarrow \sim\sim\text{[cyclobutane]}(CN)-\text{[cyclobutane]}(C=N^{\ominus})(M^{\oplus}) \xrightarrow{workup} \sim\sim\text{[cyclobutane]}(CN)-\text{[cyclobutane]}(C=O)$$

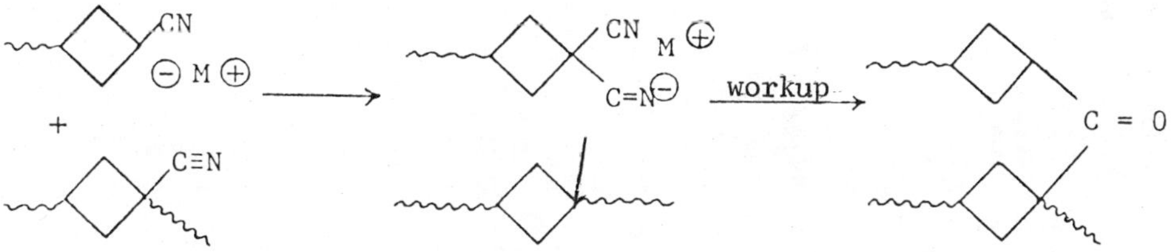

We studied a model system in an effort to determine whether or not attack of a propagating α-cyanocyclobutyl anion on nitrile groups can take place quickly enough under our reaction conditions to represent a plausible termination step. Of a number of strong bases studied, only triphenylmethylsodium cleanly abstracted the α-H of cyclobutanecarbonitrile to give the carbanion:

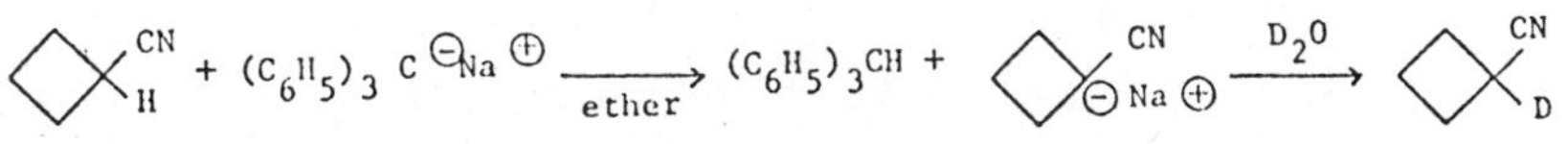

$$\text{[cyclobutane]}(CN)(H) + (C_6H_5)_3C^{\ominus}Na^{\oplus} \xrightarrow{ether} (C_6H_5)_3CH + \text{[cyclobutane]}(CN)(Na^{\oplus\ominus}) \xrightarrow{D_2O} \text{[cyclobutane]}(CN)(D)$$

Table I. Selected Polymerizations of 1-Bicyclobutanecarbonitrile

Initiator	M moles Initiator, I	M moles Monomer, M	M/I	Lewis Acid (a)	Solvent (b)	% Yield	n_{inh}(c)
$Mg(C_2H_5)(NC_5H_{10})$ (d)	1.0	13	13	-	Dioxane	95	0.41
$Mg(C_2H_5)(NC_5H_{10})$	1.0	26	26	-	Dioxane	73	0.21
$Mg(n-C_4H_9)(NC_5H_{10})$	1.0	26	26	-	Toluene	44	0.105
$Mg(n-C_4H_9)(NC_5H_{10})$	1.0	26	26	-	Dioxane	75	0.28 (e)
$Mg(NC_5H_{10})_2$, HNC_5H_{10}	1.0	13	13	-	Dioxane	88	0.23
$Mg(NC_5H_{10})_2$	2.0	26	13	-	Dioxane	51	0.086
$Mg(N-C_4H_9)_2$	1.0	13	13	$(C_2H_5)_3Al$	Dioxane	47	0.036
$Mg(S-i-C_3H_7)(NC_5H_{10})$	2.0	26	13	-	Dioxane	90	0.096
$Mg(S-i-C_3H_7)(NC_5H_{10})$	1.0	26	26	-	Dioxane-DMF (1:1)	36	0.076
$Mg(n-C_4H_9)(NC_5H_{10})$, $Mg(C_2H_5)_2$	1,1,1	13	13	-	THF	100	0.22
$n-C_4H_9Li$, $Mg(C_2H_5)_2$, HNC_5H_{10}	1,1,1	13	13	-	Dioxane	89	0.20
$n-C_4H_9Li$, $Mg(C_2H_5)_2$, $i-C_3H_7SH$	1,1,1	13	13	-	Dioxane	84	0.074
$n-C_4H_9Li$, $Mg(C_2H_5)_2$, $i-C_3H_7SH$	1,1,1	13	13	-	Dioxane-DMF (2:5)	69	0.11
$Mg(S-i-C_3H_7)_2$	1.0	26	26	-	Dioxane	64	0.20
$Mg(S-i-C_3H_7)_2$	1.0	26	26	-	Toluene	58	0.074
$Mg(S-i-C_3H_7)_2$	1.0	26	26	-	Dioxane-DMF (2:5)	20	0.15
$Mg(S-i-C_3H_7)_2$	0.25	13	52	-	Dioxane	28	0.23
$Mg(S-i-C_3H_7)_2$	0.2	13	65	-	Toluene	23	0.18

Catalyst			Cocatalyst	Solvent		
Mg(S-i-C$_3$H$_7$)$_2$	1.0	26	26	-	Dioxane-TMSO$_2$ (2:5)	35	0.051
Mg(C$_2$H$_5$)(S-i-C$_3$H$_7$)	2.0	26	13	-	Dioxane	45	0.10
Mg(C$_2$H$_5$)(S-i-C$_3$H$_7$)	1.0	26	26	-	Dioxane-DMF (2:5)	35	0.061
Mg(n-C$_4$H$_9$)(S-i-C$_3$H$_7$)	0.5	13	26	-	Dioxane	17	0.12
Mg(C$_2$H$_5$)(S-i-C$_3$H$_7$)	1.0	13	13	(C$_2$H$_5$)$_3$Al	Dioxane	12	0.077
Mg(C$_2$H$_5$)$_2$	1.0	26	26	-	Dioxane	37	0.57 (e)
Mg(C$_2$H$_5$)$_2$	0.5	26	52	-	Dioxane	18	0.71 (e)
Mg(C$_2$H$_5$)$_2$ (75°C)	0.2	13	65	-	Dioxane	20	0.31
Mg(C$_2$H$_5$)$_2$	0.2	13	65	-	Toluene	8	0.18
Mg(C$_2$H$_5$)$_2$ (70°C)	0.2	13	65	-	Toluene	22	0.33
Mg(C$_2$H$_5$)$_2$	2.0	26	13	-	TMSO$_2$	27	0.12
Mg(C$_2$H$_5$)$_2$	5.0	26	5	ZnCl$_2$	THF	25	0.15
Mg(C$_2$H$_5$)$_2$	2	26	13	(C$_2$H$_5$)$_3$Al (f)	Dioxane	87	0.084
Mg(C$_2$H$_5$)$_2$	2	26	13	(C$_2$H$_5$)$_3$Al	Toluene	10	0.075
Mg(n-C$_4$H$_9$)$_2$	1.0	26	26	-	Dioxane	27	0.53 (e)
Mg(n-C$_4$H$_9$)$_2$	2	26	13	(C$_2$H$_5$)$_3$Al	Dioxane	47	0.036
Mg(n-C$_4$H$_9$)$_2$	2	26	13	(C$_2$H$_5$)$_3$Al (f)	Dioxane	63	0.17
(n-C$_4$H$_9$)$_2$Mg, Al(C$_2$H$_5$)$_3$	1,5	26	26	-	Dioxane	45	0.11.
(n-C$_4$H$_9$)$_2$Mg, Al(C$_2$H$_5$)$_3$	2,5	26	13	-	Dioxane	49	0.12
Na⊕ ⊖ Al(C$_5$H$_5$)$_3$(S-i-C$_3$H$_7$)	0.5	13	26	-	Dioxane	23	0.243
Na⊕ ⊖ Al(C$_5$H$_5$)$_3$(S-i-C$_3$H$_7$)	0.5	13	26	-	TMSO$_2$	0	-

0.5	13	26	-	-	Dioxane	91	0.20
1	13	13			THF	23	0.39 (e)

a. 1:1 Lewis acid: Monomer ratio.
b. All runs 5 ml solvent.
c. Viscosities at 30°, 0.1%, DMF.
d. NC_5H_{10} = N⟨⟩, DMF = N,N-dimethylformamide, DMSO = dimethylsulfoxide, $TMSO_2$ = tetramethylene sulfone.
e. Viscosity done in DMSO.
f. 5 M moles Lewis acid.

This reaction was confirmed by isotopic labelling. When cyclobutanecarbonitrile was in excess, the product, even under mild conditions, was dicyclobutyl ketone ($v_{C\equiv O}$ 1700 cm^{-1}) formed by a rapid reaction of carbanion with the CN group (Thorpe-Ziegler reaction).

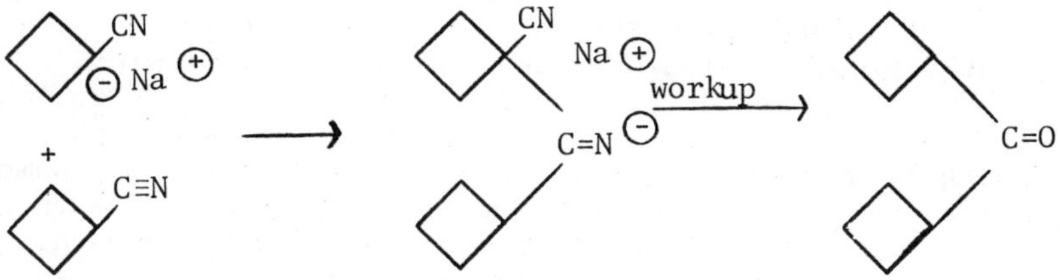

The anionic polymerization of 1a is therefore analogous to the anionic polymerization of methacrylonitrile, whose termination also involves Thorpe-Ziegler reaction with formation of carbonyl groups (17,18). For methacrylonitrile, termination involves back-biting of the growing carbanion onto a cyano group in the same polymer chain, with the formation of a six-membered ring. For bicyclobutane-1-carbonitrile, six-membered ring formation is sterically implausible and the reaction is intermolecular, as demonstrated in the model reaction. Lower yields of polymer in homogeneous solution (DMF, sulfolane), can be attributed to the higher concentrations of available CN groups. Also, more free ions, which are more reactive and less discriminating, are present in such solvents.

<u>Synthesis of a Polysubstituted Bicyclobutane</u> - To broaden our collection of bicyclobutane monomers, we have explored the utility of zwitterionic cycloaddition reactions to synthesize the required cyclobutane precursors. Such reactions of electron-rich olefins with electron-poor olefins were studied by Brannock and coworkers (19). To obtain substituents at the proper locations on the cyclobutane ring, we used trisubstituted electrophilic olefins (20) in the following reaction sequence:

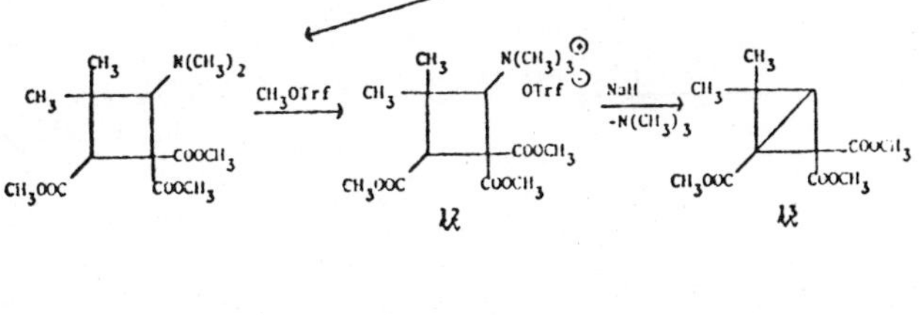

Polysubstituted bicyclobutanes

Trimethyl 4,4-dimethylbicyclobutane-1,2,2-tricarboxylate 13, a white crystalline solid obtained by this route, proved to be unreactive in either homo- or copolymerization by radical or anionic routes. Therefore too many substituents on the bicyclobutane ring render the nucleus unpolymerizable, undoubtedly due to steric hindrance.

We attempted to lessen the degree of steric hindrance by carrying out an analogous series of reactions beginning with N,N-dimethylpropenylamine or with N,N-dimethylvinylamine. The former could be carried through to trimethyl-4-methylbicyclobutane-1,2,2-tricarboxylate 14, but not in the necessary purity. N,N-dimethylvinylamine, when allowed to react with tricarbomethoxyethylene, gave no cyclobutane (21). We conclude that the ring methyl substituents which favor formation of the required cyclobutane intermediate 11 cyclobutane cause too much steric hindrance in the final monomer 13; but that, when they are absent, cyclobutane formation does not occur.

Bicyclo[2.1.0]pentane Monomers - Earlier we showed that bicyclo[2.1.0]pentane-1-carbonitrile 2 underwent anionic polymerization (3). Another monomer containing this strained structure has been synthesized recently from 4-chloro-1,2-butadiene and acrylonitrile (22).

We preferred to find another route which avoided the monovinylacetylene required to make 4-chloro-1,2-butadiene (23), and have devised a route based on 2-butyne-1,4-diol:

4-Hydroxyl-1,2-butadiene 17 (24,25) was converted to acetate 18 (26) and cycloadded to acrylonitrile (27). Hydrolysis of cycloadduct 19 gave alcohol 20, which was best purified by conversion to the trimethylsilyl derivative 21, spinning band distillation, and regeneration. It was transformed by thionyl chloride to chloride 15, which was converted to monomer 16 (22). Although longer, this synthesis can be readily scaled up and gave comparable yields.

Monomer 16 underwent free radical polymerization in 59% yield to give polymer with the rearranged structure 22.

The expected 1,5-polymerization to give structure 24 was immediately excluded because the nmr and cmr spectra of the polymer showed the presence of methyl groups on a double bond but no vinyl protons. The 1,3-polymerized structure 25 was similarly excluded. The ^{13}C-H coupling constants, as determined by gated decoupling, favored structure 22 over structure 23. Carbons 4 and 5 of 22 should have significantly different ^{13}C-H coupling constants, as found, whereas carbons 9 and 12 of 23 would be expected to show almost identical coupling constants (Table II).

Table II. Cmr Spectrum of Poly-3-Methylenebicyclo[2.1.0]pentane-1-carbonitrile 22.

Carbon Assignment	Shift (δ)	Number of Hydrogens	J^{13}C-H(±1.5Hz)
1	40.6	-	-
2	142.3	-	-
3	139.5	-	-
4	41.9	2	142
5	34.3	2	129
6	10.7	3	127
7	122.0	-	-

Our preference for 22 is also supported by the coupling constants found in the following model compounds.

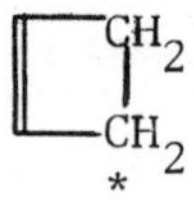

$J^{13}_{C_*H} = 140$ Hz (Ref. 28)

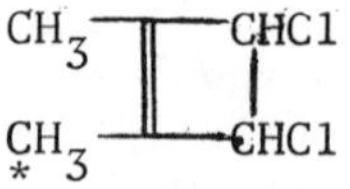

$J^{13}_{C_*H} = 127.3$ Hz (Ref. 29)

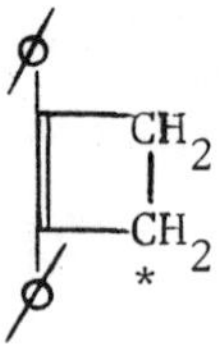

$J^{13}_{C_*H} = 138 \pm 1.5$ Hz

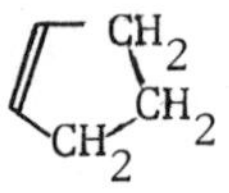

$J^{13}_{C-H} = 130 \pm 1.5$ Hz
$J^{13}_{C-H} = 132 \pm 1.5$ Hz

The difference in the chemical shifts of the methylene carbons 4 and 5 in the cmr is also more supportive of structure 22.

Free radical copolymerizations of monomer 13 with styrene gave the same isomerized structure for the incorporated monomer units. The copolymer was composed of 68% monomer 13 and 32% styrene.

Both the homo- and copolymers were soluble in DMSO, sulfolane and acetone and formed films readily when cast from acetone. Both the homo and copolymers autoxidized at room temperature and yellow rapidly when heated. The autoxidation of cyclobutene-containing polymers has been shown to be very facile (30).

The mechanism of polymerization is believed to be:

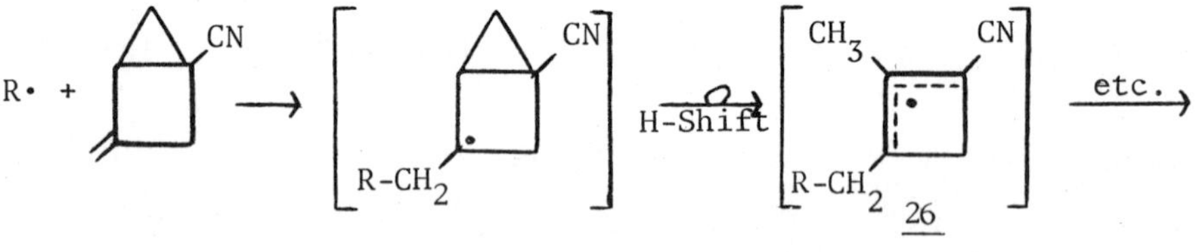

The addition of the growing radical to the double bond is analogous to the first step of 1-5 addition, but apparently conformational restraints caused by the cyclobutane ring prevents normal cyclopropyl ring opening. Instead a hydrogen shift resulting in a resonance-stabilized radical takes place. The possibility of initial thermal isomerization followed by polymerization was excluded because monomer 13 when heated in the presence of DPPH did not isomerize or polymerize.

The anionic polymerization of 13 was briefly examined. When 13 was treated with n-butyllithium, a low yield of oligomers was obtained which appeared to have mostly structure 20. This material was not isomerized to structure 18 thermally or by the presence of free radicals, an additional bit of evidence disfavoring structure 18 for the free radical polymer of 13.

Acknowledgements

Support of this research by the Asahi Electrochemical Co., U. S. Army Research Office, Standard Oil Company of Indiana, The Eastman Kodak Company and the Fulbright-Hayes Foundation is gratefully acknowledged. We also wish to thank Dr. J. C. Kauer of the duPont Co. for help with the cycloaddition reaction of 1-acetoxy-2,3-butadiene and acrylonitrile, and Dr. R. B. Bates for helpful discussions.

Experimental

The infrared spectra were obtained on a Perkin-Elmer 337 grating infrared spectrophotometer using KBr, fluorolube, HCBD, or NaCl plates. Nmr spectra were obtained on a Varian T60 spectrometer, cmr spectra were obtained on a Bruker WH-90 FT. Mass spectra data were measured on a Hitachi Perkin-Elmer RMU-6E double focusing instrument. Gas chromatograms were obtained on a Varian Aerograph 1700 instrument. Elemental analysis was done by Galbraith Laboratories, Inc., or by Chemalytics, Inc.

Reagents - Bicyclobutane-1-carbonitrile 1a (31), trimethyl 3-trimethylammonio-4,4-dimethylcyclobutane-1,2,2-tricarboxylate trifluoromethanesulfonate 12 and trimethyl 3-trimethylammonio-4-methylcyclobutane-1,2,2-tricarboxylate trifluoromethanesulfonate were prepared by literature methods (18). Cyclobutanecarbonitrile, obtained from Ash-Stevens Company, was pure as received.

Anionic Polymerization - In a typical anionic polymerization of bicyclobutane-1-carbonitrile, a 100 ml test tube was filled with nitrogen and 5 ml of solvent. Dialkylmagnesium solution, and 1 ml of co-catalyst were added and allowed to react at 28° for 0.5 hr. Bicyclobutane-1-carbonitrile, 1 ml (14 mmoles), was added to this catalyst solution. After a short while, a vigorous reaction took place and the polymer precipitated. The polymerization reaction was stopped by addition of methanol containing a small amount of hydrochloric acid. The polymer thus obtained was washed with methanol, filtered, and dried in vacuo at 28°. A white powder was always obtained; no fibrous products were observed. To obtain ir spectra free of water, KBr pellets of the polymers were made and the ir taken. The pellet was then heated to 70°C under vacuum over P_2O_5 for 24-48 hrs and the ir again taken. This was repeated until there was no change in the ir spectra. This procedure eliminated ir absorption at $\sim$1600 cm^{-1}.

These polymerization conditions were suitable for the preparation of very high molecular weight polymethacrylonitrile. Polymerization of methacrylonitrile (50 mmoles) with 1 mmole of magnesium di(isopropylmercaptide) in 5 ml of dioxane gave 69, 81 and 97% yields of polymer in separate experiments. Diethyl magnesium gave 75% yield and di-n-butylmagnesium 63% yield. Inherent viscosities ranged between 2.0 and 3.0.

Model Termination Reaction - Tritylsodium was prepared as follows: Under nitrogen, a sodium dispersion (50% in xylene), 2 g, was washed with n-hexane twice, filtered, and added with stirring to a solution of 4.2 g (0.015 moles) of trityl chloride in 80 ml of dry ether. The mixture turned red after several hours and was stirred for 16 hours at 28°. After filtration under nitrogen, the concentrations of tritylsodium was determined by acid titration to be 0.23 N.

To an ether solution of triphenylmethylsodium under nitrogen and cooled to 0°C was added the desired amount of cyclobutanecarbonitrile. Reaction took place quickly and a precipitate was formed. After 1 hour at 0°C, the reaction product was filtered under nitrogen and washed with 5 ml of ether (distilled over Na dispersion and dried over molecular sieves). This washing was repeated until the filtrate was no longer red (usually 5 times). α-Sodiocyclobutanecarbonitrile (or its mixture with the sodio-derivative of dicyclobutylketone or α-cyanodicyclobutyl ketone, in those reactions in which excess cyclobutanecarbonitrile was present) was obtained as a slightly red-colored, very air-sensitive powder. It was redissolved in THF to give a deep red solution (stable with time) and decomposed with 0.5 ml of D_2O. The products were extracted with ether, dried and evaporated. Triphenylmethane was recovered by drying and evaporating the original ether filtrate. 1-Deuterocyclobutanecarbonitrile and dicyclobutyl ketone were collected by preparative gc, and the pure fractions were analyzed by ir, nmr, and mass spectrometry. The ketone was analyzed.

(From workup with H_2O) Calc'd. for $C_9H_{14}O$: C, 78.21; H, 10.21; N, 0. Found: C, 78.36; H, 10.30; N, 0.

(From workup with D_2O) Calc'd. for $C_9H_{13}DO$: C, 77.64; H(D), 10.86; N, 0. Found: C, 77.89; H(D), 10.48; N, 0.

The mass spectrum showed the deuterated sample to be a mixture of monodeuterated (68%), dideuterated (19%), and undeuterated (13%) material. When the ratio of tritylsodium to cyclobutanecarbonitrile was changed from 1:1 to 1:2 the ratio of cyclobutanecarbonitrile to dicyclobutyl ketone went from 20:1 to 1:20.

Several other bases were tried. Tritylithium-TMEDA at 60°C also caused predominantly α-H abstraction from cyclobutanecarbonitrile. Diethylmagnesium showed some α-H abstraction but resulted in mostly addition to the nitrile. The following bases also resulted in addition products: $n-C_4H_9Li$, $n-C_4H_9Li$-TMEDA, $t-C_4H_9Li$, all at -78°C and $(i-C_3H_7)_2NLi$, $(C_5H_{10}N)_2Mg$, at 28°C. The last two also causing recovery of unreacted cyclobutanecarbonitrile. The following bases showed no reaction as determined by deuterium labelling experiments: $(C_6H_5)_3CLi$, $(C_6H_5)_3CLi$ $(C_2H_5)_2Mg$, at 60°C, $[CH_3)_3Si]_2N \ominus Na \oplus$ at -78°C, and $(i-C_3H_7S)_2Mg$ at 28°C.

<u>Trimethyl 4,4-Dimethylbicyclobutane-1,2,2-tricarboxylate</u> - A 250 ml three-necked flask fitted with a mechanical stirrer, a nitrogen inlet and a sintered glass inlet, was heated to remove moisture and cooled under nitrogen. A 57% dispersion of sodium hydride in mineral oil (1.95 g, 0.0465 mole) was washed twice with 25 ml of low boiling petroleum ether. The sodium hydride was covered with 50 ml of tetrahydrofuran dried over molecular sieves. Then 10.0 g (0.0215 mole) of trimethyl 3-trimethylammonio-4,4-dimethyl-1,2,2-cyclobutanetricarboxylate trifluoromethanesulfonate (14) was added at room temperature. The white slurry was heated to gentle reflux for 1 hour, causing evolution of 0.935 l of gas (calc'd. 0.965 l). The reaction mixture was cooled and the supernatent liquid was decanted. The remaining salts and unreacted sodium hydride were washed with a small quantity of tetrahydrofuran. To organic layers was added solid carbon dioxide. The decantation and the addition of CO_2 before the addition of the solution of sodium chloride in water were chosen to minimize the reaction of water with the excess of sodium hydride, because the base which is formed may attack the bicyclobutane already prepared or give products in which the carbomethoxy group has been replaced by a methoxy group (compounds 27 and 28). The organic layer was then poured into a mixture of ice, ether, and a saturated solution of sodium sulfate in water and shaken. The ether layer was separated, and the water layer was extracted twice with 25 ml of ether. The combined ether extracts (200 ml) were back-washed with a small amount of water, dried with stirring over magnesium sulfate during 10 minutes, filtered and rotary evaporated (bath temperature 30°C). Evaporation for 60 minutes at 0.2 mm Hg gave a yellow oil, 2.63 g, (48%) which by gas chromatography was shown to have three peaks.

Crystallization from diethyl ether at -50° gave a 34% yield of trimethyl 4,4-diemthylbicyclobutane-1,2,2-tricarboxylate as white crystals; melting point 45-46°C (lit. m.p. 44-46°C). A mixed melting point with an authentic sample was not depressed.

Anal. Calc'd for $C_{12}H_{16}O_6$: C, 56.25; H, 6.29; O, 37.46. Found: C, 56.16; H, 6.31; O, 37.61. Nmr (CDCl₃, TMS, τ) 6.30 (s, 6H, COOMe) 6.35 (s, 3H, COOMe) 6.75 (s, 1H, bridgehead) 8.55 (s, 3H, CH₃) 8.95 (s, 3H, CH₃) ir (KBr, cm⁻¹) 1710 (COOMe). Mass spectrum: parent peak at 256 (calc. = 256).

The other two compounds isolated by preparative gas chromatography, were:

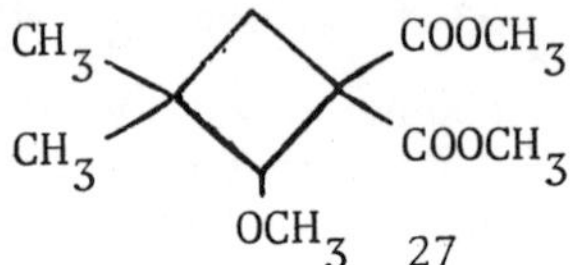

[Structure 27]

Anal. Calc'd for $C_{11}H_{18}O_5$: C, 57.38; H, 7.88; Found: C, 57.47; H, 7.70. Nmr spectrum (CDCl₃, TMS, τ) 6.25 (s, 6H, COOMe) 6.65 (s, 3H, OCH₃) 8.65 (s, 3H, Me) 6.15-6.85 (m, 3H) 9.00 (s, 3H, Me); ir (liquid, cm⁻¹) 1710 (COOMe); mass spectrum: parent peak: 231 (calc'd for $C_{11}H_{18}O_5$ = 230).

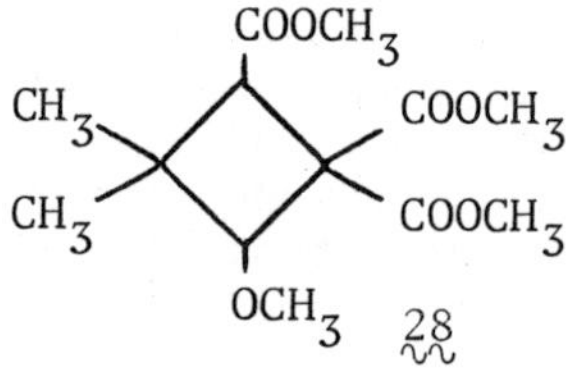

[Structure 28]

Anal. Calc'd for $C_{13}H_{20}O_7$: C, 54.16; H, 6.99; N, 0; Found: C, 54.52; H, 6.89; N trace: nmr spectrum, (CDCl₃, TMS, τ) 5.70 (s, 1H) 6.30 (t, 9H, COOMe) 5.74, 6.60 (s, 3H) 8.80 (s, 3H, Me) 6.55, 8.90 (s, 3H, Me); ir KBr, cm⁻¹) 1710 (COOMe); mass spectrum: parent peak: 288 (calc'd for $C_{13}H_{20}O_7$ = 288).

These products may result from reaction of bicyclobutene 15 with OCH₃⁻, perhaps formed by a reaction of H⊖ with COOCH₃ groups.
Similar results were obtained using 1,2-dimethoxyethane as solvent. Using N-methylpyrrolidone caused higher yields of compounds 27 and 28. Lithium hydride resulted in lower yields of 15 in both tetrahydrofuran and N-methylpyrrolidone. The use of sodium methoxide gave only compound 27 and no bicyclobutane.

Trimethyl-4-Methylbicyclobutane-1,2,2-Tricarboxylate - In a 50 ml 3-necked flask containing a stirring bar and fitted with a reflux condenser leading to a nitrogen inlet were placed, after drying, 0.42g of a 57% sodium hydride dispersion in mineral oil

(10 mmol); the hydride was washed twice with 15 ml of pentane. To the clean hydride was added a solution of trimethyl 3-trimethylammonio-4-methylcyclobutane-1,2,2-tricarboxylate trifluoromethanesulfonate (2.26g, 5 mmol) in N-methylpyrrolidone (10 ml) and the nitrogen inlet was replaced by a wet test meter. After 25 minutes 59 ml (50% of theoretical) of gas had evolved, and the reaction mixture was poured into a mixture of ice water-ether. The ether was washed twice by 40 ml of water, dried with stirring for 5 minutes over magnesium sulfate, filtered and evaporated. The remaining oil consisted mainly of the desired bicyclobutane. Nmr (CDCl$_3$) τ6.27, 6.30 and 6.32 (3s, 9, COOMe), 6.77 (d, J=1.5Hz, 1, bridgehead), 8.21 (d of q, J = 1.5Hz and 5.5Hz, 1, ring proton) and 8.63 (d, J = 5.5Hz, 3, Me). From the coupling constant between the methyl and the ring and the bridgehead protons the exo configuration is assigned to the methyl group. Attempted final purification by recrystallization, distillation and preparative gas chromatography did not succeed.

4-Chloro-2-butyne-1-ol - The procedure of Bailey and Fujiwara (24) but replacing benzene with dichloromethane, gave a 50% yield, b.p. 70°C (20 mm Hg) from technical grade 2-butyne-1,4-diol (Aldrich).

1,2-Butadien-4-ol - The procedure of Bailey and Pfeifer (25) was used with modifications. To 1800 ml of anhydrous ethyl ether was added 42 gm of lithium aluminum hydride (1.1 mole). The reaction was cooled in an ice bath with stirring and 155.24 gm of 4-chloro-2-butyne-1-ol (1.48 mole) in 300 ml of anhydrous ethyl ether was added dropwise over a period of 3 hrs. The reaction was then allowed to warm to room temperature and stirred for 18 hrs. The excess lithium aluminum hydride was destroyed by carefully adding dropwise a saturated solution of sodium sulfate in water until the lithium-aluminum salts formed white pellets. The mixture was filtered and the pellets washed with ether. The ether was rotary evaporated and the residue distilled to give a 68% yield of 1,2-butadien-4-ol (79.8 gm). b.p. 69°C (45 mm Hg).

1-Acetoxy-2,3-butadiene - The procedure of W. H. Carothers (26) was used to convert the alcohol to the acetate.

1-Acetoxy-2,3-butadiene/Acrylonitrile Cycloaddition - The procedure of H. K. Hall, Jr., and R. E. Yancy (22) was used with the convenient modification of using liquid nitrogen instead of a Dry Ice-acetone bath in the degassing procedure. For example, 9.39 gm of acrylonitrile, 4.96 gm of 1-acetoxy-2,3-butadiene, 8.40 gm of benzene, and 0.50 gm of 2,5-di-tert-butylhydroquinone were placed in an acid washed Pyrex tube, the tube was degassed and heated to 200°C at the University High Pressure Laboratory for 8 hrs. The reaction was allowed to cool, poured into 100 ml of ethyl ether and filtered. The ether was rotary evaporated and the residue distilled to give cycloadducts identical to those reported by Cripps (27). The 47.9% yield is less than that reported by Cripps, probably because of the greater difficulty in maintaining a uniform temperature and the use of benzene as a diluent, which is necessary for reasons of safety (32). The use of the glass tube does not seem to have any advantage over an autoclave.

(2,3-Butadienyloxy)trimethylsilane - Hexamethyldisilazane 25g (0.15 mole), and 21g (0.30 mole) of 1,2-butadiene-4-ol were mixed for 1 hr, during which time the temperature rose to 60°C. Chlorotrimethylsilane, 0.5 ml, was added and the mixture stirred for 0.5 hr more. The mixture was then heated to 120°C. (This was found helpful in preventing uncontrollable reactions in the cycloaddition step). The compound was then filtered and distilled at 60°C (60 mm Hg) to obtain 31.96g (96%) of product.

Cycloadditions of Substituted Allenes with Acrylonitrile - With isochloroprene, the 3-methylene-2-chloromethylcyclobutanecarbonitrile 15 and 1-chloro-1-cyclohexene-4-carbonitrile mixture was obtained in 21% yield, and 3-chloropropionitrile in 15% yield. With 1,2-butadiene-4-ol, 3-(β-hydroxyethylidene)-1-cyclobutanecarbonitrile, cis and trans-3-methylene-2-hydroxymethyl-cyclobutanecarbonitrile 20 were obtained in 29.7% yield after distillation on a spinning band column at 60-70°C (0.01 mm Hg). The two compounds can be separated by preparative gc or by silylation, followed by careful distillation.

Anal. Calc'd for C_7H_9ON: C, 68.27; H, 7.37; N, 11.37.
Found: C, 67.97; H, 7.80; N, 11.58.

In the case of (2,3-butadienyloxy)trimethylsilane, the material was distilled on a spinning band column twice to yield two main fractions in an overall yield of 47%. The first fraction w was found to be 21 by NMR, IR and mass spectra (43°C at 0.01 mm Hg).

Anal. Calc'd for $C_{10}H_{17}ONSi$: C, 61.49; H, 8.77; N, 7.17.
Found: C, 61.28; H. 8.81: N. 7.36.

The second fraction (60°C at 0.01 mm Hg) was found to be the silyl ether of 3-(β-hydroxyethylidene-1-cyclobutanecarbonitrile by nmr, ir and mass spectra. Anal. Calc'd for $C_{10}H_{17}ONSi$: C, 61.49; H, 8.77. Found: C, 61.48; H, 8.75. In two out of sixteen runs during this cycloaddition, uncontrollable exothermic reactions occurred.

The method of Cripps is not usable with isochloroprene due to the known metal-catalyzed rearrangement of the compound (23, 26). The presence of a stainless steel surface also resulted in no cycloadduct with (2,3-butadienyloxy)trimethyl silane and in lower yields with 2,3-butadien-1-ol. Silyl compounds have been reported to interact with stainless steel (33).

3-Methylene-2-chloromethylcyclobutanecarbonitrile 15 - Silyl compound 21, 2g, was hydrolyzed in 95% ethanol by the addition of a trace of base. The resulting alcohol 20 was isolated and dried over $MgSO_4$ in ether. The alcohol was then converted to the chloride 15 by the dropwise addition of purified thionyl chloride in 80% yield. This was converted to monomer 16 as before (22).

Free Radical Polymerization - 0.1483g of 3-methylene-[2.1.0]-bicyclopentane-1-carbonitrile 16 and 0.5 ml of sulfolane were placed in a dried flask along with a crystal of AIBN or benzoyl peroxide. The flask was fitted with a rubber septum, flushed with argon, and heated at 70°C for 15 hrs. The solution was then poured into methanol, containing a trace of 2,5-di-tert-butyl-hydroquinone, filtered, dissolved in acetone, reprecipitated in ethyl ether, filtered, and dried over P_2O_5 under full vacuum at room temperature, after flushing with nitrogen. The resulting polymer was obtained in 59% yield, η_{inh} = 0.95 at 30°C (0.396g/100cc in acetone). Flexible films were cast from acetone.

Anal. Calc'd for C_7H_7N: C, 79.98; H, 6.71; N, 13.32.
Found: C, 80.04; H, 6.79; N, 13.25.

The copolymerization of styrene and 3-methylene bicyclo-[2.1.0]pentane-1-carbonitrile 16 was done in the same manner using 0.3160 gm of 3-methylene-bicyclo[2.1.0]pentane-1-carbonitrile (3.006 x 10^{-3} moles), 0.3166 gm of freshly distilled styrene (3.040 x 10^{-3} moles) in 1.75 ml of sulfolane with AIBN at 70°C for 23 hrs. After workup and drying, 0.3952 gm of polymer was obtained which showed the presence of both monomer units was confirmed by ir, nmr, and cmr. η_{inh} = 0.34 at 30°C(0.445g/100cc).

Analysis: C = 81.91; H = 7.04; N = 8.95.

The analysis is low due to autoxidation. Analysis when extrapolated to 100% and nmr integration indicates a ratio of 2.2 to 1 of bicyclo monomer to styrene monomer incorporated into the polymer.

Anionic Polymerization of 3-Methylene[2.1.0]Bicyclopentane-1-Carbonitrile - The same procedure as used on [2.1.0]bicyclopentanecarbonitrile 2 (3) was used to give a 20% yield of yellow oil, which was analyzed by nmr to give the results mentioned in the Discussion. When [2.1.0]bicyclopentanecarbonitrile was used, polymer was formed in 81% yield with η_{inh} 0.24 (DMF), as described previously (3).

Literature Cited

1. Professor of Nutrition and Food Sciences, University of Ariz. Tucson, Arizona 85721.
2. Hall, Jr., H. K.; and Ykman, P.; J. Polym. Sci., Macro Reviews (1976), 11, 1.
3. Hall, Jr., H. K.; Macromol. (1971), 4, 139.
4. Closs, G. L.; Kaplan, L. R.; and Bendall, V. I.; J. Am. Chem. Soc. (1967), 89, 3376.
5. Koeberg-Telder, A.; and Cerfonfain, H.; J.C.S., Perkin II, 1974, 1206.
6. Nametkin, N. S.; Finkelshtein, E. Sh.; Yatsenko, M. S.; Portnykh, E. B.; Vdovin, V. M.; Vysokomol. Soedin.; Ser. B (1973), 15, 868. Chem. Abst., (1974), 81, 64012k.

7. Wiberg, K. B.; Upton, Jr., E. C.; and Burgmaier, G. J.; J. Amer. Chem. Soc. (1969), 91, 3372.
8. Gassman, P. G.; Topp, A.; Keller, J. W.; Tetrahedron Letters, 1969, 1093.
9. Scott, W. B.; Pincock, R. E.; J. Amer. Chem. Soc. (1973), 95, 2040.
10. Pincock, R. E.; Schmidt, J.; Scott, W. B.; Can. J. Chem. (1972), 50, 3958.
11. Takahashi, T.; J. Polym. Sci., A-1, (1968), 6, 403.
12. Lishanskii, I. S.; Linogradova, N. D.; Guliev, A. M.; Zak, A. C.; Zvyagina, A. B.; Kol'tav, A. I.; Fomina, O. S.; Khachaturov, A. S.; Sin., Str. Svoistva, pulm., 1970, 35: Chem. Abst. (1972) 76, 113671g.
13. Ohara, O.; Aso, C.; and Kunitake, T.; Nippon Kagakukuishi, 1973, 602.
14. Joh, Y.; Yoshihara, T.; Kotake, Y.; Imai, Y.; and Kurihara, S.; J. Polym. Sci., A-1, (1967), 5, 2503
15. Joh, Y.; Kurihara, S.; Sakurai, T.; Imai, Y.; Yoshihara, T.; J. Polym. Sci., A-1, (1970), 8, 377.
16. Blanchard, Jr., E. P.; and Cairncross, A.; J. Amer. Chem. Soc., (1966), 88, 487.
17. Joh, Y.; Hoshihara, T.; Kurihara, S.; Sakurai, T.; and Tomita, T.; J. Polym. Sci., A-1, (1970), 8, 1901.
18. Tsvetanov, C.; and Panazotov, I.; European Polymer J. (1975), 11, 209.
19. Brannock, K. C.; Bell, A.; Burpitt, R. D.; Kelly, C. A.; J. Org. Chem. (1964), 29, 801.
20. Hall, Jr., H. K.; Ykman, P.; J. Amer. Chem. Soc. (1975), 97, 800.
21. Glogowski, M. E., unpublished results.
22. Hall, Jr., H. K.; and Yancy, R. E.; J. Org. Chem. (1974), 39, 3862.
23. Carothers, W. H.; Berchet, G. J.; Collins, A. M.; J. Amer. Chem. Soc. (1932), 54, 4066.
24. Bailey, W. J.; and Fujiwara, F.; J. Amer. Chem. Soc. (1955), 77, 165.
25. Bailey, W. J.; and Pfeifer, C. P.; J. Org. Chem. (1955), 20, 1337.
26. Carothers, W. H.; and Berchet, G. J.; J. Amer. Chem. Soc. (1933), 55, 2807.
27. Cripps, H. N.; Williams, J. K.; and Sharkey, W. H.; J. Amer. Chem. Soc. (1959), 81, 2723.
28. Hill, E. A.; and Roberts, J. D.; J. Amer. Chem. Soc. (1967), 89, 2047.
29. Hüther, H.; and Brune, H. A., Org. Magn. Res. (1971), 3, 737.
30. Wu, C. C., and Lenz, R. W.; J. Polym. Sci: Polym. Chem. Ed. (1972), 10, 3555.
31. Hall, Jr., H. K.; Blanchard, Jr., E. P.; Cherkufsky, S. C.; Sieja, J. B.; and Sheppard, W. A.; J. Amer. Chem. Soc. (1971), 93, 110.
32. Kauer, J. C., private communication.
33. "Handbook of Silylation", p. 7, Pierce Chemical Company, Rockford, Ill., 1970.

New Polymers by Ring-Opening Polymerization of Norbornene Derivatives with Polar Substituents

S. MATSUMOTO, K. KOMATSU, and K. IGARASHI

Tokyo Research Laboratory, Japan Synthetic Rubber Co., Ltd., Kawasaki, Japan

The ring-opening polymerization of cyclic unsaturated hydrocarbons has been studied extensively(1-4), following the first report by Eleuterio(5) and subsequent discovery of homogeneous catalyst by Natta et al. (6). Typical catalyst is composed of a tungsten or a molybdenum compound and an organometallic compound. Now it is generally accepted that this polymerization involves the same type of intermediates as that of the olefin metathesis (olefin disproportionation) which was first reported by Banks et al.(7) and later extended to homogeneous system by Calderon et al.(8). The polymerization proceeds through ring-cleavage at the carbon-carbon double bond. Polymers of halogenated cyclic hydrocarbons have been prepared by this type of catalyst.

Highly strained cyclic olefins like cyclobutene or norbornene have also been polymerized in alcoholic solvents or in aqueous emulsion by using, as catalyst, ruthenium, iridium or osmium salts, which are not active for the polymerization of less strained monomers like cyclopentene or cyclooctene(9). In addition, it has been shown that these noble metal catalysts induce the ring-opening polymerization of norbornene derivatives substituted by polar groups(10-15). Monomers with ester, ether, carboxyl, hydroxyl, halogen and imide groups have been successfully polymerized. However, these catalysts were reported to be incapable of polymerizing the nitrile-substituted derivative(11).

In the course of the studies on the behavior of unsaturated compounds containing polar groups toward the metathesis catalyst(16), we have found that this type of catalyst can induce efficiently the ring-opening polymerization of norbornene derivatives with various polar substituents, including nitrile group. The polymerization of ester, nitrile, pyridyl and acid anhydride derivatives have been independently disclosed by Hepworth et al.(17) and by Ueshima et al.(18,19). The metathesis reactions of ester group-containing olefins(cyclic and acyclic) have also been reported recently(20-23).

The present paper reports the ring-opening polymerization of norbornene derivatives substituted by nitrile, amide, imide, ester, pyridyl and acid anhydride groups by the above-mentioned catalysts. The polymerization behavior of these monomers and the physical properties of the polymers of norbornenenitriles will be described.

Results and Discussion

Polymerization of Various Norbornene Derivatives.

The polymerizability of norbornene derivatives substituted by nitrile, amide, imide, ester, pyridyl, acid anhydride, ketone and aldehyde groups were examined by using the tungsten-based binary and ternary catalyst systems. The results are summarized in TABLE I.

Besides the ester and imide derivatives that are known to be polymerized by the noble metal catalysts, nitrile, amide, pyridyl and acid anhydride derivatives were found to undergo polymerization by the present catalyst systems. The ester derivatives were easily polymerized by the binary catalyst to a high conversion at the catalyst-to-monomer molar ratio as low as 10^{-4}. The nitrile derivatives were less reactive but could be polymerized to a high conversion by the ternary catalyst activated by a third component. The other derivatives were polymerized much less efficiently even by the ternary catalyst. These differences in reactivity are presumed to reflect the relative strength of the interaction of the active catalyst species with the norbornene double bond on one hand and with the polar substituent on the other.

The ketone and aldehyde derivatives, on the other hand, were not polymerized by the catalyst systems examined. As will be mentioned below, a large amount of ketones and aldehydes completely inactivates the catalyst.

All the polymers obtained were analyzed to be the products through ring-opening polymerization by means of ir and ^{1}H-nmr spectroscopy. Figure 1 shows the ^{1}H-nmr spectra of three polymers. All of them have an unresolved peak around 5.3 ppm(from tetramethylsilane), which can be assigned to the protons attached to acyclic C=C bond. The relative areas of the peaks in the spectra were found to be consistent with the structure

TABLE I. Polymerization of Various Norbornene Derivatives.

Monomer	Catalyst system	Polymer yield %	(η) dl/g	Tg (DSC) °C
⟨CN⟩	WCl_6-paralde- (1:1.5:3) hyde-Al(isoBu)$_3$	100[1]	0.66[7]	140
⟨CN, CH$_3$⟩		100[2]	0.54[7]	165
⟨COOMe⟩	WCl_6-AlEt$_3$ (1:3)	100[3]	3.16[8]	62
⟨OAc⟩	WCl_6-AlEt$_3$ (1:3)	85[4]	0.65[8]	77
⟨CONMe$_2$⟩	$W(OPh)_6$-AlEt$_3$ (1:2)	84[5]	–	114
⟨N-pyridyl⟩	WCl_6-AlEt$_3$ (1:3)	25[6]	–	–
⟨CO-O-CO (anhydride)⟩	WCl_6-MASC[9] (1:3)	10[5]	–	–
⟨CO-NPr^n-CO⟩	WCl_6-AlEt$_3$ (1:2)	7[5]	–	> 200
⟨CO-NH-CO⟩	WCl_6-AlEt$_3$ (1:2)	8[5]	–	> 250
⟨COMe⟩	WCl_6-AlEt$_3$ (1:2)	0[5]	–	–
⟨CHO⟩	WCl_6-AlEt$_3$ (1:2)	0[5]	–	–

1) Polymerization at 60°C for 4 hr. in 1,2-dichloroethane; 1 mole% of 1-hexene was added; the monomer-to-catalyst molar ratio(M/W) was 1000. 2) Polymerization at 70°C for 4 hr. with 1.4 mole% of 1-hexene. 3) Polymerization at 30°C for 17 hr. in chlorobenzene; M/W,4000. 4) Polymerization at 25°C for 6 hr. in chlorobenzene; M/W, 1000; 5 mole% of 1-heptene. 5) Polymerization at 70°C for 17 hr. in chlorobenzene; M/W, 200. 6) Polymerization at 70°C for 17 hr. in ethyl acetate; M/W, 500. 7) Measured in chloroform at 30°C. 8) Measured in toluene at 30°C. 9) MASC=$AlMe_{1.5}Cl_{1.5}$.

(I) formed through ring-opening polymerization.

$$\text{[structure with X, Y on bicyclic]} \longrightarrow \left[\underset{Y \;\; X}{\text{cyclopentane}} \text{-CH=CH-} \right]_n \quad (I)$$

The ir spectra of the polymers from imide, pyridyl and acid anhydride derivatives were similar to those of the polymers from nitrile and ester derivatives and had an absorption at 970 cm^{-1} characteristic of C=C linkage, indicating the structure (I). The polymers were resinous and hard materials. Their glass transition temperatures(Tg) are given in TABLE I.

Catalyst. Two types of catalysts were used in the present study. The one is the so-called Ziegler type catalyst composed of a tungsten, molybdenum or rhenium compound (A component) and an organometallic compound (B component). The other type of catalyst is composed of a carbonyl-carbene complex of tungsten and a Lewis acid(24).

TABLE II gives the results of the polymerization of several monomers by various catalyst systems of the first type. As the A component, tungsten compounds are more effective than molybdenum compounds. The activity of the rhenium-based catalysts was much lower than those of the tungsten- or molybdenum-based systems. Besides the compounds of the elements shown in TABLE II, organometallic compounds of lithium, sodium, magnesium, zinc, boron and germanium are effective as the B component as shown in TABLE III. Some metal hydrides are also effective. The non-organometallic system, H_2WO_4-$AlCl_3$, which had considerable activity for the metathesis reaction (polymerization) of cyclic and acyclic unsaturated hydrocarbons, has practically no activity for the present polymerization.

As in the polymerization of monomers without polar substituent, the catalytic activity was markedly enhanced by the addition of a third component (C component) such as alcohols, peroxides, hydroperoxides and epoxides. We have found that ketones, aldehydes and polymerization product of aldehydes are particularly effective as the C component for this type of catalyst. These compounds inactivate the catalyst in large amounts as shown in Figure 2. Certain metal compounds such as $Ti(OR)_4$, $FeCl_3$, chromium acetylacetonate and $Al(OR)_3$ were also effective as the C component.

TABLE IV summarizes the results of the polymerization by the second type of catalyst. The ionic (II) and non-ionic (III) complexes of tungsten were used in

TABLE II Polymerization of Norbornene Derivatives by Various Catalyst Systems: polymer yield (%). [1]

Monomer	X=CN	X=CONMe$_2$	X=COOMe	X=2-Pyridyl
W				
WCl$_6$-AlEt$_3$ (1:2)	29	71	99	16
WCl$_6$-AlEt$_3$-acetone (1:3:6)	100 [2]	–	100 [3]	–
W(OPh)$_6$-AlEt$_2$Cl (1:2)	15	5	95	0
WCl$_6$-Et$_3$SiH (1:4)	0.1	0.1	85	–
WCl$_6$-Ph$_3$Sb (1:10)	1	0	81	–
WCl$_6$-Me$_4$Sn (1:1.3)	38	1	100	0
H$_2$WO$_4$-AlCl$_3$ (1:3.2)	0	–	0	–
Mo				
MoCl$_5$-AlEtCl$_2$ (1:4)	0.2	4	49	–
Mo(OEt)$_2$Cl$_3$-AlEt$_3$ (1:2)	6	0.1	20	–
MoO$_2$(acac)$_2$-AlEtCl$_2$ (1:4)	11	–	27	–
Re				
ReCl$_5$-AlEt$_3$ (1:2)		1	0.6	–
ReCl$_5$-Me$_4$Sn (1:1.3)	–	0.1	2	–

1) Polymerizations were carried out in chlorobenzene at 70°C for 17hr: the monomer-to-catalyst molar ratio, 200. 2) Polymerization in 1,2-dichloroethane at 50°C for 4 hr: monomer-to-catalyst molar ratio, 1000. 3) Polymerization in toluene at 30°C for 15 min: monomer-to-catalyst molar ratio, 1000.

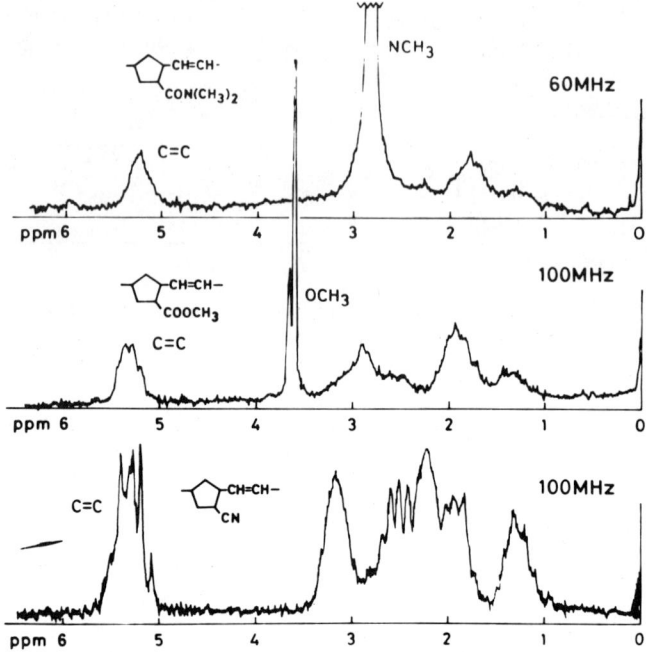

Figure 1. ¹H-NMR spectra of the polymers obtained from nitrile-, ester-, and amide-substituted norbornene derivatives: measured in $CDCl_3$ at room temperature

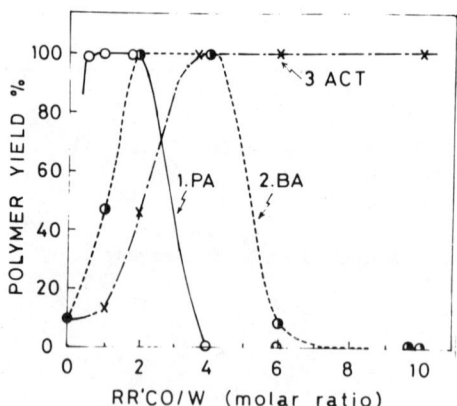

Figure 2. The effect of the addition of paraldehyde (PA), n-butyraldehyde (BA), and acetone (ACT) in the polymerization of 5-norbornene-2-nitrile by the WCl_6–Al-$(isoBu)_3$ system: polymerization in 1,2-dichloroethane at 25°C for 4 hr; the monomer-to-WCl_6 molar ratio was 1000

TABLE III Effect of Cocatalyst(B component) in the Polymerization by Binary Systems, WCl_6-Cocatalyst.

Polymer yield (%).

Cocatalyst	B/WCl_6 molar ratio	Monomer	
		⟨CN⟩	⟨COOMe⟩
n-BuLi	2	–	0.1
C_5H_5Na	2	–	6
CH_3MgBr	2	0.1	91
$ZnEt_2$	2	0.1	26
BEt_3	2	0	1
$B(n-Bu)_3$	2	0	1
$AlEt_3$	2	29	99
Et_3SiH	4	0.1	97
$Ge(CH_3)_4$	3	0	7
PEt_3	2	–	0
$LiAlH_4$	ca. 20	0.1	29
$NaBH_4$	ca. 20	0	24

Polymerizations were carried out in chlorobenzene at 70°C for 17 hr. The monomer-to-catalyst molar ratio was 200.

$[(CH_3)_4N][(CO)_5WCOPh]$ (II) $(CO)_5WC(OEt)Ph$ (III)

the present study. As the Lewis acid component, only titanium tetrahalide gave high catalytic activity. It is interesting to note that the relative effectiveness of the two complexes depends on the kind of monomer to be polymerized. For the nitrile derivative, the complex (II) is more effective than the complex (III) while the order is reversed for the ester derivatives. The reason for this phenomenon is not clear at this moment.

The enhancement of the catalytic activity by the addition of the third component was observed also with this type of catalyst. Effective third components are tertiary phosphines, sulfides, sulfoxides, quinones and N-chlorosuccinimide. Alcohols, ketones and aldehydes, on the other hand, deactivated the catalyst even in

TABLE IV Polymerization by the W Carbene Complex-Based Catalysts.[1]

Polymer yield (%).

Monomer	CN-norbornene	COOX-norbornene X = Me	COOX-norbornene X = C₆H₂Br₃
Monomer/catalyst molar ratio	1000	5000	2000[2]
[(CO)$_5$WCOPh][NMe$_4$] (II)			
(II) - TiCl$_4$	12	3	2
(II) - TiCl$_4$ - X			
X = PPh$_3$	89	–	12
PEt$_3$	50	–	–
MeSOMe	41	–	–
tBuSBut	36	–	–
Anthraquinone	48	–	–
[CO,CO]N-Cl	70	–	–
Paraldehyde	0.2	–	–
Benzoyl peroxide	5	–	–
Benzophenone	3	–	–
(CO)$_5$WC(OEt)Ph (III)			
(III) - TiCl$_4$	3(100°C,6)	100	27
(III) - TiCl$_4$ - PPh$_3$	11	100	64
[CO,CO]N-Cl	47	–	–

1) Polymerizations were carried out in bulk at 70°C for 16 hr. The Ti/W ratio was 5 and the third component/W ratio was 1.5. 2) Polymerization in dichloroethane solution.

small amounts. Furthermore, the three components had to be mixed in the absence of monomer and the order of mixing did not significantly affect the catalytic activity. In contrast, in order to attain a high activity with the first type of catalyst, the A and the C components should be reacted prior to mixing with monomer or the B component, the former being added first. These observations suggest that the mechanism of the catalyst activation by the third component in the first system is different from that in the second system.

<u>Influence of Polar Functional Groups and Acyclic Double Bond on Polymerization.</u> The polymerization can be carried out in aprotic polar solvents such as ether, tetrahydrofuran, ethyl acetate, acetonitrile and dimethylformamide as well as in hydrocarbons and halogenated hydrocarbons. Alcohols, ketones and sulfoxides strongly inhibited the reaction.

The molecular weight of the polymer product is reduced by the addition of acyclic unsaturated compounds. Olefins and non-conjugated diolefins reduced the molecular weight without appreciable effect on the rate of polymerization, terminal olefins being more effective than internal olefins. Allylic compounds with polar functional groups exerted a similar effect. Conjugated diolefins were less effective and caused reduction of catalytic activity in large amounts. Cyclopentadiene, which retards the polymerization of cyclopentene even at low concentration, did not exert significant effect in the present polymerization. Acetylenes and allenes caused reduction of molecular weight at low concentration (ca. 0.1 - 0.5 mole-% for monomer) and deactivated the catalyst at higher concentrations. Acrylic esters, acrylonitrile and maleic acid esters deactivated the catalyst without significant effect on the molecular weight.

The polymerization and the reaction with acyclic unsaturated compounds are considered to take place on the coordination sites of the active catalyst species. Therefore, the effect of various compounds described above reflect, at least in part, their relative ability of coordination to the active species. The effect of acetylenes and allenes as well as that of α,β-unsaturated esters and nitriles may reasonably be explained by their stronger power of coordination in com comparison with that of the norbornene double bond. The latter in turn coordinates more strongly than unconjugated polar group as indicated in the above results.

Inactivation of the catalyst by ketone and aldehyde groups is presumed to be due to a specific interaction of these functional groups with the active species. Deoxygenation of ketones and aldehydes by low valent tungsten species has been reported(25).

Copolymerization. Copolymers could be prepared from the mixtures of the norbornene derivatives. They could also be copolymerized with cycloolefins without polar group such as cyclopentene or cyclooctene. The formation of the copolymer was confirmed by elemental analysis, ir spectrum, thin layer chromatography(TLC) and differencial scanning calorimetry(DSC).

Furthermore, the norbornene derivatives could be polymerized in the presence of unsaturated polymers like polybutadiene or styrene-butadiene copolymer(SBR) to give block or graft copolymers. The formation of the block(graft) copolymer was substantiated by the electronmicroscopy, which revealed a two-phase structure of the products in the solid state. The block copolymerization with a unsaturated rubbery polymer was successfully utilized to improve the impact resistance of the homopolymers of norbornenenitriles.

Polymers of Norbornenenitriles. On the basis of preliminary evaluation of the properties of various polynorbornenes, the monomer availability and the polymerization behavior, the polymers of the nitrile derivatives were considered most promising as new materials. The homopolymers of 5-norbornene-2-nitrile and 2-methyl-5-norbornene-2-nitrile and a block copolymer of 5-norbornene-2-nitrile with SBR were prepared and evaluated with respect to their physical properties. The ternary system, WCl_6-paraldehyde-Al(isoBu)$_3$ was used as the catalyst, which efficiently induced the polymerization at a catalyst-to-monomer ratio as low as ca. 4×10^{-4}.

^{1}H-nmr spectrum of poly(5-norbornene-2-nitrile) showed that it is the product of ring-opening polymerization. Further detailed investigation of the structure was done by means of ^{13}C-nmr spectroscopy.

In principle, three kinds of structural isomerism are possible, i.e., cis and trans isomers of the cyano group with respect to the two 1,3-bonds of the cyclopentane ring, cis and trans isomers about the C=C bond and the head-to-tail and head-to-head (tail-to-tail) arrangements of the consecutive monomer units.

Head-to-tail

Head-to-head (tail-to-tail)

Of these, the first isomerism is determined by the monomer structure, i.e., <u>trans</u> configuration from the exo monomer and <u>cis</u> configuration from the <u>endo</u> monomer.

The ^{13}C-nmr spectrum of poly(5-norbornene-2-nitrile) had three sets of peaks at 30 - 48 ppm, about 120 ppm and 128 - 135 ppm from tetramethylsilane, which were assigned to the carbons of the cyclopentane ring, the cyano group and the C=C bond, respectively. The quantitative analyses of these peaks, using polynorbornene, poly(methyl 5-norbornene-2-carboxylate) and poly(dimethyl 5-norbornene-2,3-dicarboxylate) as references, gave the proportion of each isomer in the three structural isomerism. The results are summarized in TABLE V.

The <u>cis/trans</u> ratio of the cyano group about the cyclopentane ring is in good agreement with the <u>endo/exo</u> ratio of the starting monomer determined by gas chromatography. The content of the head-to-tail arrangement obtained shows that approximately a half of the monomer units are in the head-to-head and tail-to-tail arrangements.

The proportion of the <u>cis</u> configuration about the C=C bond varies from 20 to 80 % depending on the kind of the catalyst employed. The Tg of the polymer was not changed much in this range of the <u>cis</u> content.

Physical Properties. The physical properties of the polymers of 5-norbornene-2-nitrile and 2-methyl-5-norbornene-2-nitrile and a block(graft) copolymer of 5-norbornene-2-nitrile with SBR were evaluated. Because the molecular weight affected the processability of the polymers, polymers with the intrinsic viscosity value of 0.4 - 0.5 were prepared.

All the polymers were hard and transparent materials which were amorphous as judged by DSC analysis, which showed only the secondary transitions at 140° and 165°C for poly(5-norbornene-2-nitrile) and the methyl homologue, respectively.

TABLE VI summarizes the physical properties of the polymers. The values of the acrylonitrile-butadiene-styrene(ABS) resin and of polycarbonate are included for the sake of comparison. The norbornene polymers have good tensile and flexual properties comparable to those of the ABS resin. The heat distortion temperature (HDT) of poly(5-norbornene-2-nitrile) is raised by

TABLE V Structural Analysis of Poly(5-norbornene-2-nitrile) by ^{13}C-nmr.

Monomer	$\frac{exo}{endo} = \frac{40}{60}$	exo	endo	endo
Catalyst[1]	A	A	A	B
C≡N (ring) trans%	37	100	0	0
cis%	63	0	100	100
Head-to-tail arrangement (%)	(21)[2]	44	45	41
C=C Configuration total polymer cis%	53	–	79	30
trans%	47	–	21	70
Head-to-tail part cis%	–	88	80	19
trans%	–	12	20	81

1) Catalyst: A, WCl_6-paraldehyde-$Al(isoBu)_3$; B, WCl_6-acetal-$AlEt_2Cl$. 2) The value showed considerable fluctuation because of the complexity of the spectrum.

about 10°C in the methyl derivative and these values lie between the HDT's of the ABS resin and polycarbonate. Outstanding properties of the homopolymers are transparency, high creep resistance and high abrasion resistance that are comparable to those of engineering plastics. The Izod impact strength of the homopolymer is lower than those of the ABS resin and polycarbonate, but this can be greatly improved by block copolymerization with a small amount of SBR as is evident from TABLE VII.

Experimental Section

Materials. The monomers were prepared by the Diels-Alder reaction of cyclopentadiene with the corresponding vinyl and maleic compounds. The norbornene imides were prepared by the reaction of norbornene dicarboxylic acid anhydride with the corresponding amines. Liquid monomers were purified by distillation under reduced nitrogen pressure and freed from residual water

TABLE VI Properties of Polynorbornenenitriles.

Polymer	Poly-[1] NN	NN/SBR Block copolymer	Poly-[2] MNN	Polycarbonate	ABS Resin
SBR content(wt-%)	0	10	0	–	–
$[\eta]$ (dl/g)	0.37	0.5	0.54	–	–
Density (25°C)	1.03	1.07	–	1.22	1.08
HDT (246 psi; °C) [3]	115	111	127	139	87
Tensile strength at yield [4] (kg/cm²) at break [4]	531 374	480 380	– –	667 796	479 382
Elongation (%) at yield [4] at break [4]	13 270	– 145	– –	14 250	10 50
Flexual strength(kg/cm²)	898	820	790	1170	831
Flexual modulus(kg/cm²) [4],[5]	2.1×10⁴	2.2×10⁴	2.1×10⁴	2.9×10⁴	2.5×10⁴
Izod impact strength [6] (kg·cm/cm; at 25°C)	3.2	110	2.8	15	13
Rockwell hardness(R scale)	120	109	122	123	107
Taber abrasion resistance [7] (mg at 1000g; 25°C)	17	52	–	15	76
Creep(% at 75°C; 100 kg/cm²) [8]	1.9	0.8	–	1.3	broken

1) NN = 5-Norbornene-2-nitrile. 2) MNN = 2-Methyl-5-norbornene-2-nitrile. 3) ASTM D648, without annealing. 4) At 25°C. 5) ASTM D790. 6) ASTM D256, notched sample. 7) ASTM D1044. 8) 100 hr.

by treating with molecular sieves 4A. Solid monomers were recrystallized from appropriate solvents and dried in vacuo at room temperature.

The W-carbene complexes were prepared by the method reported by Fischer et al.(26). Other catalyst components were commercial reagents which were used without further purification.

Solvents were commercial reagents of appropriate purity and purified by usual methods. They were freed from water by treatment with molecular sieves. Other materials were commercial reagents and purified by usual techniques.

Polymerization. Polymerication runs were carried out by usual methods using sealed glass ampoules. The solvent-to-monomer weight ratio was 3 - 5. Polymers were recovered by precipitation in methanol or, in the case of amide and pyridyl derivatives, in petroleum ether. For the evaluation of properties, polymer samples were prepared by using a 14 l stainless-steel autoclave and recovered by steam distillation of the volatile materials. The products were further purified by reprecipitation with chloroform-methanol when necessary.

Measurements. The 'H-nmr and ^{13}C-nmr spectra were recorded on a Nihon Denshi MH 60 nmr spectrometer and on a Nihon Denshi PS 100 nmr spectrometer, respectively. Ir spectra were taken on a Nihon Bunko IRA-1 grating infrared spectrophotometer. Differencial scanning calorimetry was carried out with a Thermoflex 8001 differencial scanning calorimeter(Rigaku Denki).

For the measurement of the physical properties, polymers were injection-molded to pieces of specified size. Measurements were made by the standard procedures. Polycarbonate (Panlite 1225, Teijin Co.) and the ABS resin (JSR ABS#55, Japan Synthetic Rubber Co.) were also evaluated as the references.

Acknowledgements. The authors wish to express their gratitude to Dr. F. Imaizumi for his active collaborations and helpful discussions. Their grateful thanks are also due to Mr. M. Ikeyama for nmr analyses and to Mr. M. Nagata for the evaluation of the physical properties. The authors are much indebted to Japan Synthetic Rubber Co., Ltd. for generous permission to publish this work.

Literature Cited

(1) Calderon,N., J. Macromol. Sci., Revs. Macromol. Chem., (1972),C7, 105.
(2) Günther,P., Haas,F., Marwede,G., Nützel,K., Oberkirch,W., Pampus,G., Shön,N., Witte,J., Angew. Makromol. Chem., (1971),16/17, 27.
(3) Scott,K.W., Polymer Preprints, (1972),13(2), 874.
(4) Calderon,N., Acc. Chem. Res., (1972),5, 127.
(5) Eleuterio,H.C., US, 3,074,918(1963).
(6) Natta,G., Dall'Asta,G., Mazzanti,G., Angew.Chem. (1964), 76,765.
(7) Banks,R.L., Bailey,G.C., Ind. Eng. Chem., Prod. Res. Develop., (1964),3,170.
(8) Calderon,N., Ofstead,E.A., Ward,J.P., Judy,W.A., Scott,K.W., J. Am. Chem. Soc., (1968), 90, 4133.
(9) Natta,G., Dall'Asta,G., Mortani,G., Makromol. Chem., (1963), 69, 163.
(10) Michelotti,F.W., Carter,J.H., Polymer Preprints, (1965),6,224.
(11) Michelotti,F.W., Keaveney,W.P., J. Polym. Sci., Part A, (1965),3, 895.
(12) Rinehart,R.E., Smith,H.P., J. Polym. Sci., Part B, (1965),3, 1049.
(13) Charbonnage de France, Fr, 1,594,943(1970).
(14) Charbonnage de France, Fr, 1,543,497(1968).
(15) Porri,L., Rossi,R., Diversi,P., Lucherini,A., Polymer Preprints, (1972),13(2), 897.
(16) Nakamura,R., Matsumoto,S., Echigoya,E., Chem. Lett., (1976), 1019.
(17) Hepworth,P.(to ICI), Ger. Offen., 2,231,995(1973).
(18) Ueshima,T., Kobayashi,S., Matsuoka,M.(Showa Denko), Ger. Offen., 2,316,087(1973).
(19) Ueshima,T., Kobayashi,S., Japan Plastics, (1974), 11.
(20) Van Dam,P.B., Mittelmeijer,M.C., Boelhouwer,C., Chem. Commun., (1972), 1221.
(21) Van Dam,P.B., Mittelmeijer,M.C., Boelhouwer,C., Fette. Seifen Anstrichm, (1974),76, 264.
(22) Ast,W., Rheinwald,G., Kerber,R., Makromol. Chem., (1976), 177, 1341.
(23) Ast,W., Rheinwald, G., Kerber,R., Makromol. Chem., (1976), 177, 1349.
(24) Kroll,W.R., Doyle,G., Chem. Commun., (1971), 839.
(25) Sharpless,K.B., Umbreit,M.A., Nieh,M.T., Flood, T.C., J. Am. Chem. Soc., (1972), 94, 6538.
(26) Fischer,E.O., Maasbol,A., Chem. Ber., (1967), 100, 2445.

22

Polymerization of Aryl Cyclic Sulfonium Zwitterions

D. L. SCHMIDT

The Dow Chemical Co., Midland, MI 48640

Trialkyl sulfonium salts react with nucleophiles by displacement at a carbon atom adjacent to the trivalent sulfur (1). Dialkyl sulfide is eliminated and the nucleophile becomes alkylated [eq. (1)].

$$\text{Nuc}^- + R_3S^+ \longrightarrow R\text{-Nuc} + R_2S \qquad (1)$$

Since sulfonium groups are hydrophilic, water soluble polymers may be prepared by incorporating enough of this functionality onto the polymer chains. Subsequent removal of the water and thermal curing destroys the sulfonium groups and yields water-insensitive polymers (2)(3)(4).

Benzylmethylsulfonium salts are much more labile than simple trialkyl sulfoniums, and they react with almost exclusive cleavage of the benzyl to sulfur bond (5). Because of this reactivity, polymers may be prepared at moderate temperatures by reacting [arylenebis(methylene)]bis(dimethylsulfonium) salts with difunctional nucleophiles such as dicarboxylates or diphenolates (6)(7).

Hatch (8) utilized the unique chemistry of sulfoniums by designing a zwitterionic structure that has a cyclic sulfonium attached through the sulfur to a phenolic, aromatic ring. Upon heating, these "monomers" polymerize by a mechanism involving nucleophilic attack by the phenolic anion upon the ring carbon α to the sulfonium sulfur [eq. (2)].

$$(\overparen{CH_2)_x {+}S}\text{-}\bigcirc\text{-}O^- \longrightarrow [(CH_2)_x\text{-}S\text{-}\bigcirc\text{-}O]_n \qquad (2)$$

$$x = 4 \text{ or } 5$$

Polymerization proceeds by ring-opening and loss of charge to yield a nonionic polymer (9). There are no sulfide by-products from this reaction since no sulfur-phenyl bond cleavage occurs.

Preparation of Monomers

Five-membered cyclic sulfonium zwitterions may be prepared by three different methods. The first method (1) involves the reaction of a phenolic compound with tetrahydrothiopene 1-oxide and hydrogen chloride (10). A similar method requires the use of tetrahydrothiophene and chlorine or sulfuryl chloride with a phenolic (method 2) (11)(12).

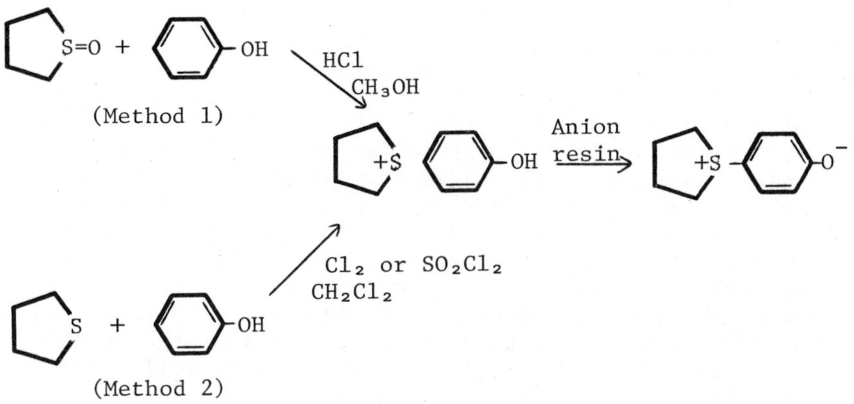

(Method 1)

(Method 2)

Reagents: HCl / CH_3OH ; Cl_2 or SO_2Cl_2 / CH_2Cl_2 ; Anion resin

In both methods 1 and 2, the hydrochloride salts are converted to the zwitterion with anion exchange resin (OH^- form). The tetrahydrothiophene substitution goes essentially all para to the phenolic hydroxy. If the para position is blocked, the tetrahydrothiophene substitutes in the ortho position; no meta substitution has been observed. Phenolic compounds that have electron-withdrawing groups (such as Cl) will not react under conditions of either method 1 or 2. Chlorine-containing derivatives may be prepared by chlorination of the appropriate zwitterion hydrochloride (9). Six membered cyclic sulfonium zwitterions can be prepared in only very low yields by methods 1 and 2. These compounds may be obtained by the reaction of 1,5-dibromopentanes with 4-(methylthio)phenol in refluxing chlorobenzene to yield the sulfonium salt and methyl bromide (method 3)(9). Method 3 is also applicable to the preparation of meta substituted derivatives (13).

$$Br(CH_2)_x Br + CH_3-S-\underset{}{\bigcirc}-OH$$

(Method 3)

$$\downarrow$$

$$\underset{\overline{Br}}{(CH_2)_x {}^+S}-\underset{}{\bigcirc}-OH + CH_3Br$$

Properties of Monomers

Sulfur to phenyl bond cleavage has not been observed with aryl cyclic sulfonium compounds. Presumably, this is because the sulfonium sulfur to phenyl bond is much stronger than the sulfur to alkyl bond (14). Cleavage of the sulfur to alkyl bond in di-alkyl aryl sulfonium zwitterions is much more difficult than with five-membered cyclic sulfonium zwitterions. This is probably due to the strain in the five-membered ring. The cyclic sulfonium ring probably gains considerable stability due to the contribution of the quinoid form of the resonance hybrid [eq. (3)]. The resulting system would be delocalized and thus the phenolate ion would have decreased nucleophilicity and the cyclic sulfonium would have decreased positive charge density and increased stability toward nucleophiles. This is consistent with the

$$\underset{}{\bigcirc}{}^+S-\underset{}{\bigcirc}-O^- \longleftrightarrow \underset{}{\bigcirc}S=\underset{}{\bigcirc}=O \qquad (3)$$

observed large difference in stability between zwitterion monomers with the sulfonium group ortho or para and those meta to the phenolic oxygen. Doorakian et al. (13) demonstrated that the meta isomers will polymerize much faster and at lower temperatures (10 min. at 40°C) than the ortho and para isomers. Also consistent with resonance stabilization of the zwitterion is the observation that methylation of the phenolic oxygen greatly increases the reactivity of the resulting cyclic sulfonium (15).

Zwitterion monomers that are not substituted with an electron-withdrawing group, such as chlorine, may be isolated only as crystalline hydrates. Attempts to remove the water of hydration in all cases causes polymerization of the monomer. Stability of the crystalline, hydrated zwitterions is probably due to their high energies of interaction with the polar water molecules in the crystal lattice. Introducing enough energy to remove the water of hydration from the crystalline solid is sufficient to cause polymerization. In solution, the stability of the monomers increases with increasing polarity of the solvent. For example, the rates of polymerization of various zwitterions in methanol are between 3 and 46 times faster than the rates measured in water.

Zwitterion monomers that are substituted with chlorine have markedly different properties than the unsubstituted monomers. Monomers I and II are examples of these two classes of monomer and their physical properties, and polymerizations have been studied in some detail.

The thermal stability of 1-(4-hydroxy-3-methylphenyl)tetrahydrothiophenium hydroxide inner salt (dihydrate), monomer I is much lower than 1-(3,5-dichloro-4-hydroxyphenyl)tetrahydrothiophenium hydroxide inner salt, monomer II. Differential thermal analysis of I indicates polymerization begins at 75°C, but monomer II does not begin polymerization until 150°C (9).

Monomer I (2H$_2$O)   Monomer II

The stability of II can be rationalized by two arguments: first, the nucleophilicity of the phenoxide group is greatly decreased due to the ortho electron-withdrawing chlorine atoms; second, monomer II does not contain water of hydration and is a stable, crystalline material. Zwitterion molecules should gain stability by arrangement in a crystal lattice. Monomer I probably gains less stability from crystal energy than II because of the ease with which water may be removed. Monomer I kept in a sealed container at room temperature polymerizes only about 0.7% per year but will polymerize significantly under vacuum in 24 hours.

Polymerization

The mechanism of polymerization of aryl cyclic sulfonium zwitterions involves initiation by displacement of the sulfonium moiety of one monomer by the phenolate ion of another. The resulting activated, linear dimer has both a more reactive cyclic sulfonium and a more nucleophilic phenolate ion than the initial monomer. The sulfonium moiety has increased reactivity because of the disappearance of negative charge attached to the aromatic ring and the loss of the stabilization contribution of the quinoid resonance hybrid. The phenolate anion of the linear dimer has both increased nucleophilicity and basicity due to the loss of the electron-withdrawing sulfonium on the aromatic ring [eq. (4)]. Thus, after initiation, a zwitterionic, bifunctional propagating species is produced which can grow by reaction with monomer or linear dimer from either or both ends. In theory, termination of only one end of the growing polymer chain should

$$\text{[+S-C}_6\text{H}_4\text{-O}^-\text{]} \longrightarrow \text{[+S-C}_6\text{H}_4\text{-O-(CH}_2)_4\text{-S-C}_6\text{H}_4\text{-O}^-\text{]} \quad (4)$$

Activated, Linear Dimer

Initiation

retard but not stop polymerization. A monofunctional chain can react with monomers, dimers or bifunctional chains, but the monofunctional units cannot react with each other. If initiation is slow, chains could grow to moderate molecular weight by reaction of growing, dimer-initiated chains with monomer. Later, with depletion of monomer, polymerization should proceed by combination of dimers and growing chains. If initiation is fast, then chains would reach moderate molecular weight only by the combination of short, growing chains.

Because both ends are reactive, the linear dimer or subsequent short, growing chains can react intramolecularly to form cyclics. Because of the attraction of the opposite charged ends of these molecules, cyclization of zwitterionic species should be much easier than other nonionic bifunctional systems. Cyclization can proceed until the growing chains become long enough that the probability of the ends coming in contact is small or until termination of either or both ends occurs.

Polymerization studies of crystalline monomers I and II have been reported earlier (9). In the crystalline forms, zwitterionic monomers should arrange themselves, due to electrostatic interaction, with alternate positive and negative sites in chain patterns similar to the arrangement of the atoms in the polymer. This orientation should favor rapid polymerization and is probably one reason for the fast polymerization (1-2 min.) of crystalline zwitterions at elevated temperatures.

Monomer I easily polymerizes merely by removal of the water of hydration to give polymer A, poly[oxytetramethylenethio(3-methyl-1,4-phenylene)]. A gel permeation chromatograph (GPC) of polymer A, prepared by heating (105°C for 35 min.) recrystallized I, is shown in Figure 1. This analysis indicates that besides polymer, cyclic dimer and trimer are present. Mass spectrometric analysis of the latter material gave m/e peaks at 388 (dimer) and 582(trimer). The dimer may be isolated by sublimation upon heating polymer A under vacuum. There was little change in the chromatographs of samples polymerized under vacuum or over a wide range of temperatures and times.

Recrystallized samples of monomer I were mixed with various nucleophiles by dissolving them in methanol containing the additive followed by vacuum removal of the solvent. Amines, sodium iodide and sodium methoxide were among the nucleophiles added. Changing the nucleophile (sodium methoxide) to monomer I mole ratio between 1/500 and 1/1000 had little effect on the gel permeation chromatographs of polymer A samples. When the additive to monomer ratio was increased to 1/100, the cyclics to polymer ratio was decreased and the molecular weight was lowered (Figure 1).

Polymerization of crystalline monomer II, to obtain polymer B, requires a considerably higher optimum polymerization temperature (170°C) than monomer I. Polymer B, poly[oxytetramethylene-thio(3,5-dichloro-1,4-phenylene)], generally has a higher number average molecular weight ($\overline{M}_n$) than polymer A. Unlike monomer I, the polymerization of II is very dependent upon purity. A series of nucleophiles were incorporated into samples of recrystallized monomer II and then polymerized (9). Table I (16) illustrates the effect of additives upon the specific viscosity (ηsp) in chlorobenzene of the resulting polymers. It is evident that most

Table I

EFFECT OF ADDITIVES ON POLYMERIZATION OF MONOMER II
(Initial ηsp = 0.22)

Additive	Additive in II, mole-%	ηsp After Polymerization
NaCl	0.10-0.50	0.25-0.27
NaBr	0.10-0.50	0.31-0.35
NaI	0.10-0.50	0.34-0.38
$\overset{+}{N}(CH_3)_4OH$	0.10-0.50	0.34-0.37
N,N,N,'N,'-Tetramethyl-ethylenediamine	0.10-0.50	0.40-0.43
N,N'-Dimethyl-1,6-hexane-diamine	0.10-1.00	0.30-0.38
NaOCH$_3$	0.10-2.00	0.39.0.51
Monomer I	0.10-3.00	0.22-0.23

nucleophiles introduced at an additive to monomer ratio between 1/500 and 1/1000 increased the ηsp. Typical gel permeation chromatographs of polymer B prepared from both pure monomer II and II with added nucleophile are shown in Figure 2 (16). Although part of the high molecular weight portion of the polymer samples is insoluble in the tetrahydrofuran solvent, these GPC studies clearly indicate that less cyclic material is formed when a nucleophile is present during polymerization. Solvent extraction of polymer B samples, prepared with and without added nucleophile, also shows that these additives significantly lower the amount of cyclics. Most of the cyclics may be removed by dissolving polymer B in hot chlorobenzene and then cooling and collecting the

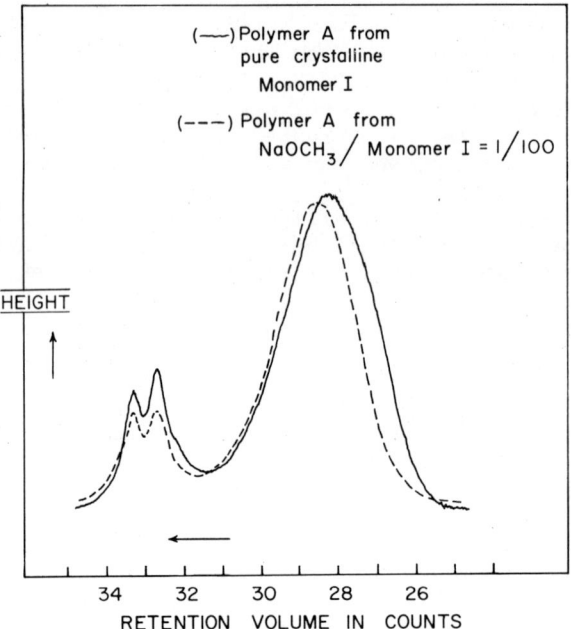

Figure 1. Gel permeation chromatographs of polymer A with and without $NaOCH_3$ present during solid state polymerization (high $\overline{M}_n$ column H)

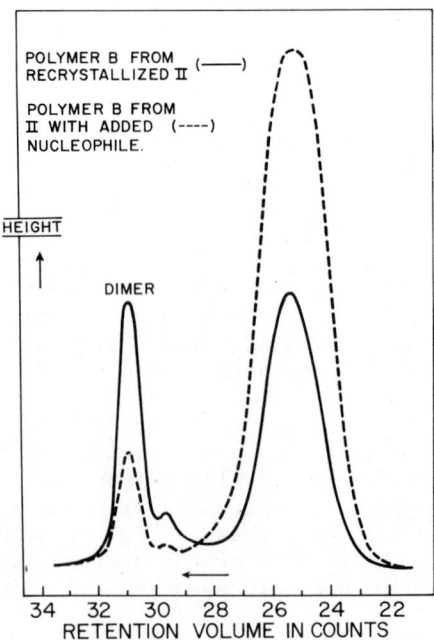

Figure 2. Gel permeation chromatographs of polymer B with and without nucleophile present during solid state polymerization (high $\overline{M}_n$ column D)

precipitated polymer. These purified polymers have increased ηsp values but the samples obtained by polymerizing II with traces of nucleophiles still have considerably higher ηsp values than polymers from pure II. Thus, the addition of small amounts of nucleophiles to II not only gives less cyclic material but also yields polymers with higher molecular weights.

A sample of polymer B prepared by using sodium methoxide as an additive had a ηsp of 0.5 and a $\overline{M}_n$ of 46,000 (9). The glass transition temperature was between 4 and 10°C and the crystalline melting point was between 140 and 160°C. The specific gravities of monomer II and polymer B are respectively 1.528 and 1.483; this would indicate an expansion upon polymerization. The mechanical properties of this sample of polymer B are shown in Table II (16).

Table II

MECHANICAL PROPERTIES OF POLYMER B

Mechanical Test	Unannealed Polymer	Annealed Polymer[a]
Ultimate tensile, psi (ASTM D638-68)	2960	4100
Tensile modulus X 10^5, psi (ASTM D638-68)	0.91	1.28
Elongation, % (ASTM D638-68)	>260	>260
Heat distortion, °C (ASTM D648-61)	34	79
Impact strength, ft-lb/in. of notch (ASTM D256-61)	16+	1.84

[a] Annealed 48 hr at 110°C, then 7 days at 70°C.

Obtaining a solvent applicable for studying the polymerization of zwitterion monomers is difficult because of the large difference in solubility between the ionic monomer and the non-ionic polymers. Monomers I and II and their subsequent polymers may be kept in solution by using a special solvent consisting of 2% dipropylene glycol, 13% diethylene glycol and 85% chlorobenzene. Studies of both monomer systems were carried out in sealed tubes at 1% (weight solids) solution. The solutions were heated at 65°C for five days and the resulting polymers isolated and analyzed by GPC.

Monomer I was polymerized in solution both as a purified monomer and with a sodium methoxide to monomer mole ratio of 1/100. The gel permeation chromatographs of the resulting polymer A samples are shown in Figure 3. It is evident that the nucleophile decreases the cyclic dimer to polymer ratio but the relative amount of dimer is still high. There is no apparent change in molecular weight with the addition of nucleophile. The molecular weight of this polymer is much lower than those obtained by solid state polymerization.

Monomer II was polymerized in the solvent system with varying amounts of sodium methoxide. The chromatographs of the polymer B samples, obtained by using nucleophile to monomer ratios of 1/50 and 1/300, are shown in Figure 4. It is evident that the addition of nucleophile decreases the cyclic to polymer ratio but does not affect the polymer molecular weight. The $\bar{M}_n$ of polymer B, from solvent polymerization of II, is much lower than that obtained by solid state polymerization, but it is higher than the $\bar{M}_n$ of polymer A obtained from solvent polymerization of I. The polydispersities ($\bar{M}_w/\bar{M}_n$) of polymer B samples, prepared in solution, have realatively low values of about 1.2. After polymer B was isolated from the 1/50 monomer to nucleophile experiment, it was allowed to react with more monomer II under the same polymerization conditions. The resulting polymer had a higher molecular weight and could not be completely analyzed by GPC since it was partially insoluble in the GPC solvent. The low molecular weight fraction gave a $\bar{M}_w/\bar{M}_n$ of 1.5.

From the limited data available, one can only speculate on the differences in the polymerization mechanisms of monomers I and II. One possible theory requires three assumptions: first, the monomers are more stable to polymerization than linear dimers and zwitterion chains and this difference in stability is greatest with the monomer II system; second, zwitterion dimers and chains can terminate by β-elimination to give a terminal double bond, $-S(CH_2)_2CH=CH_2$; third, zwitterion monomers are more stable towards β-elimination than the corresponding sulfonium end groups of the linear zwitterion dimers and chains. The monomers have both lowered positive charge on the sulfurs and less basic phenolic oxygens than the linear dimers. Both of these effects should stabilize the monomer cyclic sulfonium moieties toward elimination (1). The stability of I and II is born out by the observation that reaction of the monomers with excess sodium methoxide yields mostly methoxy derivatives rather than β-elimination. In the case of zwitterion dimers and chains, the higher positive charge density and the presence of a more basic phenolate anion should promote elimination [eq. (5)].

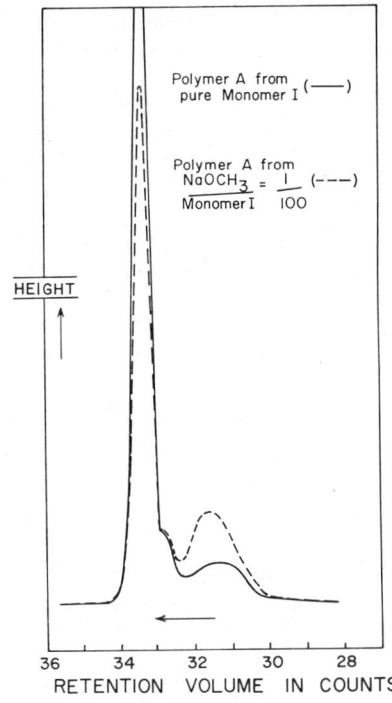

Figure 3. Gel permeation chromatographs of polymer A with and without $NaOCH_3$ present during solution polymerization (high $\overline{M}_n$ column H)

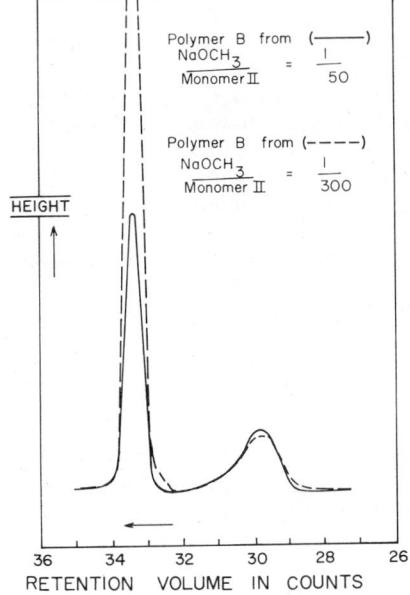

Figure 4. Gel permeation chromatographs of polymer B with and without $NaOCH_3$ present during the solution polymerization (high $\overline{M}_n$ column H)

Solid monomer I undergoes fast initiation. The monomer quickly disappears and polymerization proceeds by combination of dimers and small chains [eq. (6)]. Termination can occur by cyclization and β-elimination at the sulfonium-containing end groups of the growing chains. Without termination, these chains would continue to grow by combination to attain much higher $\bar{M}_n$ than has been observed for polymer A. A sample of polymer A exhibited a

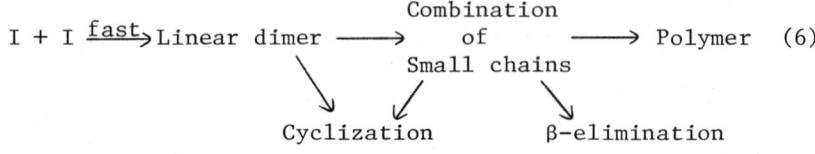

weak infrared band at 912 cm^{-1} which could be due to the CH_2 wagging vibration of a terminal double bond. This is consistent with the proposed termination by β-elimination.

Small amounts of nucleophile have little effect on the polymerization of I. Initiation by a nucleophile is probably not much faster than monomer to monomer initiation. Larger amounts of nucleophile (Figure 1) show an effect by destroying more sulfonium end groups, thus preventing cyclization, and also by terminating the end of growing chains. This decreases the amount of cyclic material and also lowers the molecular weight of polymer A.

In solution, polymerization of I yields high amounts of cyclics and low $\bar{M}_n$ polymer A. The low polarity solvent system speeds initiation and greatly promotes cyclization due to the inability of the charged species to extend themselves and obtain the charge separation needed for polymerization. The small amount of polymer probably results both from chains that attain a noncyclizable size and from sulfonium end group termination. The chains then grow to limited size by reaction with what little monomer is left. Addition of nucleophile gives an equivalent amount of low molecular weight polymer by preventing cyclization.

If monomer II is considerably more stable than its dimer, then slow initiation will be followed by a fast reaction of initiated species with monomers to form growing chains. This mechanism [eq. (7)] will lead to high molecular weight and should not be very sensitive to termination by β-elimination. The concen-

$$\text{II} + \text{II} \xrightarrow{\text{slow}} \text{Dimer} + \text{II} \xrightarrow{\text{fast}} \text{Polymer B} \qquad (7)$$
$$\downarrow$$
$$\text{Cyclic dimer}$$

tration of low molecular weight chains capable of termination would remain low. When monomer becomes depleted, then growth by combination of chains would predominate. Because of the decreased basicity of the chlorine-flanked phenolate anion of II, termination by β-elimination should be less likely during the polymerization of monomer II than I.

Added nucleophile can react with II at temperatures well below the polymerization of pure II (9). Nucleophile would produce an end-capped, initiated species and propagation would proceed chiefly by reaction of monomer with growing chains [eq. (8)]. This would result in both an increase in molecular weight and less cyclic material.

$$Nuc^- + II \xrightarrow{fast} Nuc\text{-}(CH_2)_4\text{-}S\text{-}C_6H_3Cl_2\text{-}O^- + II \longrightarrow Polymer\ B \quad (8)$$
$$\downarrow Cyclization$$

In solution, as in the case of monomer I, cyclization should predominate. The polymer obtained should have been generated from those chains that gained a noncyclizable size or became end-capped. The polymer B produced has a larger $\bar{M}_n$ than polymer A obtained under the same reaction condition. Presumably, this is because the higher concentration of the more stable II allows longer chains to form. In solution, the addition of nucleophile also prevents cyclization and thus increases the amount of polymer B. The relatively low $\bar{M}_w/\bar{M}_n$ of 1.2 would be consistent with most of the initiation occurring at the beginning of the reaction followed by reaction of monomer with growing chains. Addition of II to previously prepared polymer B gives higher molecular weight polymer by growth of previously formed polymer with monomer. The $\bar{M}_w/\bar{M}_n$ increases in this case because, besides growth due to previously formed chains, new initiation occurrs to give a lower molecular weight polymer fraction.

<u>Utility</u>

Linear polymers from monomers similar to I and II do not appear to have very much commercial potential because of their high production costs and poor properties. By incorporating two or more zwitterions in the same molecule, monomers are obtained that yield highly crosslinked polymers (17) [eq. (9)]. These

$$[\text{structure with three phenolate-sulfonium units bridged by X}]_n \xrightarrow{heat} Crosslinked\ polymer \quad (9)$$

X = bridging group

($-CH_2-$, $O(CH_2)_2O$, or $-C(CH_3)_2-$

monomers appear promising in water-based coating applications. The polyfunctional monomers give high-solids, low-viscosity, aqueous solutions which cure at low temperatures to yield highly crosslinked coatings. Polyfunctional monomers may also be used as an efficient method of crosslinking water-soluble or water-dispersible carboxy-containing polymers (18). The carboxy groups readily open sulfonium rings to yield esters. The adhesion of carboxy-containing, latex based films is greatly increased by the addition of as little as 1 to 5 weight % of polyfunctional monomers (19)(20).

Acknowledgment

I am indebted to J. R. Runyon and J. E. Jones for the gel-permeation chromatograph studies and to R. A. Nyquist for the interpretation of infrared spectra. I am grateful for the experimental work carried out by D. Urchick and for valuable discussions with T. Alfrey, G. D. Jones, T. C. Klingler, W. C. Meyer R. A. Wessling, J. W. Rakshys and R. A. Kirchhoff.

Literature Cited

1. Ingold, C. K., "Structure and Mechanism in Organic Chemistry" 2nd ed., Cornell Univ. Press, Ithaca, N.Y., 1969, Chapter VII and IX.
2. Hatch, M. J. and McMaster, E. L., U.S. Patent 3,078,259 (1963); Chem. Abstr. 58, 10327a (1963).
3. Wessling, R. A. and Zimmerman, R. G., U.S. Patent 3,401,152 (1968); Chem. Abstr. 69, 87735 (1968).
4. Fang, J. C., U.S. Patent 3,310,540 (1967); Chem. Abstr. 66, 106007 (1967).
5. Swain, C. G. Burrows, W. D. and Schowen, B. J., J. Org. Chem. 33, 2534 (1968).
6. Hatch, M. J., U.S. Patent 3,502,710 (1970); Chem. Abstr. 72, 121180 (1970).
7. Alfrey, T., paper presented in part at the Biannual Polymer Symposium, Ann Arbor, Michigan, June 13, 1972.
8. Hatch, M. J., Yoshimine, M., Schmidt, D. L. and Smith, H. B., J. Amer. Chem. Soc., 93, 4617 (1971).
9. Schmidt, D. L., Smith, H. B., Yoshimine, M. and Hatch, M. J., J. Polym. Sci., A-1, 10, 2951 (1972).
10. Goethals, E. and deRadzitsky, P., Bull. Soc. Chem. Belg., 73, 546 (1964).
11. Cisney, M. E. and Camas, M., U.S. Patent 3,259,660 (1966).
12. Klingler, T. C., Schmidt, D. L., Jensen, W. J. and Urchick, D., U.S. Patent applied for.
13. Doorakian, G. A. and Schmidt, D. L., U.S. Patent applied for.
14. Price, C. C. and Oae, S., "Sulfur Bonding", Ronald Press, New York, N.Y., 1962, pp. 151-158.

15. Schmidt, D. L., Smith, H. B., Hatch, M. J. and Broxterman, W. E., U.S. Patent 3,898,247 (1975); Chem. Abstr., 84, 5824 (1976).
16. Used by permission of John Wiley and Sons, Inc., Publishers, 605 Third Ave., New York, N.Y. see Reference 9.
17. Schmidt, D. L., Smith, H. B. and Broxterman, W. E., J. Paint Technol. 46, 41 (1974).
18. Julier, R. M., The Dow Chemical Company, personal communication, 1976.
19. Plueddemann, E. P., paper presented at Symposium on Polyelectrolytes and Their Applications, California Institute of Technology, May 23-25, (1973).
20. Bergman, R. W. and Schmidt, D. L., The Dow Chemical Company, private communication, 1977.

23

Spontaneous Alternating Copolymerization of Cyclic Phosphorus Compounds via Phosphonium Zwitterion Intermediates

TAKEO SAEGUSA, SHIRO KOBAYASHI, YOSHIHARU KIMURA, and TSUNENORI YOKOYAMA

Department of Synthetic Chemistry, Faculty of Engineering, Kyoto University, Kyoto 606, Japan

Induction of polymerization reaction usually requires initiator (or catalyst) or radiation. Recently we have explored a new type of copolymerization which takes place spontaneously without any added catalyst ($\underline{1}$). In this copolymerization, the reactivity characters of two monomers are very important, i.e., one is of nucleophilic reactivity (M_N) and the other is of electrophilic reactivity (M_E). Reaction occurs between two monomers to produce a zwitterion **1** (Eq 1), which is responsible for initiation as well as for propagation.

$$M_N + M_E \longrightarrow \underset{\mathbf{1}}{{}^+M_N - M_E^-} \qquad (1)$$

$$\mathbf{1} + \mathbf{1} \longrightarrow \underset{\mathbf{2}}{{}^+M_N - M_E M_N - M_E^-} \qquad (2)$$

$$\mathbf{2} + \mathbf{1} \times n \longrightarrow \underset{\mathbf{3}}{{}^+M_N \!-\!\!\left(\!M_E M_N\!\right)_{\!\overline{n+1}} M_E^-} \qquad (3)$$

The two moles of the "genetic zwitterion" **1** react with each other to produce its dimer **2** which is the smallest species of propagation (Eq 2). Then, the propagation species grows by its reaction with **1** (Eq 3). Zwitterions **2** and **3** ($n \geq 1$) are called "macro zwitterion", which are differenciated from genetic zwitterion **1**. The intermolecular reaction between two moles of macrozwitterion and the intramolecular cyclization of macro zwitterion are also possible, although the contribution of each process depends upon the natures of monomers, reaction conditions and the extent of monomers' conversion.

In addition, free monomers sometimes take part in propagation, i.e., M_N and M_E may react respectively with the cationic and anionic sites of zwitterion (genetic or macro)(Eqs 4 and 5).

$$^+M_N\!\!-\!\!(M_EM_N)_{\overline{n}}\, M_E^- \longrightarrow \begin{cases} \xrightarrow{M_N} {}^+M_NM_N\!\!-\!\!(M_EM_N)_{\overline{n}}\, M_E^- \quad (4) \\ \\ \xrightarrow{M_E} {}^+M_N\!\!-\!\!(M_EM_N)_{\overline{n}}\, M_EM_E^- \quad (5) \end{cases}$$

$(n \geq 1)$

When the processes of Eqs 1-3 occur exclusively throughout the course of polymerization, alternating copolymer $-(M_NM_E)_{\overline{n}}$ is produced. In many combinations of M_N and M_E, the so-called spontaneous alternating copolymerizations have been realized. In some cases, however, one of homo-propagations of Eqs 4 and 5 takes place to result in copolymers of biased compositions.

Illustrative Example of Spontaneous Alternating Copolymerization

Alternating copolymerization between 2-oxazoline and β-propiolactione **5** (BPL) occurs at room temperature through the genetic zwitterion **6**, where **4** acts as M_N and **5** behaves as M_E (2, 3).

The interaction between two zwitterions (genetic zwitterion or macro zwitterion) occurs via the opening of the oxazolinium ring of the cationic site (Eq 6).

Scope of Spontaneous Copolymerization

On the basis of the above-mentioned principle, many copolymerizations have been explored by the combinations of several M_N and M_E monomers. Table I shows the structures of six M_N monomers and seven M_E monomers as well as the cationic and anionic species which are derived respectively from these two groups of monomers. Among 42 possible combinations of copolymerization, 24 essential ones have been examined, which are shown by the respective references.

Alternating Copolymerizations of Cyclic Phosphorus Compounds

The main topic of the present paper is the copolymerization of cyclic phosphorus compounds, 2-phenyl-1,3,2-dioxaphospholane **11** and 2-phenoxy-1,3,2-dioxaphospholane **20** as M_N.

20

Copolymerizations of 2-Phenyl-1,3,2-dioxaphospholane **11**

Without any added catalyst, **11** has successfully been copolymerized with several M_E monomers such as β-propiolactone **13** (_13_), 3-hydroxypropanesulfonic acid lactone (propanesultone) **15** (_14_), acrylic acid **16** (_13_), acrylamide **17** (_13_), ethylenesulfonamide **19** (_14_), acrylic ester (_16_), vinyl ketone (_16_) and α-keto acid (_17_). In every case, copolymerization occurs at temperatures above 100°C to produce alternating copolymer.

The two copolymers of the combinations of **11**-BPL and **11**-acrylic acid have the identical structure of **23** which is derived from the common zwitterion **22**. In the case with acrylic acid, an unstable carbanion intermediate is first formed, which is then transformed into **22** by hydrogen transfer.

Table I
Nucleophilic and Electrophilic Monomers and Ionic Groups Derived from Them

M_N	M_N^+-	M_E	$-M_E^-$
7 (2-9) cyclic iminoether (5-ring)	cyclic iminoether cation	**13** β-propiolactone (2,3 / 10-13 / 15)	$-CH_2CH_2CO_2^-$
8 (7,8,10) cyclic iminoether (6-ring)	cyclic iminoether cation (6-ring)	**14** succinic anhydride (4,15)	$-\underset{\underset{O}{\|}}{C}CH_2CH_2CO_2^-$
9 (11) pyrrolidone	pyrrolidinium	**15** sultone (5,14)	$-(CH_2)_3SO_3^-$
10 (12) Me,Me-N-Me azetidine	Me,Me-N,Me azetidinium	**16** $CH_2=CH-CO_2H$ (6,10-13,15)	$-CH_2CH_2CO_2^-$
11 (13,14) cyclic phosphite (P-Ph)	cyclic phosphonium (P-Ph)	**17** $CH_2=CH-CONH_2$ (7,13)	$-CH_2CH_2C\!\!\begin{array}{c}\nearrow NH\\ \searrow O\end{array}$
12 (15) Ph-CH=N-Ph	Ph-CH⋯N⁺-Ph	**18** $CH_2=CH-\underset{\underset{O}{\|}}{C}OCH_2CH_2OH$ (8)	$-CH_2CH_2CO_2^-$ / $-CH_2CH_2O^-$
		19 $CH_2=CH-SO_2NH_2$ (9,14)	$-CH_2CH_2\underset{\underset{NH}{\|}}{\overset{\overset{O}{\|}}{S}}=O$

The reaction mode of the opening of the phosphonium ring of **22** is understood on the basis of the scheme of the Arbuzov reaction. Here it is to be noted that the trivalent phosphorus in **11** is oxidized to pentavalent state during copolymerization.

In the copolymerizations of **11** with acrylamide **17** and ethylenesulfonamide **19**, hydrogen transfer occurs similarly in the first-formed carbanion zwitterions **24** and **26**, respectively.

$$11 + 17 \longrightarrow \left[\begin{array}{c} O \\ O \end{array} \!\!\! P \!\! \begin{array}{c} +/Ph \\ \bar{} \\ CH_2\bar{C}HCONH_2 \end{array} \right] \longrightarrow \begin{array}{c} O \\ O \end{array} \!\!\! P \!\! \begin{array}{c} +/Ph \\ \bar{} \\ CH_2CH_2C\bar{N}H \\ \parallel \\ O \end{array}$$
$$\quad\quad\quad\quad\quad\quad\quad\quad\quad\quad \mathbf{24} \quad\quad\quad\quad\quad\quad\quad\quad\quad \mathbf{25}$$

$$11 + 19 \longrightarrow \left[\begin{array}{c} O \\ O \end{array} \!\!\! P \!\! \begin{array}{c} +/Ph \\ \bar{} \\ CH_2\bar{C}HSO_2NH_2 \end{array} \right] \longrightarrow \begin{array}{c} O \\ O \end{array} \!\!\! P \!\! \begin{array}{c} +/Ph \\ \bar{} \\ CH_2CH_2SO_2\bar{N}H \end{array}$$
$$\quad\quad\quad\quad\quad\quad\quad\quad\quad\quad \mathbf{26} \quad\quad\quad\quad\quad\quad\quad\quad\quad \mathbf{27}$$

The reaction sites of the anionic part of ambident nature in **25** and **27** are different from each other, i.e., the reaction occurs at nitrogen atom in **25** and oxygen atom in **27**. The structures of the two alternating copolymers are expressed by **28** and **29**, respectively. The regiospecificity of the ambident anions was quite high. The understanding of the difference of the site of nucleophilic reaction between the two ambident anions requires further studies.

$$\mathbf{25} \rightleftharpoons \underset{\mathbf{28}}{-\!\!\!\left(CH_2CH_2O\underset{\underset{O}{\parallel}}{\overset{\overset{Ph}{|}}{P}}-CH_2CH_2CONH \right)_{\!p}\!-}$$

$$\mathbf{27} \rightleftharpoons \underset{\mathbf{29}}{-\!\!\!\left(CH_2CH_2O\underset{\underset{O}{\parallel}}{\overset{\overset{Ph}{|}}{P}}-CH_2CH_2\underset{NH}{\overset{\overset{O}{\parallel}}{S}}-O \right)_{\!p}\!-}$$

In the reactions of **11** with **16** (or **13**) and with **17** at lower temperatures (e.g., room temperature — 50°C), pentacovalent phosphorus compounds of spiro ring system, **30** and **31** were produced in high yields, which were isolated in crystalline form (18). The two compounds of **30** and **31** are a new class of pentacovalent organophosphorus compounds, which are derived by the intramolecular ring closures in the zwitterions **22** and **25**. On heating at temperatures above 120°C, each of **30** and **31** was polymerized to produce **23** and **28**, respectively. At high temperatures, P-O

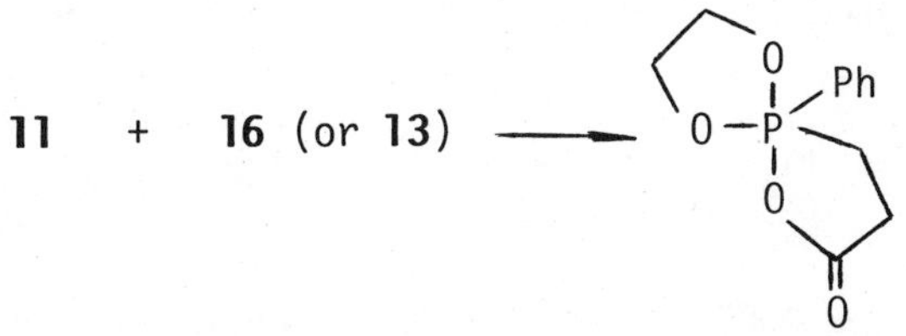

11 + 16 (or 13) ⟶ **30** (mp, 80°C)

11 + 17 ⟶ **31** (mp, 158°C)

bond of **30** and P-N bond of **31** are broken in a heterolytic way to generate zwitterions **22** and **25**, and the polymerizations proceed via these zwitterions. Thus, the homopolymerizations of **30** and **31** constitute a novel pattern of the "thermally induced ring-opening polymerization via zwitterion."

Alternating copolymerizations of **11** with acrylate (**32a**) and with vinyl ketones (**32b** and **32c**) occurred at 130°C to give lower molecular weight polymers (mol. wt. 700-1600) (Eq 7) (16).

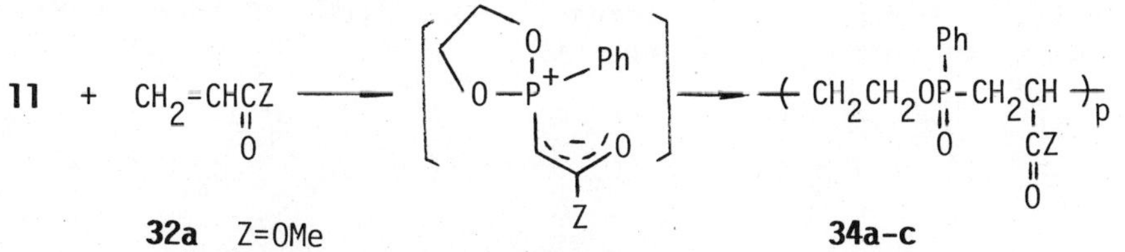

$$11 + CH_2=CHCZ \longrightarrow [33] \longrightarrow 34a\text{-}c \quad (7)$$

32a Z=OMe
  b Z=Me
  c Z=Ph

At lower temperatures (room temperature to 50°C), pentacovalent phosphorus compounds of a spiro structure, **35a** and **35b** were obtained in fairly good yields. They were formed by the combination of cationic site (phosphorus atom) and anionic site (oxygen of enolate).

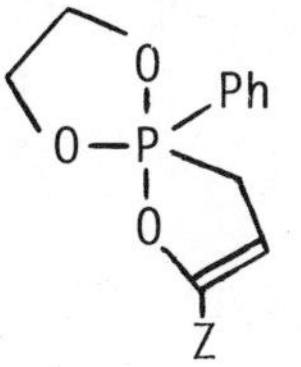

35a Z=OMe (bp, 107-110°C/0.04 mmHg)
  b Z=Me (mp, 131°C)

Heating of isolated samples of **35a** and **35b** in bulk at 200°C gave polymers **34a** and **34b**, respectively.

Extension of the copolymerization of Eq 7 has brought about a very interesting finding of the 1:1:1 alternating terpolymerization of **11**, **32a** and CO_2 (Eq 8) (16).

$$\mathbf{11} \ + \ \mathbf{32a} \ + \ CO_2 \ \longrightarrow \ {\Large(}CH_2CH_2O\underset{Ph}{\overset{\overset{O}{\|}}{P}}-CH_2\underset{CO_2Me}{CH}-\overset{\overset{}{}}{\underset{\underset{O}{\|}}{C}O}{\Large)}_p$$
$$\mathbf{36} \qquad\qquad (8)$$

Thus, CO_2 was introduced at atmospheric pressure to an equimolar mixture of **11** and **32a** at 50°C. A zwitterion **38** is assumed to be the key intermediate which is derived by the addition of CO_2 onto the carbanion of enolate anion of **37**.

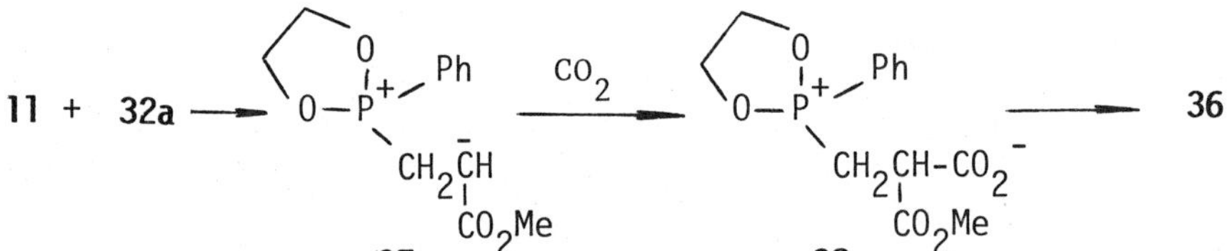

$$\mathbf{11 + 32a} \longrightarrow \underset{\mathbf{37}}{\begin{array}{c}\text{[cyclic structure]}\\ \text{with } CH_2\bar{C}H\text{-}CO_2Me\end{array}} \xrightarrow{CO_2} \underset{\mathbf{38}}{\begin{array}{c}\text{[cyclic structure]}\\ \text{with } CH_2CH\text{-}CO_2^-\text{, }CO_2Me\end{array}} \longrightarrow \mathbf{36}$$

In the above terpolymerization, **32a** can be replaced by acrylonitrile, i.e., a 1:1:1 alternating terpolymer of **39** was successfully prepared under atmospheric pressure of CO_2 (Eq 9).

$$\mathbf{11} \ + \ CH_2{=}CHCN \ + \ CO_2 \ \longrightarrow \ {\Large(}CH_2CH_2O\underset{Ph}{\overset{\overset{O}{\|}}{P}}-CH_2\underset{CN}{CH}-\underset{O}{\overset{\|}{C}}O{\Large)}_p \qquad (9)$$
$$\mathbf{39}$$

Copolymerization of **11** with 3-hydroxypropanesulfonic acid sultone **15** proceeds at 140°C in benzonitrile to produce the alternating copolymer of **41** (14). The following scheme may well be assumed.

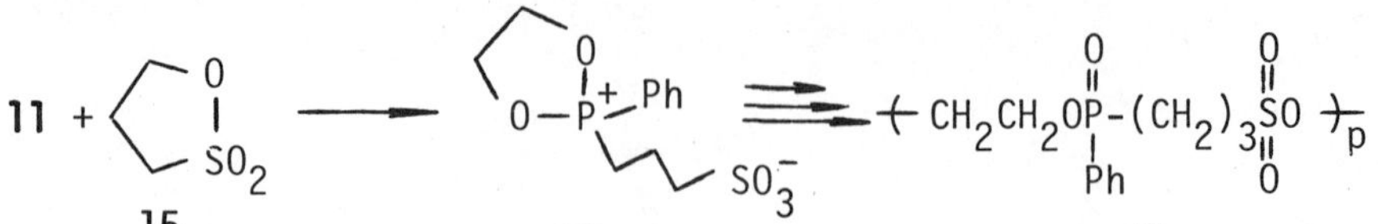

$$\mathbf{11} + \underset{\mathbf{15}}{\begin{array}{c}\text{[cyclic }SO_2\text{]}\end{array}} \longrightarrow \underset{\mathbf{40}}{\begin{array}{c}\text{[cyclic P}^+\text{-Ph}\\ \text{with } SO_3^-\text{]}\end{array}} \rightleftharpoons {\Large(}CH_2CH_2O\underset{Ph}{\overset{\overset{O}{\|}}{P}}-(CH_2)_3\underset{O}{\overset{\overset{O}{\|}}{S}}O{\Large)}_p$$
$$\mathbf{41}$$

Another interesting copolymerization of **11** was discovered with 4-carboxybenzaldehyde **42**. At 130-140°C, alternating copolymer **43** was produced (20).

$$\textbf{11} + OHC-\langle\underline{}\rangle-CO_2H \longrightarrow -(CH_2CH_2O\overset{O}{\underset{Ph}{P}}-OCH_2-\langle\underline{}\rangle-CO_2)_p-$$

$$\textbf{42} \qquad\qquad\qquad\qquad\qquad \textbf{43}$$

A zwitterion **44** is assumed to be the key intermediate.

$$\textbf{11} + \textbf{42} \longrightarrow \left[\underset{\textbf{44}}{\overset{O}{\underset{O\bar{C}H-\langle\underline{}\rangle-CO_2H}{\underset{|}{O-P^+\diagup Ph}}}} \right] \xrightarrow{\text{hydrogen transfer}} \underset{\textbf{45}}{\overset{O}{\underset{OCH_2-\langle\underline{}\rangle-CO_2^-}{\underset{|}{O-P^+\diagup Ph}}}}$$

These copolymerizations are closely related to the alternating copolymerization of cyclic phosphite **20** with α-keto acid which is discussed in the following section.

Copolymerization of 2-Phenoxy-1,3,2-dioxaphospholane 20

The second cyclic phosphorus compound which was successfully adopted as M_N is 2-phenoxy-1,3,2-dioxaphospholane **20**, a cyclic phosphite. The copolymerization of **20** with α-keto acid **46** is of big interest (19).

$$\textbf{20} + R\underset{O}{\overset{\parallel}{C}}CO_2H \longrightarrow -(CH_2CH_2O\overset{O}{\underset{|}{P}}-O\underset{|}{C}HCO_2)_p-$$
$$\qquad\qquad\qquad\qquad\qquad Ph \qquad R$$

46a R=Me **47a**
  b R=Ph   **b**

The above alternating copolymerization occured at 120°C. At 0°C, an equimolar mixture of **20** and **46** in ether produces acyl pentaoxyphosphorane derivatives **48** (21). This finding has opened a new synthetic method of pentaoxyphosphorane.

20 + 46 ⟶ [structure 48]

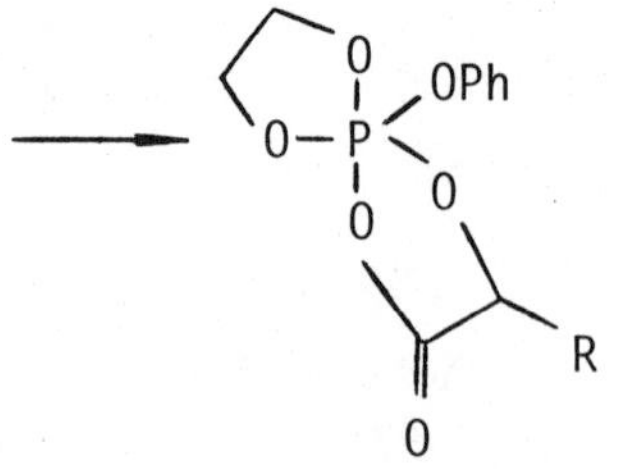

48a R=Me (mp, 87°C)
  b R=Ph (mp, 60–62°C)

Heating of **48a** in bulk at 120°C produced its polymer **47a**.

The following scheme of reactions explains the alternating copolymerization as well as the formation of the acyl pentaoxyphosphorane. A zwitterion **50**, which is generated by the hydrogen transfer in a transient zwitterion **49**, acts as a key intermediate in these two interesting reactions.

20+46 ⟶ [**49**] ⟶ [**50**] ⇌ **47**
                                ⇌ **48** (heat)

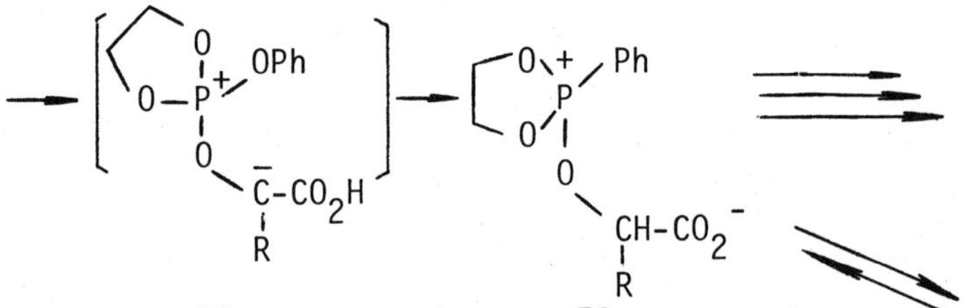

Two significant points are to be pointed out from the viewpoint of polymerization chemistry. Thus, α-keto acid has been copolymerized for the first time. In the copolymerization, one monomer is oxidized and the other is reduced. It may be well called "Redox Copolymerization", which is entirely different from the so-called "Redox—initiated Polymerization" where a free radical is produced by the redox reaction between an oxidizing and a reducing components. From organic synthesis, the above findings are also significant, i.e., α-hydroxy acid is readily obtained by the hydrolysis of **48**. In practical synthesis, the isolation of **48** is not necessary. Reduction of α-keto acid to α-hydroxy acid is readily performed in a single batch process at room temperature, i.e., admixing of **20** with **46** followed by the hydrolysis of the reaction mixture (22).

Copolymerization of a six-membered cyclic phosphite **51** with **46** also occurred without added catalyst (23). However, the product copolymer did not possess the 1:1 alternating structure as expressed by **52** (Eq 11).

$$\underset{51}{\bigg\langle\!\!\!\underset{O}{\overset{O}{\diagdown}}\!\!P\!\!\underset{OPh}{\diagup}} + 46 \longrightarrow$$

$$\left[-(CH_2)_3 O\overset{O}{\underset{OPh}{\overset{\|}{P}}}O - (\overset{R}{\underset{O}{\overset{|}{C}HC\overset{\|}{O}}})_n - \right]_p + (n-1) \underset{O}{\bigg\langle\!\!\!\overset{O}{\diagdown}\!\!P\!\!\underset{OPh}{\overset{O}{\diagup}}}$$

$$\underset{52\ (n=1.5-3.0)}{} \qquad (11)$$

Production of excess amount of the unit of α-hydroxy acid ester is due to the nucleophilic attack of carboxylate group at the growing species onto the α-carbon atom of carboxylate or carboxylic group in a zwitterion **53**.

53

In the case of five-membered ring, the ring-opening path occurred exclusively to produce the alternating copolymer. The difference may be attributed to the difference of the ring-opening reactivity between five- and six-membered cyclic phosphonium rings. Furthermore, the relative contribution of each path in **53** is dependent upon the nature of R and the reaction conditions such as temperature and solvent.

Summary

Spontaneous alternating copolymerizations of two cyclic phosphorus compounds, 2-phenyl-1,3,2-dioxaphospholane **11** and 2-phenoxy-1,3,2-dioxaphospholane **20**, with several electrophilic monomers were discussed. As the electrophilic monomers, β-propiolactone **13**, 3-hydroxypropanesulfonic acid lactone (propanesultone) **15**, acrylic acid **16**, acrylamide **17**, acrylate were sucessfully copolymerized with **11**. At temperatures lower than those of copolymerization, the combinations of **11** — **13**, **11** — **16**, **11** — **17** and **11** — acrylate produced trioxyphosphorane derivatives having spiro-ring structures, which were significant in two respects, i.e., a strong support to the polymerization mechanism as well as a new family of organophosphorus compounds. These spiro-ring P^V compounds were polymerized on heating. These polymerizations constitute a new type of ring-opening polymerization, which is induced thermally and proceed through zwitterion intermediate. Another cyclic phosphite of **20** was successfully copolymerized with α-keto acids such as pyruvic acid **46a** and phenylglyoxilic acid **46b** without any added catalyst. This copolymerization may well be called "Redox Copolymerization" in which **20** was oxidized and α-keto acid was reduced.

"Literature Cited"

1. Review articles on "Spontaneous Alternating Copolymerization via Zwitterion Intermediate".
    a) Saegusa, T., Chem. Tech. (Amer. Chem. Soc.), (1975) **5**, 295;
    b) Saegusa, T., Kobayashi, S., Kimura, Y. and Ikeda, H., J. Macromol. Sci. Chem., (1975) **A-9**, 641;
    c) Saegusa, T., Kobayashi, S. and Kimura, Y., Pure and Appl. Chem., in press.
    d) Saegusa, T., Angew. Chem., in press.
2. Saegusa, T., Ikeda, H. and Fujii, H.,Macromolecules, (1972) **5**, 354.
3. Saegusa, T., Kobayashi, S. and Kimura, Y., Macromolecules, (1974) **7**, 1.
4. Saegusa, T., Kimura Y. and Kobayashi, S.,Presented at 28th Annual Meeting of Soc. Polymer Sci., Japan, April, 1973.
5. Saegusa, T., Ikeda, H., Hirayanagi, S. and Kobayashi, S., Macromolecules, (1975) **8**, 259.
6. Saegusa, T., Kobayashi, S. and Kimura Y., Macromolecules, (1974) **7**, 139.
7. Saegusa, T., Kobayashi, S. and Kimura, Y., Macromolecules, (1975) **8**, 374.
8. Saegusa, T., Kimura Y. and Kobayashi, S., Macromolecuels, (1977) **10**, in press.
9. Saegusa, T., Kobayashi, S. and Furukawa, J., Macromolecules, (1976) **9**, 728.
10. Saegusa, T., Kimura, Y. and Kobayashi, S., Macromolecuels, (1977) **10**, in press.
11. Saegusa, T., Kimura, Y. and Sawada, S. and Kobayashi, S., Macromolecules (1974) **7**, 956.
12. Saegusa, T., Kimura, Y., Sawada, S. and Kobayashi, S., Macromolecules (1974) **7**, 956.
13. Saegusa, T., Kimura, Y., Ishikawa, N. and Kobayashi, S., Macromolecules (1976) 9, 724.
14. Saegusa, T. , Kobayashi, S. and Furukawa, J., Macromolecules, (1977) **10**, in press.
15. Saegusa, T., Kobayashi, S. and Furukawa, J., Macromolecules, (1975) **8**, 703.

16. Saegusa, T., Kobayashi, S. and Kimura, Y., Macromolecules, (1977) **10**, in press.
17. Saegusa, T., Yokoyama, T. and Kobayashi, S., to be published.
18. Saegusa, T., Kobayashi, S. and Kimura, Y., J. Chem. Soc., Chem. Commun., **1976**, 443.
19. Saegusa, T., Yokoyama, T., Kimura, Y. and Kobayashi, S., to be published.
20. Saegusa, T., Yokoyama, T., Kimura, Y. and Kobayashi, S., Presented at 26th Annual Meeting of Soc. Polymer Sci., Japan, May 1976; Submitted to Macromolecules.
21. Saegusa, T., Kobayashi, S., Kimura, Y. and Yokoyama, T., J. Amer. Chem. Soc., (1976) **98**, 7843.
22. Saegusa, T., Kobayashi, S., Kimura, Y. and Yokoyama, T., Submitted to J. Org. Chem.
23. Saegusa, T., Yokoyama, T. and Kobayashi, S., Unpublished results.

INDEX

A

Absorbance, effect of composition on	257
Acetals	60, 65, 77
Acetone	308
1-Acetoxy-2,3-butadiene	299
Achiral enantiogenic process	206
Acid anhydride, dibasic	146, 147
Acrylamide	334, 336
Acrylic acid	334
Acrylic ester	334
Acrylonitrile	300
cycloaddition, 1-acetoxy-2,3-butadiene/	299
Acylamidine salts used as model compounds	138
Acylamidinium ions	136
Acyl fluoride, difunctional	277
Additives on polymerization, effect of	203, 323
AgX precipitation	30
Aldehyde syntheses	116
Aldehydes, polymerization of	112
Alkali metal counterion	223
Alkoxide initiator, trimer	277
Alkylimino group	146
Allenes, cycloadditions of substituted	300
Allyl chloride	26
Allyl halide	27, 32
Amides, protonation of	130
Amidine HCl groups	139
Amines, cyclic	1
β-(4-Aminophenyl) propionic acid	251
Anion, α-cyanobutyl	287
Anionic polymerization (see Polymerization, anionic)	
Antisteric choice	196
Antisteric process	196
Asymmetric polymer synthesis	204
N-(Azacyclohepten-(1)-yl-(2))-caprolactam	138
Azetidines	9
Aziridine, N-substituted	147

B

Benzimidazole	282
Benzocyclobutene	285
Benzocyclopropenecarbonitrile	285
Benzonate end-groups	68
1-Benzyl aziridine (BA)	6, 9

1-Benzyl-2,2-dimethyl aziridine (BDMA)	9
1-Benzyl-2-methyl aziridine (BMA)	9
Benzylmethylsulfonium salts	318
Bicyclobutane-1-carbonitrile	286
Bicyclobutanes	285, 291
Bicyclopentane-[1.1.0]carbonitrile	285
Bicyclo[2.1.0]pentane monomers	292
Bimolecular reactions	247
Binary systems, polymerization by	309
Binding model	217
Bisoxonium salt, monomer bisester and trimer	21
Bisphenol-A diglycidyl ether	51
Block copolymerization	5, 171
Bond	
acyclic double	311
angle distortion	249
carbonyl double	113
Bromal	120
Bulk polymerization of perhaloacetaldehydes	118
Bulky substituent	178
1,2-Butadien-4-ol	299
(2,3-Butadienyloxy)trimethylsilane	300
1-tert-Butylaziridine	5, 6
(R)-tert-Butyloxirane	180, 182, 186, 189
P((RS)-tert-Butyloxirane)	188
Butyl rubbers, halogenated	33, 34
tert-Butylthiirane	203
n-Butyraldehyde (BA)	308

C

C–C bond opening, ring-opening polymerization via	285
11-CF-4, polymerization of	100, 107
CM7, conversion data for	243
Cadmium allyl thiolates	194
Cadmium salt catalysts	193
Caprolactam	
copolymeriaztion with organometallic compounds	149
epsilon (ϵ-caprolactams)	63, 148
with β-(3,4-diaminophenyl) propionic acid, copolymerization of	251
–LiCl, phase diagram of	223
polymerization of	222

345

Caprolactam (*Continued*)
 –LiCl mixtures, polycaproamide
  synthesized from 229
 polymers, cationic 134
Caprolactone copolymerization with
 organometallic compounds 149
Carprolactone, epsilon
 (ε-caprolactone) 148, 152
 polymerization of 152
Carbene complex-based catalysts, W 310
Carbocation 109
Carbonyl double bond, polarization
 of the 113
4-Carboxy-6-alkyl-2-piperidone 236
4-Carboxy-5-alkyl-2-pyrrolidone 236
4-Carboxy-6-ethyl-2-piperidone 234
4-Carboxy-5-ethyl-2-pyrrolidone 234
Carboxylic groups, concentration of .. 139
α-Carboxymethyl caprolactam 234, 236
β-Carboxymethyl caprolactam 234, 236
Carboxymethyl lactams 234
4-Carboxymethyl-2-piperidone 234
4-Carboxy-6-methyl-2-piperidone 234
β-Carboxymethyl-2-piperidone 236
4-Carboxymethyl-2-pyrrolidone 234, 236
4-Carboxy-5-methyl-2-pyrrolidone 234
4-Carboxy-2-piperidone 234, 236
4-Carboxy-2-pyrrolidone 234, 236
Catalysis, electrophilic 246
Catalyst(s)
 cadmium salt 193
 coordination 161, 165
 lithium tetraalkylaluminate 149
 metathesis 303
 ring-opening coordination 165
 site control 166
 W carbene complex-based 310
 Ziegler type 306
Catalytic
 activity, enhancement of the 309
 behavior in homopolymerization 168
 properties 167
Cationic oligomerization of
 (R)-*tert*-butyloxirane 186
Cationic polymerization
 (*see* Polymerization, cationic)
Ceiling temperature of perhaloacetaldehyde polymerization 124
1-(2-Ceproethyl)aziridine 6
Cesium fluoride 270
Chain(s)
 aromatic, uncyclized 258
 configuration of polymer 238
 end control 166
 ethylene glycols and cyclic formals,
  open 92
 extended polymers 280, 283
 growth 243

Chain(s) (*Continued*)
 propagation 132
 transfer 157
Chiral β-lactones 211
Chiral sites 207
Chloral 112, 114
Chlorobromoacetaldehydes 119
Chlorobutyl rubber in isooctane 35
4-Chloro-2-butyne-1-ol 299
Chlorodibromoacetaldehyde (CDBA) 119
Claisen-type condensation 225
Co-initiator, hydroxyl containing 161
Collagen, melting temperature of 217
Composition on absorbance, effect of 257
Condensation, Claisen-type 225
Condensation process 166
Configurational choice 195
Conidine 10
Consumption equation, first order 198
Consumption equation, second order 199
Conversion data for CM7 243
Coordination catalysts, ring-opening .. 165
Coordination type 154
Copolymers(s)
 block- 5
 nylon.6-polybutadiene-nylon.6
  triblock 174
 phase transition behavior 150
 sequenced 78
Copolymerization(s) 312
 block 171
 of 11-CF-4 with styrene 107
 of ε-caprolactam with β-(3,4-
  diaminophenyl) propionic acid 251
 caprolactone and caprolactam 149
 of cyclic compounds containing
  O and N atoms 145
 of cyclic phosphorus compounds 322, 334
 by oxonium ion, graft 24
 of perhaloacetaldehydes 121, 125
 N-phenylaziridine and phthalic
  anhydride 148
 of 2-phenoxy-1,3,2-dioxa-
  phospholane 334, 339
 R,S- 178
 relative reactivity ratios in 169
Correlation time 180
Counterion, alkali metal 223
Counterion, reaction with 27
Crystalline fraction 192
Crystalline properties of the racemic
 polymer 212
Crystallinity properties 258
Crystallinity relationships, structure– 210
α-Cyanocyclobutyl anion 287
1,(2-Cyanoethyl)-2-methyl aziridine
 (CEMA) 8

INDEX

Cyclic compounds containing O and
  H atoms, copolymerization of 145
Cyclization 233, 328, 329
Cycloaddition, 1-acetoxy-2,3-
  butadiene/acrylonitrile 299
Cycloadditions of substituted allenes
  with acrylonitrile 300
Cyclohexane rings 42
Cyclohexanone 152
Cyclohexene sulfide (CS) 204

D

Dainton's equation 102
Decomposition temperatures on
  polypyrrolidone 217
1,3-Dehydroadamantane 285
Depolymerization of the polymer 107
Diamines 281
β-(3,4-Diaminophenyl) propionic
  acid, copolymerization of
  ϵ-caprolactam with 251
Dichlorobromoacetaldehyde (DCBA) 119
1-(3,5-Dichloro-4-hydroxyphenyl)-
  tetrahydrothiophenium hydroxide
  inner salt 321
1,1-Dichloro-2-vinylcyclopropane 49
Diethylzinc-(+) 3,3 dimethyl 2
  butanol initiator system 197
Difluorobromoacetaldehyde (DFBA) 117
Difluorochloroacetaldehyde (DFCA) 115
1,1-Difluoro-2,2-dibromoethylene 117
Diisotactic structures 205
Dimer, cyclic 107, 248
(−)3,3 Dimethyl 1,2 butane diol
  (DMBD) 198
6,6-Dimethyl-4-carboxy-2-
  piperidone 234, 236
5,5-Dimethyl-4-carboxy-2-pyrrolidone 236
5,5-Dimethyl-4-carboxyl-2-pyrrolidone 234
Dimethyl cyclobutene-1,2-
  dicarboxylate 58
4,4-Dimethyl-Diox 63
cis-4,5-Dimethyl-Diox 63
trans-4,5-Dimethyl-Diox 63
Dimethyl sulfoxide (DMSO) 213
cis-2,3 Dimethyl thiirane (DMT) 204
3,9-Dimethylene-1,5,7,11-tetra-
  oxaspiro[5.5]undecane 49
Diols 281
1,3-Dioxacycloacycloalkanes ..103, 106, 107
1,3-Dioxacyclooctane 108
1,3-Dioxacycloundecane (octanediol
  formal) 95
1,3-Dioxlan (Diox) 60, 64
Dioxolan formation 82
Dioxolans, polymerizability of 62
α,α-Disubstituted-β-propiolactones 210

Disyndiotactic structures 205
Di-*tert*-butyl peroxide 50

E

Electrophilic monomers 335
β-Elimination 326, 328
Enantiomer consumption 198
Enantiomeric composition of the
  monomer 200
Enantiomeric purity of the monomer 201
End-groups
  benzonoate 68
  formed in cationic lactam
    polymerizations 137
  in polyacetals 67
Epoxides 146
Ester interchange 157
Ethers, heat of polymerization of
  cyclic 104
Ethylene glycols, open chain 92
Ethylenesulfonamide 334, 336
Ethylenimines, N-substituted 3
Expansion during polymerization 47
Expansion in volume, ring-opening
  polymerization with 38

F

^{19}F NMR 14
Flip-flop mechanism 168
Flory parameter 61
Fluoral 112, 115
Fluorobromoacetaldehydes 117
Fluorocarbon epoxides 269
Fluorochloroacetaldehydes 115, 116
Formals, macrocyclic 99
Formals, polymerization of cyclic ..104, 108
Free radical polymerization 301
Functional groups, influence of polar 311

G

Gauche interactions 64
Global rate constants 199
Glycols, open chain ethylene 92
GPC 16
Graft copolymerization by oxonium
  ion mechanism 24
Grafting from halogenated butyl
  rubbers 34

H

Halide, allyl 32
Halide precipitated from halogenated
  butyl rubbers 33
Halogenated polymers 31
Hexafluoropropylene (HPF) 274

Hexafluoropropylene epoxide
  (HFPO) 269
HO group 146
1.3.6.9.12.15-Hexaoxacyclohepta-
  decane (pentaethylene glycol
  formal) 95
Homopolymerization 168, 170
Homosteric correlations 196
HX from allyl halide, elimination of .. 27
Hydrogen abstraction from
  2-methyltetrahydrofuran 27
Hydroxyl containing co-initiator 161
4-Hydroxyl-1,2-butadiene 293
1-(4-Hydroxy-3-methylphenyl)-
  tetrahydrothiophenium hydroxide
  inner salt 321
3-Hydroxypropanesulfonic acid
  lactone (propanesultone) 334

I

Imide linkage 221, 242
Imides, cyclic 146
Initiation 13, 27, 156, 130
                                 220, 287, 322
  efficiency of 25
  mechanisms of 159
  triflic anhydride 13
Initiator(s)
  diethylzinc-(+) 3,3 dimethyl 2
    butanol 197
  difunctional 279
  influence of the nature of the 195
  monofunctional and difunctional .. 271
  pyridine 122
  triethyloxonium tetrafluoroborate .. 2, 6
  trimer alkoxide 277
  on the two-stage polymerization,
    effect of 103
Isolation of silver halide 29, 31
Isomerization polymerization of
  lactams 233
Isomers, rotational 189
Isooctane, chlorobutyl rubber 35
IsoSyn mechanism 184

J

Jacobson–Stockmayer theory 102

K

k_p/k_t, determination of 2
Killing agent 72
Kinetic considerations 156
Kinetics 242

L

Lactam(s)
  carboxymethyl 234
  isomerization polymerization of 233
  –lithium chloride interactions ... 216, 219
  polymerizations, anionic 142, 216, 228
  polymerizations, cationic 129, 137, 142
  substituted 234
Lactone 145
  polymerization 170, 211
β-Lactones, chiral 211
Ligand 196
Lithium chloride (LiCl)
  in the anionic polymerization of
    lactams 216
  –lactam interactions 219
  on the melting and decomposition
    temperatures, effect of 217
  mixtures, polycaproamide syn-
    thesized from caprolactam– 229
  phase diagram of ε-caprolactam– .. 223
Lithium tetraalkylaluminate 149
Living
  macromolecules 70
  mechanism 158
  tetrahydrofuran polymers 13

M

Macrocyclic formals, addition of 101
Macroester and macroion, equilibra-
  tion between 13
Macroion/macroester ratio 15
Macromolecules, living 70
Markov chain mechanism, first order 185
Mechanical properties 241
Melt viscosity for polymerization of
  ε-caprolactone 159
Melting
  behavior of polycaprolactam ... 223, 229
  point depression of the polymers ... 219
  temperature of collagen 217
  temperatures of polypyrrolidone 217
Metathesis catalyst 303
1-Methylazetidine 10
α-Methyl-α-butyl-β-propiolactone 213
Methyl β-(3,4-diaminophenyl)
  propionate 253
3-Methylene[2.1.0]bicyclopentane-
  1-carbonitrile, anionic polymerization of 301
2-Methylene-1,5,7,11-tetraoxaspiro-
  [5.5]undecane 51
Methylene groups 218
2-Methyl-5-norbornene-2-nitrile 312
Methyloxirane 184
2-Methyltetrahydrofuran 26, 27, 32
Methylthiirane 197, 202

Model compounds, acylamidine
  salts used as 138
Model studies, isolation of silver
  halide from 29
Molecular weight distributions 162, 275
Molecular weight–viscosity relations 241
Monofilament, oriented 238
Monomer 196
  bisester salt, mechanisms of
    formation of 21
  -to-catalyst molar ratio 309
  enantiomeric composition of the 200
  enantiomeric purity of the 201
  influence of the nature of the 198
  nucleophilic and electrophilic 335
  preparation of 115, 319
  properties of 320
  R-content in unchanged 183
  ring-opening polymerization of
    heterocyclic 165
  structure 2
Morphology controls in polymer
  blends 174

N

N-substituted
  aziridine 147
  ethylenimines 3
  propylenimines 7
N-terminal groups 138, 139
Nitrogen atoms, copolymerization of
  cyclic compounds containing 145
NMR data for PCM7 and PC6 240
NMR studies 26
Norbornene derivatives 303, 305
5-Norbornene-2-nitrile 308, 312
Norbornenenitriles 312
Nucleophilic monomers 335
Nylon.6-polybutadiene-nylon.6
  triblock copolymer 174

O

Octafluoroisobutylene epoxide
  (OFIBO) 269
Octanediol formal,
  (1.3-dioxacycloundecane) 95
1,4,6,10,12,15,16,19-Octaoxatrispiro
  [4.2.2.4.2.2] nonadecane 55
Oligomerization of (R)-tert-
  butyloxirane 186
Oligomers, cyclic 87, 102, 162
Oligomers, structure of 132
Optical purity 197
Optical rotatory dispersion spectra
  of poly(alkyloxirane) 181

Organometallic compounds, caprolactone and caprolactam
  copolymerization with 149
1,3,4 Oxadiazoles 282
2-Oxazoline 333
Oxiranes 198
  polymerization 168, 191, 206
μ-Oxo-alkoxides 165, 167
Oxonium ion 109
  mechanism, graft copolymerization
    by 24
Oxygen atoms, ring-opening
  copolymerization of cyclic
  compounds containing 145
Oxymethylene and oxyethylene units 78

P

Paraformaldehyde 100
Paraldehyde (PA) 308
PC6 240
PCM7 .... 238, 240
Pentaethylene glycol formal
  (1.3.6.8.12.15-hexaoxacyclo-
  heptadecane) 95
1.3.6.9.12-Pentaoxacyclotetradecane
  (tetraethylene glycol formal) 94
Penultimate effect 184
Perhaloacetaldehyde
  polymerization 111, 118, 121, 122, 125
Perhaloaldehyde polymerization 113
Phase diagram of ϵ-caprolactam–
  LiCl 223
Phase transition behavior of
  copolymers 150
2-Phenoxy-1,3,2-dioxaphospholane 334, 339
N-Phenylaziridine and phthalic
  anhydride copolymerization 148
Phenyl ether-terminated products 17
1-(2-Phenylethyl)aziridine 6
1-(2-Phenylethyl)-2-methyl
  aziridine (PEMA) 8
α-Phenyl-α-ethyl-β-propiolactone 214
Phenylisocyanate, copolymerization
  of perhaloacetaldehydes with 125
Phosphonium zwitterion
  intermediates 332
Phosphorus compounds, cyclic 332, 334
Phthalic anhydride copolymerization,
  N-phenylaziridine and 148
Poisson molecular weight
  distribution 158
Polar substituents, polymerization
  of norbornene derivatives with .. 303
Polarization of the carbonyl double
  bond 113
Polyacetals, end-groups in 67

Poly(alkyloxirane) 178, 181
Polyamides 281
Poly(R-tert-butyloxirane) 179
Poly(tert-butyl thiirane) 200
Polycaproamide, melting behavior of 229
Polycaprolactam, melting behavior
  of 223
Poly(caprolactone-b-isocyanates) ... 171
Poly(caprolactone-b-oxiranes) 171
Poly(caprolactone-b-styrene) 172–174
Poly(caprolactone-b-thiiranes) 171
Poly(cyclohexene sulfide) 205
Poly-Diox 70
Poly-1,3-dioxolan 67
Polyesters 281
Poly(2,5-ethylene benzimidazole) ... 251
Polyethyleneglycol 100
Poly(R-isopropyloxirane) 179
Polymer(s)
  blends, morphology controls in ... 174
  chain extended 283
  chains, configuration of 238
  characterization 275
  crosslinked 329
  crystalline properties of the
    racemic 212
  depolymerization of the 107
  difunctional 275
  halogenated 31
  living tetrahydrofuran 13
  melting point depression of the .. 219
  mole fraction difunctional 276
  monofunctional 275
  of norbornenenitriles 312
  preparation of 115
  –salt interactions 216
  structure of 132
  synthesis, asymmetric 204
Polymerization
  of acetals 60, 65, 77
  addition 39
  of aldehydes 112
  anionic 295
    of bicyclobutane-1-carbonitrile .. 286
    coordinated 203
    of α,α-disubstituted-β-
      propiolactones 210
    of fluorocarbon epoxides 269
    lactam 142, 216, 228
    of 3-methylene [2.1.0] bicyclo-
      pentane-1-carbonitrile 301
    ring-opening 220
  of aryl cyclic sulfonium
    zwitterions 318
  by binary systems 309
  bulk 118
  of tert-butyloxirane, selectivity in
    the 182

Polymerization (Continued)
  of tert-butylthiirane, stereoelec-
    tivity in the 203
  of ε-caprolactam 152, 222
  of ε-caprolactone 159
  cationic
    of 11-CF-4 100
    of chloral 114
    of cyclic amines 1
    of cyclic formals 108
    of lactams 129, 137
  of cyclic
    acetals 60, 65, 67, 77
    amines 1
    dimer 107
    ethers 104
    formals 108
  degree of (DP) 271
  of 1,3-dioxacycloalkanes 106, 107
  of 1,3-dioxacyclooctane 108
  effect of additives on 323
  effect of initiators and solvents on
    the two-stage 103
  expansion during 47
  influence of polar functional groups
    and acyclic double bond on 311
  isomerization of lactams 233
  of lactams, anionic 216, 228
  of lactams, isomerization 233
  lactones 170
  mechanisms 159
  of 3-methylene[2.1.0]bicyclo-
    pentane-1-carbonitrile 301
  of norbornene derivatives ... 303, 305
  oxiranes 168, 191, 206
  perhaloacetaldehyde ... 111, 118, 121, 124
  for perhaloaldehydes 113
  of α-phenyl-α-ethyl-β-propiolactone 214
  rate of 2, 31, 121
  ring opening
    anionic 220
    of tert-butyloxirane 180
    via C–C bond opening 285
    with expansion in volume 38
    of heterocyclic monomers 165
    of macrycyclic acetals 77
    of norbornene derivatives 303
    shrinkages for 40
  solid state 324
  solution 237
  stereoelective 191, 194
  stereoselective 191
  tetrahydrofuran 31
  thermodynamics of 61
  of thiiranes 191, 206
  by the W carbene complex-based
    catalysts 309

Poly-3-methylenebicyclo[2.1.0]-
  pentane-1-carbonitrile 293
Poly-α-methyl-α-propyl-β-
  propiolactone 210
Polymethyl thiiranes 194
Poly(5-norbornene-2-nitrile) 314
Polynorbornenenitriles 314
Poly[oxy(1-alkyl)ethylene] 178
Polyoxycarbonate 48, 52
Poly(R-oxypropylene) 178, 179
Poly-α-phenyl-α-ethyl-β-
  propiolactone 211
Polypivalolactone 212
Polypyrrolidone 217
Poly(styrene-b-aziridine) 5
Polytetrahydrofuran (PTHF) 25, 35
Poly(THF-b-aziridine) 5
Poly-1,3,5-trioxan 67
Preequilibrium 156
Product distribution 18
Propagation 13, 157, 221, 287
  kinetics of 74
  reaction 1, 133
Propanesultone (3-hydroxypropane-
  sulfonic acid lactone) 334
β-Propiolactone 333, 334
Propylene oxide 192
Propylenimines, N-substituted 7
Protonation of amides 130
Pyridine initiator 122

R

Racemic polymer, crystalline
  properties of the 212
Rate
  constants for the polymerization of
    α-phenyl-α-ethyl-β-propio-
    lactone 214
  enhancement by addition of
    macrocyclic formals 101
  investigations 213
  polymerization 2, 31, 121
R-content in unchanged monomer 183
Reactivity, electrophilic and
  nucleophilic 332
Reactivity ratios in copolymeriza-
  tions, relative 169
Regioselectivity in the ring-opening
  polymerization 180
Relaxation time, spin-lattice 186
Ring
  expansion assumption 74
  number, heat of polymerization as
    a function of 104

Ring (Continued)
  opening
    coordination catalysts 165
    copolymerization of cyclic com-
      pounds containing O and
      N atoms 145
    polymerization (see Polymeriza-
      tion, ring-opening)
Rotational isomers 189
Rubber in isooctane, chlorobutyl 35
Rubbers, halogenated butyl 33, 34

S

Salt(s)
  benzylmethylsulfonium 318
  on the melting temperature of
    collagen, effect of 217
  –polymer interactions 216
  trialkyl sulfonium 318
Selectivity in the polymerization of
  tert-butyloxirane 182
Shrinkages for addition
  polymerization 39
Shrinkages for ring-opening
  polymerization 40
Side chain size in perhalo-
  acetaldehyde polymerization 111
Side reactions and irregular
  structures 225
Silver
  halide, isolation of 29, 31, 32
  hexafluorophosphate 26
  triflate 25
Solid state polymerization 324
Solution polymerization 327
Solvents, effect of 103, 203
Spin-lattice relaxation time 186
[2,4]-Spiroheptadine 286
Spiro ortho carbonate 48, 52
Spiro-o-xylylene 49
Spontaneous alternating copolymeri-
  zation of cyclic phosphorus
  compounds 332
Stereoelective polymerization 191, 194
Stereoelectivity, effect of 197, 198, 203
Stereoelectivity ratio 200
Stereoregularity in perhaloacetalde-
  hyde polymerization 111
Stereoregularity of polymethyl
  thiiranes 194
Stereoselective polymerization of
  oxiranes and thiiranes 191
Stereoselectivity 184, 200
Stereospecific polymerization of
  oxiranes and thiiranes 206
Stress–strain curves for
  poly(CL-b-St) 173

Structure–crystallinity relationships .. 210
Structures, side reactions and
  irregular 225
Styrene, copolymerization of
  11-CF-4 with 107
Succinimide 146
Sulfonium zwitterions, polymerization
  of aryl cyclic 318
Synthetic methods 166

T

Temperature of perhaloacetaldehyde
  polymerization, ceiling 124
Tensile properties 242, 264
Termination reaction 1, 157, 287, 296
Tetracyclooctanes 285
Tetraethylene glycol formal
  (1.3.6.9.12-pentaoxacyclotetra-
  decane) 94
Tetraglyme 270
Tetrahydrofuran (THF) 213
Tetrahydrofuran polymerizations ... 13, 31
Tetrahydropyran 32
Tetramer of (R)-tert-butyloxirane 186, 189
1.3.6.9-Tetraoxacycloundecane
  (triethylene glycol formal) 85
Thermal properties 242, 258, 264
Thermodynamics of polymerization ... 61
Thiiranes, stereoselecitve and stereo-
  elective polymerization of 191, 198, 206
Thorpe–Ziegler reaction 291
Titration of cationic caprolactam
  polymers, potentiometric 134
Torsion modulus of poly-
  (caprolactone-b-styrene) 173
Trialkyl sulfonium salts 318
s-Triazines 283
1,3,4 Triazole crosslink 282
Triethylene glycol formal, (1.3.6.9-
  tetraoxacycloundecane) 85, 87
Triethyloxoniumhexafluorophosphate 67
Triethyloxonium tetrafluoroborate
  initiator 2, 6
Triflic anhydride initiation 13

Trifluoromethane sulfonic anhydride
  initiator 13
Triisobutylaluminum 147
Trimer alkoxide initiator 277
Trimer bisoxonium salt 21
1,3,3-Trimethylazetidine 9
Trimethyl 4,4-dimethylbicyclo-
  butane-1,2,2-tricarboxylate 292, 297
Trimethyl-4-methylbicyclobutane-
  1,2,2-tricarboxylate 292, 298
2,6,7-Trioxabicyclo[2.2.2]octanes 45
1.3.5-Trioxacycloheptane (trioxepane) 79
1.3.6-Trioxacyclooctane (trioxocane) 83
1,4,6-Trioxaspiro[4,4]nonane 40
Trioxepane, polymer of 81
Triphenylmethylsodium 87

V

δ-Valerolactam 233
van der Waals' distance 39
1-Vinylbicyclobutane 286
Vinylcyclopropanes 49, 286
2-Vinyl-1,3-dioxane 65
2-Vinyl-1,3-dioxolan 65
Vinyl ketone 334
Viscosity–molecular weight relations 241

W

Water structure role 217

X

m-Xylylenediammonium isophthalate 251

Z

Ziegler type catalyst 306
Zinc allyl thiolates 194
Zwitterion
  aryl cyclic sulfonium 318
  genetic 332
  intermediates, phosphonium 332
  macro 332